At the Size Limit—Effects of Miniaturization in Insects

Alexey A. Polilov

At the Size Limit—Effects of Miniaturization in Insects

Springer

Alexey A. Polilov
Department of Entomology,
Faculty of Biology
Lomonosov Moscow State University
Moscow
Russia

ISBN 978-3-319-39497-8 ISBN 978-3-319-39499-2 (eBook)
DOI 10.1007/978-3-319-39499-2

Library of Congress Control Number: 2016952897

Printed on acid-free paper

This Springer imprint is published by Springer Nature
The registered company is Springer International Publishing AG
The registered company address is: Gewerbestrasse 11, 6330 Cham, Switzerland

Preface

Miniaturization is a widespread trend in animal evolution and one of the principal directions of evolution in insects. It is known that body size, especially when extremely small, is an important characteristic that largely determines the morphology, physiology, and biology of animals. Pattern of miniaturization-related structural changes have been described for many vertebrates, but until recently remained largely unknown for insects. Over the last few decades, extremely interesting finds about the structure of the smallest insects have attracted the interest of many researchers to the problem of miniaturization in insects. This book summarizes those finds.

Not only does the study of the structure of the smallest insects, the length of which is only several tenths of a millimeter, considerably expand our notions about the phenomenon of miniaturization in the animal kingdom, it is also of exceptional theoretical value and general fundamental importance: insects of such sizes live in a peculiar 'microworld,' in which surface tension of liquids and capillary and electrostatic forces are greater than the weight of an organism.

Furthermore, insects, displaying a huge range of sizes (the largest and smallest insects differ in body length by a factor of more than 2000, which is considerably greater than the range of sizes in any class of vertebrates), provide unique material for such a topical field of modern biology and biotechnology as the study of scaling in biological structures and processes. Moreover, the study of the scaling of insect organs provides novel material for discussing the factors that limit body size in animals in general.

Miniaturization is more than a principal trend in insect evolution: it is also an important field of modern engineering. The data that have been obtained and patterns that have been revealed about the structure of microinsects can be used in microbotics, nano-optics, and modeling of neural networks and sensory systems. The unique phenomenon of anucleate neurons may greatly change our notions about the structure and function of the central nervous system of non-human animals and humans and prove a key point in the study of memory storage and neuron regeneration.

I am deeply grateful to my teachers, R.D. Zhantiev—for the idea to study the smallest beetles—and S.Y. Chaika—for introducing me to the wonderful world of histology and electron microscopy. I am also grateful to members of the interdepartmental Electron Microscopy Laboratory, Lomonosov Moscow State University, for their constant technical support, to E.L. Mockford, R.G. Beutel, and V.V. Grebennikov for providing material, and to A.A. Makarova, M.I. Yavorskaya, E.V. Vortsepneva, Y.M. Marusik, and V.S. Polilova for their support of my studies of microinsects and help in preparing the manuscript of this book. Finally, I thank P.N. Petrov for translating the manuscript of this book into English.

Those various projects of studying microinsects that have provided the foundation for this book were supported by the Russian Foundation for Basic Research and Russian Science Foundation, by grants of the President of the Russian Federation, and by the Alexander von Humboldt Foundation. Work on this book was supported by the Russian Science Foundation (project no. 14-14-00208).

Moscow, Russia — Alexey A. Polilov

Contents

Chapter 1
Introduction

1.1 Introduction

In his review on the subject—the monograph *Scaling: Why Is Animal Size So Important*—K. Schmidt-Nielsen wrote: "Size is one of the most important aspects of an animal's endowment, and yet size differences are so obvious that often we give no further thought to them" (Schmidt-Nielsen 1984). Indeed, size, especially extremely small, is a characteristic of major importance and largely determines the morphology, physiology, and biology of animals. Miniaturization, or body size diminution, is a widespread trend of animal evolution (Hanken and Wake 1993) and one of the principal directions of evolution in insects (Chetverikov 1920), some of which have become comparable in size to unicellular organisms. The study of the smallest insects is also of exceptional theoretical value and general fundamental importance: insects of such sizes live in a peculiar "microworld," in which surface tension of liquids and capillary and electrostatic forces are greater than the weight of an organism. The study of the effects of miniaturization on the structure of insects has recently become a prominent field of modern entomology.

1.2 The Phenomenon of Miniaturization in the Animal Kingdom

Several terms are used in the literature to refer to the evolutionary diminution of body size, e.g., *pumilization* or *nannization* (Gorodkov 1984), but *miniaturization* is the most commonly used term. It follows from Hanken and Wake's definition of this term (Hanken and Wake 1993) that miniaturization is evolution toward extremely small body size.

Peculiar features of structure related to body size diminution have been described in many animals (McMahon and Bonner 1983; Schmidt-Nielsen 1984;

A.A. Polilov, *At the Size Limit—Effects of Miniaturization in Insects*,
DOI 10.1007/978-3-319-39499-2_1

Hanken and Wake 1993). Most of the available studies on the effects of miniaturization on structure in animals treat vertebrates, in spite of the fact that they are not so small that minute size could have fundamental effects on their structural organization (Hanken and Wake 1993).

Although invertebrates have gone much farther than vertebrates in the evolutionary trend of size diminution, there are only a few studies that specifically investigate the effects of miniaturization of their structure (Laurin and Garcia-Joral 1990; Mooi 1990; Edlinger 1998; Rundell and Leander 2010) and rather scant studies on the structure of miniature annelids (Westheide 1984; Worsaae and Rouse 2008, 2010), mollusks (Thorson 1965; Snyder and Bretsky 1971; Turner and Yakovlev 1983; Strong and Glaubrecht 2008; Brenzinger et al. 2013), crustaceans (Serban 1960; Hartmann 1973; Noodt 1974; Petrunina and Kolbasov 2012), tardigrades (Kristensen 1976, 1978, 1979; Schmidt-Rhaesa and Kulessa 2007; Zantke et al. 2008; Halberg et al. 2009; Schulze and Schmidt-Rhaesa 2011), micrognathozoans (Kristensen and Funch 2000; Kristensen 2002), rotifers (Martini 1912; Ware and Lopresti 1975; Hochberg 2007, 2009; Hochberg and Ablak Gurbuz 2008; Wilts et al. 2009), loriciferans (Kristensen 2002; Gad 2005), various interstitial animals (Swedmark 1964; Kirsteurer 1972; Westheide 1987), spiders (Quesada et al. 2011), Acari (Sylvere and Stein-Margolina 1976), and hemichordates (Worsaae et al. 2012).

Until recently, the miniaturization of insects was almost never discussed or specially studied. Particular studies that touched upon miniaturization-related features of structure in insects treated relatively large insects (over 2 mm long) (Rensch 1948), described only particular aspects of larval morphology (Beutel and Haas 1998; Grebennikov and Beutel 2002), or were purely speculative rather than fact-based (Gorodkov 1984). However, the publication of the first special studies on the subject of miniaturization in insects (Polilov 2004, 2005; Beutel et al. 2005) was followed by rapid development of this field of study, and over a dozen publications that touch upon miniaturization in insects are published annually in recent years (Polilov 2015a).

1.3 Body Size of Insects

The number of miniature insects is extremely great. There are at least 245 families of the class Insecta that include representatives of body length smaller than 2 mm and at least 106 families of this class that include representatives 1 mm long or smaller (Fig. 1.1). About 52 % of all families that include insects of 2 mm long or smaller represent either the order Coleoptera or the order Hymenoptera. The percentage of these orders among families of insects less than 1 mm long is as high as 69 %, and the size class of insects smaller than 0.5 mm is represented exclusively by species of these two orders. For convenience of discussion, insects of adult body length smaller than 2 mm are termed microinsects. The boundary of 2 mm

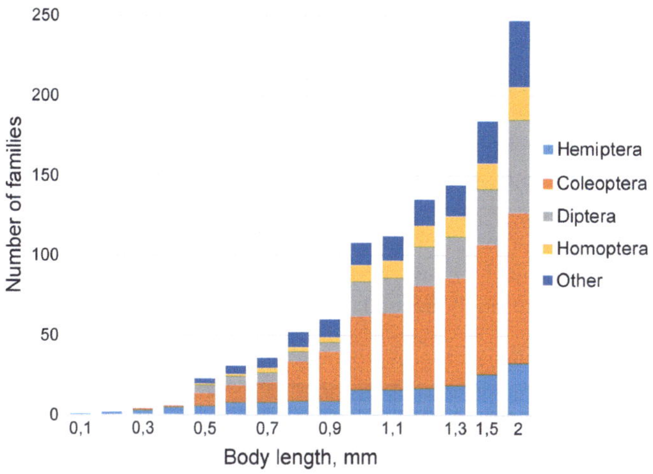

Fig. 1.1 Number of insect families that include representatives of particular body lengths

corresponds to the upper threshold of pumilistic degeneration recognized by Gorodkov (1984) and is similar to the size characteristics of the meiofauna (Rundell and Leander 2010).

The range of sizes among insects is more than twice as great as among any class of vertebrates (Table 1.1). Among representatives of different orders of insects, the greatest range is found in beetles, in which this parameter is considerably greater than in any other insect order and tens of times as great as in mammal orders.

Table 1.1 Size characteristics of some taxa

Class	Order	Superfamily	Size range	Min–max body size
Insecta			**2200**	0.14–310 mm
	Coleoptera		**660**	0.3–200 mm
		Staphylinoidea	200	0.25–50 mm
		Chrysomeloidea	250	0.8–200 mm
		Curculionoidea	158	0.6–95 mm
	Hymenoptera		430	0.14–60 mm
		Chalcidoidea	140	0.14–20 mm
	Hemiptera		130	0.5–65 mm
	Diptera		108	0.5–54 mm
Osteichthyes			**1000**	0.007–7 m
Mammalia			**825**	0.04–33 m
	Cetacea		30	1.1–33 m
	Carnivora		27	11–300 cm
	Rodentia		26	5–130 cm
Reptilia			500	0.02–10 m

Only three families of insects (excluding Entognatha) include representatives less than 0.5 mm long: Mymaridae, Trichogrammatidae (both Hymenoptera), and Ptiliidae (Coleoptera). These insects are nearly as small as the smallest metazoans.

The smallest insects (Fig. 1.2) are the wingless males of the egg parasitoid *Dicopomorpha* (Mymaridae); the smallest flying insects are wasps of the genera *Kikiki* (Mymaridae) and *Megaphragma* (Trichogrammatidae); and the smallest nonparasitic insects are beetles of the genus *Scydosella* (Ptiliidae). The orders Hymenoptera and Coleoptera, to which these genera belong, include the majority of miniature insects (Table 1.2) and display the greatest size range (Table 1.1). Therefore, greatest attention in the study of insect miniaturization should be paid to these two orders.

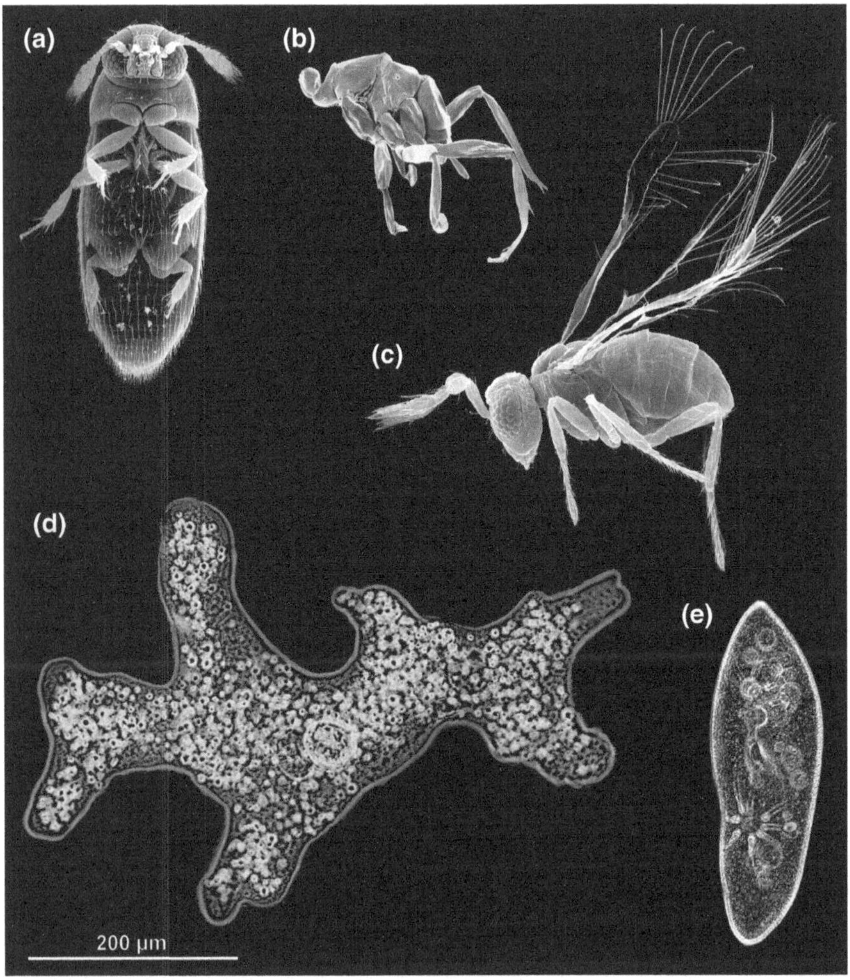

Fig. 1.2 Relative body sizes of smallest insects and some unicellular organisms: **a** *Nanosella* sp., **b** *Dicopomorpha echmepterygis* (modified from Huber and Landry 1999), **c** *Megaphragma mymaripenne*, **d** *Amoeba proteus*, **e** *Paramecium caudatum*

Table 1.2 The smallest insects (body length 300 μm or less)

Species	Body size (μm)	Fam.*	Source
Dicopomorpha echmepterygis Mockford 1997	139 (male)	Mym.	Mockford (1997)
Kikiki huna Huber & Beardsley 2000	158 (female)	Mym.	Huber and Noyes (2013)
Megaphragma caribea Delvare 1993	170	Tri.	Delvare (1993)
Alaptus borinquensis Dozier 1932	200 (male)	Mym.	Huber and Noyes (2013)
Tinkerbella nana Huber & Noyes 2013	210 (male)	Mym.	Huber and Noyes (2013)
Alaptus magnanimus Anandale 1909	210	Mym.	Borror et al. (1981)
Megaphragma amalphitanum Viggiani 1997	225	Tri.	Viggiani and Bernardo (1997)
Megaphragma longiciliatum Subba Rao 1969	250	Tri.	Subba Rao (1969)
Megaphragma mymaripenne Timberlake 1924	252**	Tri.	Timberlake (1924)
Megaphragma deflectum Lin 1992	260	Tri.	Lin (1992)
Megaphragma striatum Viggiani 1997	270 (male)	Tri.	Viggiani (1997)
Megaphragma decochaetum Lin 1992	280	Tri.	Lin (1992)
Megaphragma stenopterum Lin 1992	280	Tri.	Lin (1992)
Megaphragma ghesquierei Ghesquière 1939	290	Tri.	Ghesquière (1939)
Scydosella musawasensis Hall 1999***	300	Pti.	Hall (1999)
Megaphragma macrostigmum (Lin 1992)	300	Tri.	Lin (1992)

*Mym., Mymaridae (Hymenoptera); Tri., Trichogrammatidae (Hymenoptera); Pti., Ptiliidae (Coleoptera)
***M. mymaripenne* was reportedly measured to be 172 μm (Dozier 1932), but the species actually measured was probably *M. caribea*
***According to the newest measurements, the minimum length of *Scydosella musawasensis* is 325 μm (Polilov 2015b)

References

Beutel, R. G., & Haas, A. (1998). Larval head morphology of *Hydroscapha natans* LeConte 1874 (Coleoptera, Myxophaga, Hydroscaphidae) with special reference to miniaturization. *Zoomorphology, 118*(2), 103–116.

Beutel, R. G., Pohl, H., & Hunefeld, F. (2005). Strepsipteran brain and effect of miniaturization (Insecta). *Arthropod Structure & Development, 34*(3), 301–313.

Borror, D. J., DeLong, D. M., & Triplehorn, C. A. (1981). *An Introduction to the study of insects.* Philadelphia: Saunders.

Brenzinger, B., Haszprunar, G., & Schrodl, M. (2013). At the limits of a successful body plan - 3D microanatomy, histology and evolution of Helminthope (Mollusca: Heterobranchia: Rhodopemorpha), the most worm-like gastropod. *Frontiers in Zoology, 10*(1), 37.

Chetverikov, S. S. (1920). *The Fundamental Factor of Insect Evolution*. Washington, DC: Gov. Print. Off.

Delvare, G. (1993). Guadeloupe avec la description d'une espèce nouvelle (Hymenoptera, Trichogrammatidae). *Revue Française d'Entomologie, 15*, 149–152.

Dozier, H. L. (1932). Descriptions of new trichogrammatid (Hymenoptera) egg parasites from the West Indies. *Proceedings of the Entomological Society of Washington, 34*, 29–37.

Edlinger, K. (1998). Konstruktionsprobleme der Miniaturisierung im Tierreich. Zur Unumkehrbarkeit konstruktiven Wandels. *Natur und Museum, 128*(5), 139–151.

Gad, G. (2005). Successive reduction of the last instar larva of Loricifera, as evidenced by two new species of Pliciloricus from the Great Meteor Seamount (Atlantic Ocean). *Zoologischer Anzeiger, 243*, 239–271.

Ghesquière, J. (1939). Contribution a l'étude des Hyménoptères du Congo Belge. VI. Déscription d'un Mymaride nouveau et remarques sur le Gn. *Megaphragma* Timb. (Trichogrammatidae). *Revue de Zoologie et de Botanique Africaines, 35*, 33–41.

Gorodkov, K. B. (1984). Олигомеризация и эволюция систем морфологических структур: 2. Олигомеризация и уменьшение размеров тела (Oligomerization and evolution of the morphological structure systems. 2. Oligomerization and body size decrease. *Zoologicheskii Zhurnal, 63*, 1765–1778.

Grebennikov, V. V., & Beutel, R. G. (2002). Morphology of the minute larva of *Ptinella tenella*, with special reference to effects of miniaturisation and the systematic position of Ptiliidae (Coleoptera: Staphylinoidea). *Arthropod Structure & Development, 31*(2), 157–172.

Halberg, K. A., Persson, D., Mobjerg, Wanninger N. A., & Kristensen, R. M. (2009). Myoanatomy of the marine tardigrade *Halobiotus crispae* (Eutardigrada: Hypsibiidae). *Journal of Morphology, 270*, 996–1013.

Hall, W. E. (1999). Generic revision of the tribe Nanosellini (Coleoptera: Ptiliidae: Ptiliinae). *Transactions of the American Entomological Society, 125*(1/2), 39–126.

Hanken, J., & Wake, D. B. (1993). Miniaturization of body size: organismal consequences and evolutionary significance. *Annual Reviews of Ecology and Systematics, 24*, 501–519.

Hartmann, G. (1973). Zum gegenwartigen Stand der Erforschung der Ostracoden interstitieller Systeme. *Annales de Spéléologie, 28*, 417–426.

Hochberg, R. (2007). Topology of the nervous system of *Notommata copeus* (Rotifera: Monogononta) revealed with anti-FMRFamide, -SCPb, and -serotonin (5-HT) immunohistochemistry. *Invertebrate Biology, 126*(3), 247–256.

Hochberg, R. (2009). Three-dimensional reconstruction and neural map of the serotonergic brain of *Asplanchna brightwellii* (Rotifera, Monogononta). *Journal of Morphology, 270*(4), 430–441.

Hochberg, R., & Ablak Gurbuz, O. (2008). Comparative morphology of the somatic musculature in species of *Hexarthra* and *Polyarthra* (Rotifera, Monogononta): Its function in appendage movement and escape behavior. *Zoologischer Anzeiger, 247*(3), 233–248.

Huber, J. T., & Landry, J.-F. (1999). Cutio nanissimus incredibilis. *Nouv'ailes: Bulletin de nouvelles de l'Association des entomologistes amateurs du Québec 9*(3), 11.

Huber, J. T., & Noyes, J. (2013). A new genus and species of fairyfly, *Tinkerbella nana* (Hymenoptera, Mymaridae), with comments on its sister genus *Kikiki*, and discussion on small size limits in arthropods. *Journal of Hymenoptera Research, 32*, 17–44.

Kirsteurer, E. (1972). The interstitial nemertean fauna of marine sand. *Smithsonian Contributions to Zoology, 76*, 17–19.

Kristensen, R. M. (1976). On the fine structure of *Batillipes noerrevangi* Kristensen 1976, 1. Tegument and moulting cycle. *Zoologischer Anzeiger, 197*(3/4), 129–150.

Kristensen, R. M. (1978). On the structure of *Batillipes noerrevangi* Kristensen 1978. 2. The muscle-attachments and the true cross-striated muscles. *Zoologischer Anzeiger, 200*(3/4), 173–184.

Kristensen, R. M. (1979). On the fine structure of *Batillipes noerrevangi* Kristensen, 1978 (Heterotardigrada). 3. Spermiogenesis. *zeszyty Naukowe Uniwersytetu Jagiellonskiego. Prace Zoologiczne, 25*, 97–105.

Kristensen, R. M. (2002). An introduction to Loricifera, Cycliophora and Micrognathozoa. *Integrative and Comparative Biology, 42*(3), 641–651.

Kristensen, R. M., & Funch, P. (2000). Micrognathozoa: A new class with complicated jaws like those of Rotifera and Gnathostomulida. *Journal of Morphology, 246*(1), 1–49.

Laurin, B., & Garcia-Joral, F. (1990). Miniaturization and heterochrony in *Homoeorhynchia meridionalis* and *H. cynocephala* (Brachiopoda, Rhynchonellidae) from the Jurassic of the Iberian range, Spain. *Paleobiology, 16*(1), 62–76.

Lin, N. (1992). Descriptions of five new species of *Megaphragma* and *Paramegaphragma* gen. nov. (Hymenoptera: Trichogrammatidae) from China. *Entomotaxonomia, 14*(2), 129–138.

Martini, E. (1912). Studien über die Konstanz histologischer Elemente. III. Hydatina senta. *Zeitschrift für wissenschaftliche Zoologie, 102*, 425–645.

McMahon, T., & Bonner, J. (1983). *On size and life*. New York: Scientific American Books.

Mockford, E. L. (1997). A new species of *Dicopomorpha* (Hymenoptera: Mymaridae) with diminutive, apterous males. *Annals of the Entomological Society of America, 90*(2), 115–120.

Mooi, R. (1990). Progenetic miniaturization in the sand dollar *Sinaechinocyamus*: Implications for clypeasteroid phylogeny. In C. De Ridder, P. Dubois, M.-C. Lahaye, et al. (Eds.), *Echinoderm research* (pp. 137–143). Rotterdam: Balkema.

Noodt, W. (1974). Anpassung an interstitielle Bedingungen: Ein Faktor in der Evolution höherer Taxa der Crustacea. *Faunistisch-Oekologische Mitteilungen, 4*, 445–452.

Petrunina, A., & Kolbasov, G. (2012). Morphology and ultrastructure of definitive males of *Arcticotantulus pertzovi* and *Microdajus tchesunovi* (Crustacea; Tantulocarida). *Zoologischer Anzeiger, 251*(3), 223–236.

Polilov, A. A. (2004). Анатомические пределы миниатюризации насекомых на примере мельчайших жесткокрылых (Anatomical limits of miniaturization in insects by the example of the smallest Coleoptera). *Zeiss Today, 25*, 2.

Polilov, A. A. (2005). Anatomy of the feather-winged beetles *Acrotrichis montandoni* and *Ptilium myrmecophilum* (Coleoptera, Ptiliidae). *Entomological Review, 85*(5), 467–475.

Polilov, A. A. (2015a). Small is beautiful: Features of the smallest insects and limits to miniaturization. *Annual Review of Entomology, 60*, 103–121.

Polilov, A. A. (2015b). How small is the smallest? New record and remeasuring of *Scydosella musawasensis* Hall, 1999 (Coleoptera, Ptiliidae), the smallest known free-living insect. *ZooKeys, 526*, 61–64.

Quesada, R., Triana, E., Vargas, G., Douglass, J. K., Seid, M. A., Niven, J. E., et al. (2011). The allometry of CNS size and consequences of miniaturization in orb-weaving and cleptoparasitic spiders. *Arthropod Structure & Development, 40*(6), 521–529.

Rensch, B. (1948). Histological changes correlated with evolutionary changes in body size. *Evolution, 2*, 218–230.

Rundell, R. J., & Leander, B. S. (2010). Masters of miniaturization: Convergent evolution among interstitial eukaryotes. *BioEssays, 32*(5), 430–437.

Schmidt-Nielsen, K. (1984). *Scaling: Why is animal size so important?*. Cambridge: Cambridge Univ. Press.

Schmidt-Rhaesa, A., & Kulessa, J. (2007). Muscular architecture of *Milnesium tardigradum* and *Hypsibius* sp. (Eutardigrada, Tardigrada) with some data on *Ramazottius oberhaeuseri*. *Zoomorphology, 126*, 265–281.

Schulze, C., & Schmidt-Rhaesa, A. (2011). Organisation of the musculature of *Batillipes pennaki* (Arthrotardigrada, Tardigrada). *Meiofauna Marina, 19*, 195–207.

Serban, M. (1960). La neotenie et le probleme de la taille chez les Copepodes. *Crustaceana, 1*, 77–83.

Sylvere, A. P., & Stein-Margolina, V. V. (1976). Tetrapodili — четырехногие клещи: Электронномикроскопическая анатомия, проблемы эволюции и взаимоотношения с возбудителями болезней растений (Tetrapodili, Four-legged Mites: Electron Microscopic Anatomy, Evolution Problems, and Mutual Relations with Plant Pathogenic Organisms) Valgus, Tallinn.

Snyder, J., & Bretsky, P. W. (1971). Life habits of diminutive bivalve mollusks in the Maquoketa formation (Upper Ordovician). *American Journal of Science, 271*(3), 227–251.

Strong, E. E., & Glaubrecht, M. (2008). Anatomy and systematics of the minute syrnolopsine gastropods from Lake Tanganyika (Caenogastropoda, Cerithioidea, Paludomidae). *Acta Zoologica, 89*(4), 289–310.

Subba Rao, B. R. (1969). A new species of *Megaphragma* (Hymenoptera: Trichogrammatidae) from India. *Proceedings of the Royal Society of London B 38*(7/8), 114–116.

Swedmark, B. (1964). The interstitial fauna of marine sand. *Biological Reviews, 39*, 1–42.

Thorson, G. (1965). A neotenous dwarf form of *Capulus ungaricus* (L.) (Gastropoda, Prosobranchia) commensalistie on *Turritella communis* risso. *Ophelia, 2*, 175–210.

Timberlake, P. H. (1924). Descriptions of new chalcid-flies from Hawaii and Mexico. *Proceedings of the Hawaiian Entomological Society, 5*, 395–417.

Turner, R. D., & Yakovlev, Y. (1983). Dwarf males in the Teredinidae (Bivalvia, Pholodacea). *Science, 219*(4584), 1077–1078.

Viggiani, G. (1997). Notes on the type of *Megaphragma* Timberlake (Hymenoptera: Trichogrammatidae), with description of a new species. *Bollettino del Laboratorio di Entomologia Agraria Filippo Silvestri, 53*, 117–119.

Viggiani, G., & Bernardo, U. (1997). Two species of *Megaphragma* (Hymenoptera Trichogrammatidae), egg-parasitoids of Heliothrips haemorrhoidalis Bouché (Thysanoptera) in southern Italy, with description of a new species. *Bollettino di Zoologia Agraria e di Bachicoltura, 291*, 51–55.

Ware, R. W., & Lopresti, V. (1975). Three-dimensional reconstruction from serial sections. *International Review of Cytology, 40*, 325–340.

Westheide, W. (1984). The concept of reproduction in polychaetes with small body size; adaptation in interstitial species. *Fortschritte der Zoologie, 29*, 265–287.

Westheide, W. (1987). Progenesis as a principle in meiofauna evolution. *Journal of Natural History, 21*(4), 843–854.

Wilts, E. F., Ahlrichs, W. H., & Martínez Arbizu, P. (2009). The somatic musculature of *Bryceella stylata* (Milne, 1886) (Rotifera: Proalidae) as revealed by confocal laser scanning microscopy with additional new data on its trophi and overall morphology. *Zoologischer Anzeiger, 248*(3), 161–175.

Worsaae, K., & Rouse, G. W. (2008). Is Diurodrilus an annelid? *Journal of Morphology, 269*(12), 1426–1455.

Worsaae, K., & Rouse, G. W. (2010). The simplicity of males: Dwarf males of four species of *Osedax* (Siboglinidae; Annelida) investigated by confocal laser scanning microscopy. *Journal of Morphology, 271*(2), 127–142.

Worsaae, K., Sterrer, W., Kaul-Strehlow, S., Hay-Schmidt, A., & Giribet, G. (2012). An anatomical description of a miniaturized acorn worm (Hemichordata, Enteropneusta) with asexual reproduction by paratomy. *PLoS ONE, 7*(11), e48529.

Zantke, J., Wolff, C., & Scholtz, G. (2008). Three-dimensional reconstruction of the central nervous system of Macrobiotus hufelandi (Eutardigrada, Parachela): implications for the phylogenetic position of Tardigrada. *Zoomorphology, 127*(1), 21–36.

Chapter 2
Methods of Collecting and Studying Microinsects

2.1 Introduction

Microinsects are one of the least thoroughly studied groups of insects and quite objective reasons are largely associated with their microscopic body size, difficulty in collecting the material, technical and methodological difficulties of studying samples, and others.

The current renaissance of insect morphology is due to new methods that have become available to researchers (Friedrich et al. 2014). Though many of the new approaches provide good results in studying the microinsects, such studies often require modification in methods for adjusting them to the specifics of microinsects. In the past few years, morphology of the microinsects is the the best-studied field of knowledge (Polilov 2015), but other aspects of the biology of microinsects remain almost unstudied.

2.2 Collecting Methods

Microinsects have extremely diverse modes of life and occur almost ubiquitously. Some miniature insects are free-living at all the stages of their development, e.g., beetles of the families Ptiliidae and Corylophidae, which livein all kinds of substrates such as forest litter, decaying wood, polypore, or feces of various animals in both adult and larval stages, while some are parasites of plants, e.g., many thrips. The smallest insects are called as parasitoids, where their larvae develop inside eggs of other insects. Some microinsects are nidicolous pests of stored products, e.g., psocopterans of the family Liposcelididae. Based on different modes of life found in microinsects, the methods used to collect them also strongly differ depending on the target group of insects.

A.A. Polilov, *At the Size Limit—Effects of Miniaturization in Insects*,
DOI 10.1007/978-3-319-39499-2_2

Small coleopterans are usually collected by sifting the substrate through a soil sieve with a mesh size of 0.7–2 mm, and processing the substrate in eclectors or windowpane trapping (for more details, refer Suter 1966). Miniature hymenopterans are most successfully collected by the following methods: aerial sweeping, yellow traps, pyrethrum spraying, Malaise traps, and hatching from infested hosts (for more details, refer Noyes 1982). Thrips that live on plants and other parasites of plants can be collected by sweeping and shaking off from plants (for more details, refer Stannard 1968). The other thrips and small psocopterans are collected by the methods identical to those used to collect microcoleopterans.

2.3 Fixation and Storage of Material

A lot of methods are used to fix biological samples and many of them are applicable to microinsects. The choice of the fixative depends on the intended further usage of the material and expected methods for studying the samples.

For faunistic or taxonomic purposes, microinsects are usually fixed and preserved in 70–80 % ethanol, in which they are immersed immediately once captured. The traditional method of killing and subsequent storage on cotton beds is impractical for such small insects, because they often crumple and can be strongly damaged during manipulations. For the purpose of studying the DNA of the collected samples, the material is stored in 100 % ethanol at −20 °C.

External morphology can also be studied using the fixed material as described above; this method of fixation is good for preserving most structures. Finer work, especially in larvae with soft integuments, is preferably done using histological fixatives described as follows.

Two groups of fixatives are used for studying anatomy (internal morphology), one for histological purposes and the other for immunolabeling. In spite of their small size, the integument of microinsects is poorly permeable or not permeable to fixatives; therefore, it is preferable to perforate such samples prior to fixation and increase the duration of fixation. The most widely used histological fixatives are Brodsky's fixative (FAE: 80 % ethanol: 37 % formaldehyde [PFA]: acetic acid = 2:1:0.3) and the Duboscq-Brazil fixative (Alcoholic Bouin : 95 % ethanol : saturated water solution of picric acid : 37 % formaldehyde : acetic acid = 4:3:2.2:0.8). The duration of fixation is 2–12 h, depending on the size of the specimen and intactness of its integument. After fixation, the material can be stored in 70–80 % ethanol. For immunolabeling, buffered formalin is usually used (4 % PFA, 0.1 M phosphate buffer pH 7.2–7.4), with an admixture of a detergent (usually 0.05–0.3 % Triton X-100) added to accelerate the penetration of the fixative into the specimen and increase the wettability of the integument. The material is fixed for 3 h at room temperature or overnight at 4 °C. After fixation, the samples are stored in the same buffer with 0.05 % sodium azide. Another fixative quite suitable for fixation of microinsects for immunolabeling is Dent's fixative (methanol : dimethylsulfoxide = 4:1); the material is usually fixed overnight at 4 °C. After fixation using Dent's fixative, the material can be preserved for a long time in methanol at −20 °C.

Fixation for studying ulstrastructural organization is often complicated by the slow penetration of fixatives through the integument (or even holes in it) and by the nonwettability of the integument in the majority of microinsects. The most widely used method is fixation in buffered glutaraldehyde (2.5 % GA, 0.1 M sodium cacodylate buffer pH 7.2) for 2–10 h at 4 °C and subsequent postfixation in osmium oxide solution (2 % OsO_4, 0.1 M sodium cacodylate buffer pH 7.2) for 2–10 h at 4 °C. Prior to fixation, the integument must be perforated. This method of fixation is suitable for many but not for all microinsects and purposes; if the result is poor, it is better to use simultaneous fixation in 1 % OsO_4 and 1 % GA in 0.1 M sodium cacodylate buffer, pH 7.2, for 1 h at 4 °C or fixation in a mixture of 2 % GA, 2 % PFA, and 2 % DMSO in the same buffer for 2–10 h at 4 °C and postfixation in 2 % OsO_4 in the same buffer. The material can be stored in the buffer or, after dehydration, in 100 % ethanol, at 4 °C.

2.4 Methods of Studying the Structure of Microinsects

2.4.1 Optical Microscopy

It is most convenient to study external morphology for morphological, diagnostic, or taxonomic purposes using total preparations or preparations of particular parts of bodies under transmission light microscopes. Prior to making the preparations, the samples are cleared with a solution of an alkali, hydrogen peroxide, or lactic acid. The concentration, duration, and temperature of exposure to clearing liquids are adjusted depending on the size of the sample and pigmentation of the integument. After clearing, temporary preparations in glycerol or permanent preparations in a permanent mounting medium are made. Larvae are often studied using phase-contrast or differential interference contrast (DIC).

Because of their extremely small size, microinsects are seldom dissected for studying their anatomy using histological methods, confocal microscopy, and tomography.

Because of small size and relatively strong integument of microinsects, classical histology methods based on paraffin embedding are not used in their anatomical studies. Sections are obtained from specimens (fixed and dehydrated in a series of alcohols of increasing concentrations) by embedding them in special mounting media (usually Araldite or Epon). The blocks obtained by this procedure are cut into sections of about 0.5–2 μm thick using a diamond or glass knife. Sections can be stained, without removing the mounting medium, with toluidine blue and pyronin, azure and eosin, methylene blue, or other universal histological stains.

Confocal microscopy is rarely used for studying the structure of microinsects, because its usage is hindered by the nonpermeability for large molecules and strong autofluorescence of their cuticle. The study of particular organs and body parts is complicated by their minute size. It is also impossible to obtain sections of microinsects by freezing or using soft mounting media, which are usually used in confocal microscopy, because insect integument is too hard. One exception in using the confocal microscopy is the study of skeletal structures and musculature, because of the strong autofluorescence of these structures; fixed material is usually decolorized with hydrogen peroxide, dehydrated, cleared in benzyl alcohol/benzyl benzoate (Murray's clear) or methyl salicylate, and finally autofluorescence is studied at wavelengths of 488–546 nm (for more details, refer Zucker 2006; Smolla et al. 2014).

2.4.2 Electron Microscopy

Scanning electron microscopy (SEM) is a very convenient and efficient method for studying the external morphology. For SEM studies, fixed samples have to be dehydrated, critical point-dried or hexamethyldisilazane-dried, glued to metal mounts with glue or double-sided tape, and sprayed with metal (for more details, refer Bolte 1996).

The ultrastructural organization of cells and tissues in microinsects is successfully studied using transmission electron microscopy (TEM). On the one hand, small size complicates the preparation and fixation of samples, on the other hand entire samples can fit into ultrathin sections. Samples are usually embedded by the standard method: fixed samples are dehydrated and embedded on a mounting medium (usually Epon), and the blocks obtained by this procedure are cut into ultrathin sections with an ultratome and transferred to special blends; the sections are usually additionally contrasted with uranyl acetate or lead citrate by the standard method.

Methods of 3-D studies of ultrastructural organization of cells and tissues have been actively developed over the past few years. Until recently, the only chance to study the spatial organization at the cellular and subcellular levels was to analyze series of ultrathin sections under a transmission electron microscope, but now a whole range of new methods are available; these include using focused ion beam and scanning electron microscopes (FIB-SEM), serial block-face SEM, nanotomography, and others. The key problem in the study of 3-D ultrastructural organization of biological systems is the preparation of samples, which provides quality for fixation, contrasting, and preservation of the sample up to its deepest layers. Protocols are currently developed that allow studying an entire mouse brain (Mikula and Denk 2015) or fruitfly brain (Hayworth et al. 2015) at the subcellular level, but these protocols either cannot be used for an entire organism or for an entire head detached from the body, because they cannot provide quality in a sample that consists of heterogeneous tissues. Three-dimensional electron microscopy is also very promising for studying structure and ultrastructure of the microinsects, because in the near future it will allow to obtain comprehensive data in their 3-D organization at the cellular and subcellular levels.

2.4.3 Tomography

Tomography (μ-ct) is one of the most efficient and widely used modern methods for studying the anatomy of insects, because it allows obtaining comprehensive 3-D data on the external and internal morphology of samples without destroying them (Friedrich et al. 2014). Samples for tomography are critical point-dried or immersed in a liquid (usually ethanol); if required, samples can be additionally contrasted with heavy metals (e.g., osmium) or iodine (Metscher 2009). Though tomography yields good results for microinsects, modern desktop tomographs have insufficient resolution (effective resolution about 1 μm per pixel), while synchrotron-based tomographs are very expensive to use and still only approaching the resolution of ordinary light microscopes.

2.4.4 3-D Computer Modeling

3-D modeling has long become an integral part of morphological studies; it is also widely used in studies of microinsects. A wide range of 3-D modeling software is currently available including both free pakages such as Reconstruct or IMOD, and large commercial packages such as Bitplate Imaris or Visage Amira, which provide everything that is needed for imaging, segmentation, and analysis of large sets of morphological data.

References

Bolte, K. B. (1996). Techniques for obtaining scanning electron micrographs of minute arthropods. *Proceedings of the Entomological Society of Ontario, 127*, 67–87.

Friedrich, F., Matsumura, Y., Pohl, H., Bai, M., Hörnschemeyer, T., & Beutel, R. G. (2014). Insect morphology in the age of phylogenomics: innovative techniques and its future role in systematics. *Entomological Science, 17*(1), 1–24.

Hayworth, K. J., Xu, C. S., Lu, Z., Knott, G. W., Fetter, R. D., Tapia, J. C., et al. (2015). Ultrastructurally smooth thick partitioning and volume stitching for large-scale connectomics. *Nature Methods, 12*(4), 319–322.

Metscher, B. D. (2009). MicroCT for comparative morphology: simple staining methods allow high-contrast 3D imaging of diverse non-mineralized animal tissues. *BMC Physiology, 9*(1), 1–14.

Mikula, S., & Denk, W. (2015). High-resolution whole-brain staining for electron microscopic circuit reconstruction. *Nature Methods, 12*(6), 541–546.

Noyes, J. S. (1982). Collecting and preserving chalcid wasps (Hymenoptera: Chalcidoidea). *Journal of Natural History, 16*(3), 315–334.

Polilov, A. A. (2015). Small is beautiful: Features of the smallest insects and limits to miniaturization. *Annual Review of Entomology, 60*, 103–121.

Smolla, M., Ruchty, M., Nagel, M., & Kleineidam, C. J. (2014). Clearing pigmented insect cuticle to investigate small insects' organs in situ using confocal laser-scanning microscopy (CLSM). *Arthropod Structure & Development, 43*(2), 175–181.

Stannard, L. J. (1968). The thrips, or Thysanoptera of Illinois. *Illinois National History Survey Bulletin, 29*(4), 215–552.

Suter, W. R. (1966). Techniques for the Collection of Microcoleoptera of the Families Pselaphidae, Ptiliidae, and Scydmaenidae. *Coleopterists Bulletin, 20*(2), 33–38.

Zucker, R. M. (2006). Whole insect and mammalian embryo imaging with confocal microscopy: Morphology and apoptosis. *Cytometry Part A, 69A*(11), 1143–1152.

Chapter 3
Structure of the Principal Groups of Microinsects. I. Booklice (Psocoptera: Liposcelididae)

3.1 Introduction

The smallest psocopterans and some of the smallest hemimetabolous insects belong to the family of Liposcelididae. They are ubiquitous; many of them are synanthropic or nidicolous. The morphology of Liposcelididae, as well as that of many other families of Psocoptera, is known only fragmentarily. Some aspects of their external morphology are touched upon in studies on psocopterans in general (Badonnel 1951; Yoshizawa 2005) and on the morphology of antennae (Jentsch 1940; Seeger 1975; Hu et al. 2009) and of receptors of the musculature of the mandibles (Paulus and Honomichl 1989). As for internal morphology, the only known aspects are topography of the principal systems of organs (Noland 1924), structure of the foregut (Chapman 2003), and structure of the reproductive system (Goss 1954; Wong and Thornton 1968; Vyashnikova 1970; Biining and Sohst 1990; Emeljanov et al. 2001; Golub 2003; Yoshizawa and Johnson 2006). Only fragmentary information on the nymphal morphology of Liposcelididae is available in studies on pests of stored products (Kucerova et al. 2009). The anatomy of adults and first instar nymphs of *Liposcelis bostrychophila* was the subject of a recent study (Polilov 2016). Large representatives of the order have been studied much better (Yoshizawa 2005). The external and internal structures of adult psocopterans have been described in detail for Stenopsocidae (Badonnel 1934; Finlayson 1949).

3.2 Brief Description of the Structure of Liposcelididae

Small wingless insects from 0.8 to 2 mm long (on average about 1 mm). Their body is oval or elongate oval, flattened dorsoventrally. Coloration of integument in different representatives of this family varies from white to brown often with thoracic segments darker than other parts of body. Whole body covered with sparse

A.A. Polilov, *At the Size Limit—Effects of Miniaturization in Insects*,
DOI 10.1007/978-3-319-39499-2_3

short hairs and also with isolated long hairs. Integument has no pronounced sculpture, but has fine punctate microsculpture. Head with large prognathous, mobile. Cranium solid, almost without sutures, except for frontoclypeal suture and gular sutures. Eyes lateral, consisting of 4–6 ommatidia each. Ocelli absent. Antennae filiform, 15-segmented; mouthparts consisting of well-developed mandibles, maxillae, labrum, and labium. Thorax consists of detached prothorax and fused meso- and metathorax. Wings are absent or strongly reduced; legs ambulatorial, with normal division into coxa, trochanter (metatrochanter fused with femur), femur, tibia, and tarsus. Tarsi is 3-segmented with two simple claws. Abdomen consists of nine segments. Number of visible sternites 5 because sternite 1 reduced and several distal ones modified and sunk into abdomen. Number of visible tergites 7–8. External male genitalia are represented by penis and parameres. Females with short ovipositor consists three pairs of valves. External sexual dimorphism weakly pronounced.

3.3 Morphology of Adult *Liposcelis bostrychophila*

3.3.1 External Morphology

Body elongate oval, flattened dorsoventrally (Fig. 3.1). Size of body 890–960 μm (M = 928, n = 10).

Structure of head

Cranium almost without sutures, except for frontoclypeal suture and gular sutures (Fig. 3.2). Clypeus (postclypeus) well developed, markedly convex. Anteclypeus weakly developed and appears as a narrow fold. Head connected with prothorax via articular membrane and cervical sclerites. Tentorium of *Liposcelis* morphologically identical to that of the most psocopterans and consists of corporotentorium and anterior, dorsal, and posterior arms. Anterior arms well-developed and oval in section; dorsal arms short and slender; posterior arms over much of their length fused with posterior margin cranium.

Eyes consist of 4–6 ommatidia each (Mazokhin-Porshnyakov and Kazyakina 1981: prosommatidia). **Ocelli** absent.

Antennae filiform, 15-segmented, attached laterally in front of eyes. Scape and pedicel markedly dilated, segments of flagellum with ring-shaped microsculpture. Antennae with four types of sensilla: microtrichia, trichoid sensilla, basiconical sensilla, and shortened basiconical sensilla (Hu et al. 2009: Bohm bristles).

Mouthparts typically morphological to those of other psocopterans (Fig. 3.3) **Labrum** movably joined with cranium and appears as convex plate bearing about 40 short setae. Internal surface of labrum and upper surface of oral cavity with complex epipharynx bear several groups of cuticular teeth. **Mandibles** well developed, symmetrical or almost symmetrical, short with wide base. Anterior and

Fig. 3.1 External morphology of *Liposcelis bostrychophila*, SEM: **a–c** adult; **d–f** first instar nymph; **a**, **d** dorsal view; **b**, **e** lateral view; **c**, **f** ventral view

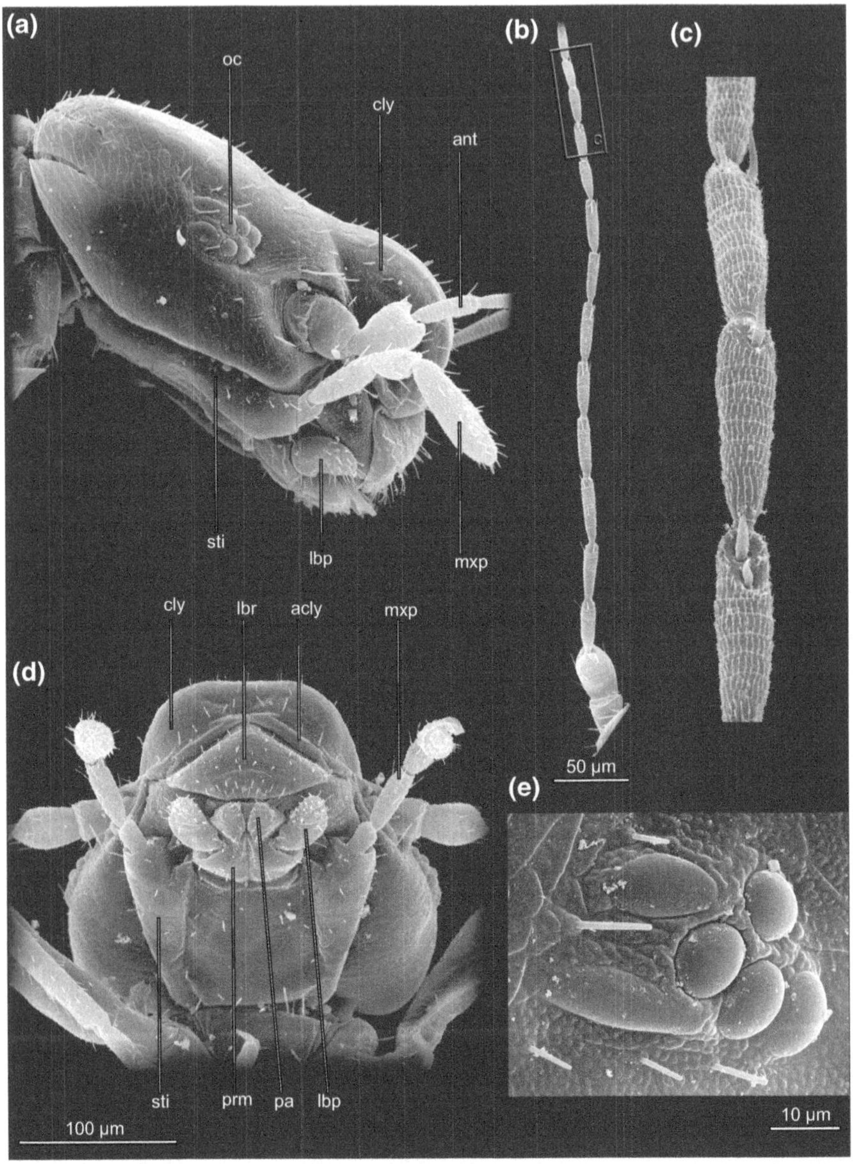

Fig. 3.2 External morphology of head adult *Liposcelis bostrychophila*, SEM: **a** head, lateral view; **b**, **c** antenna; **d** head ventral view; **e** eye; *acly* anteclypeus, *ant* antenna, *cly* clypeus, *lbp* labial palp, *lbr* labrum, *mxp* maxillary palp, *oc* eye, *pa* paraglossae, *prm* prementum, *sti* stipes

dorsal margin bears several large teeth. Molar process bears several rows of blunt teeth. Medial surface of mandibles with membranous prostheca. **Maxillae** well-developed. Cardo and palpiger fused with stipes. Maxillary palp, 4-segmented,

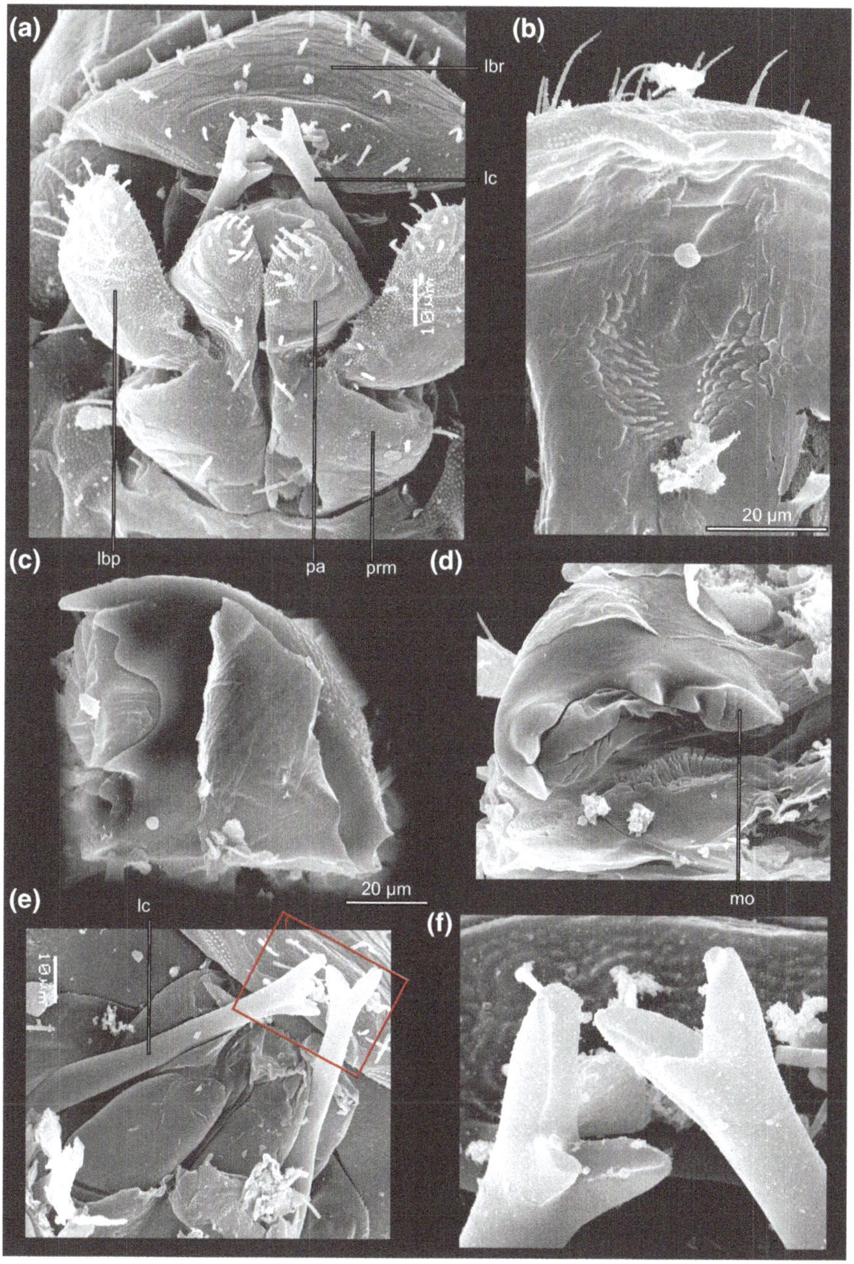

Fig. 3.3 Mouthparts of adult *Liposcelis bostrychophila*, SEM: **a** mouthparts, frontal view; **b** epipharynx; **c** mandible, ventral view; **d** mandible, medial view; **e**, **f** lacinia; *lbp* labial palp, *lbr* labrum, *lc* lacinia, *mo* mola, *pa* paraglossae, *prm* prementum, *sti* stipes

covered with microtrichia and bears trichoid and basiconical sensilla. Galea is short and weakly sclerotized. Lacinia of peculiar structure typical of most psocopterans is represented by long rod-shaped cuticular structure (sometimes termed as maxillary fork), with base sunk deep into cranium and apex protruding beyond mouthparts and bearing several teeth (three in *Liposcelis*). **Labium** strongly reduced, consisting of prementum dissected into two halves by longitudinal fold, 1-segmented palps, and paraglossa. Internal surface of labium bears intricately structured hypopharynx. Glossal sclerites connected with pharyngeal sclerites by bifurcating cord.

Structure of thorax

Thorax of Liposcelididae consists of detached prothorax and fused meso- and metathorax (Fig. 3.4).

Prothorax is represented by narrow convex pronotum. Pleurites rather small, without visible division into episternum and epimeron, on internal surface with well-developed pleural apodeme (pleural apophysis, pleural arm, episternal endophragme). Basisternum rather small, square. Profurca (furca of prothorax) appears as paired tube directed anteriad. Prodorsum (dorsum of prothorax) well developed.

Mesothorax and **metathorax** fused. Notal part represented by common shield of fused meso- and metanotum without additional differentiation. Pleurite of mesothorax divided into episternum and epimeron, pre-episternum not separated from episternum. Pleurite of metathorax without visible division into episternum, pre-episternum, and epimeron, but internally episternum and epimeron divided by straight pleural ridge (crista pleuralis). Basisterna of meso- and metathorax fused into one sclerite. Meso- and metafurca simple, slightly curved. Mesothorax with spiracles of tracheal system.

Most representatives of this family completely lack **wings** or wing primordia, some species have wing primordia with reduced venation. **Legs** ambulatorial, consisting of coxa, trochanter, femur, tibia, tarsus, and pretarsus. Metatrochanter fused with femur. Metafemur strongly dilated. Tarsi 3-segmented. Apical tarsomere bears two claws with indented internal margins.

Structure of abdomen

Number of visible sternites 6, because sternite 1 reduced, and several distal ones modified and sunk into abdomen. Number of visible tergites 7–8 (Fig. 3.5). Pleurites not distinguished. Abdominal apex bears anal valves: epiproct and two paraprocts. Spiracles usually present on tergites 1–6. Stigmata simple, ring-shaped, without filter apparatus. Females with short **ovipositor**, formed by three pairs of valves: ventral valves (stiletto shaped), dorsal valves, and external valves. **Male genitalia** of Liposcelididae consist of penis and parameres. Hypandrium simple and flap-shaped.

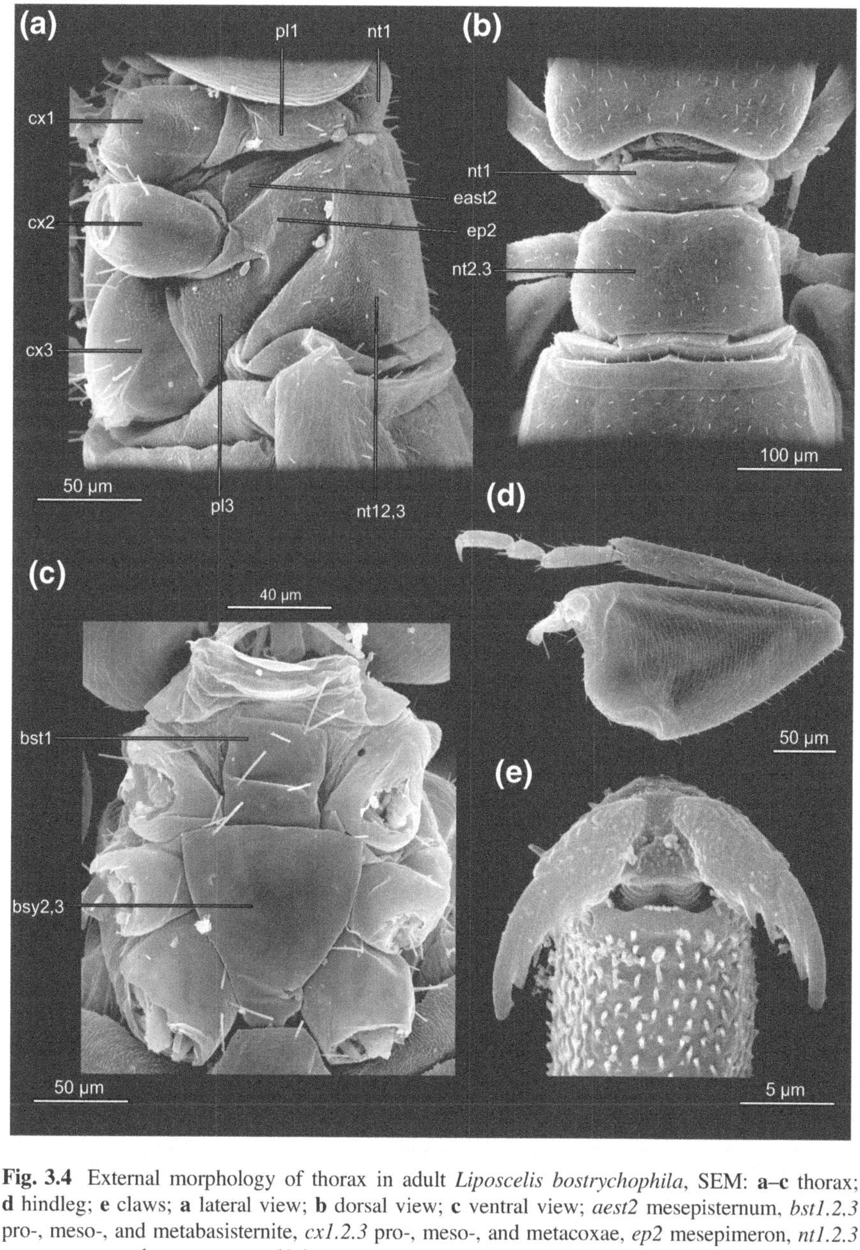

Fig. 3.4 External morphology of thorax in adult *Liposcelis bostrychophila*, SEM: **a–c** thorax; **d** hindleg; **e** claws; **a** lateral view; **b** dorsal view; **c** ventral view; *aest2* mesepisternum, *bst1.2.3* pro-, meso-, and metabasisternite, *cx1.2.3* pro-, meso-, and metacoxae, *ep2* mesepimeron, *nt1.2.3* pro-, meso-, and metanotum, *pl1.3* pro- and metapleurite

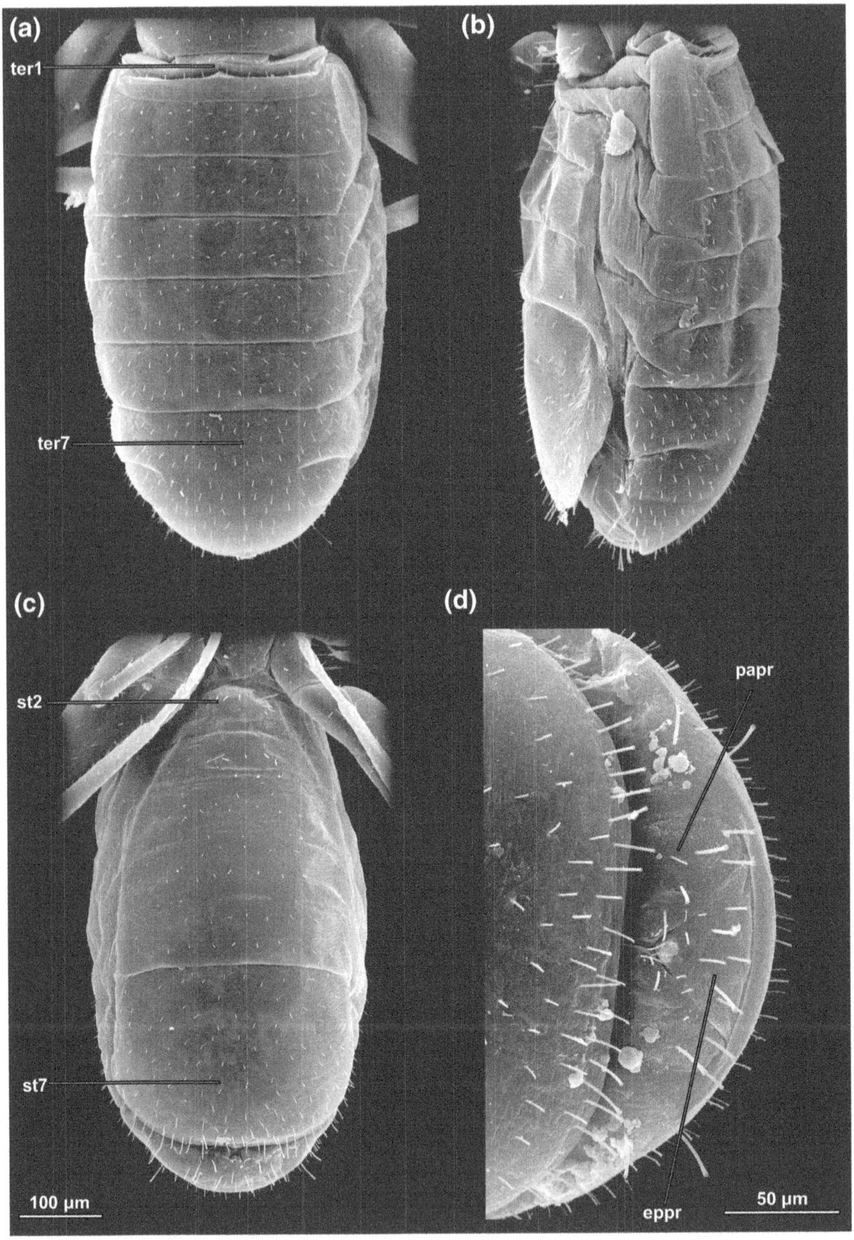

Fig. 3.5 External morphology of abdomen in adult *Liposcelis bostrychophila*, SEM: **a** dorsal view, **b** lateral view, **c** ventral view, **d** apex of abdomen, posterior view; *eppr* epiproct, *papr* paraproct, *st* sternite, *ter* tergite

3.3.2 Internal Morphology

General plan of internal structure: intestine medial, central nervous system markedly oligomerized and compact, considerable part of thorax occupied by musculature, reproductive system occupies considerable part of abdomen (Fig. 3.6).

Particular organ systems are described in more detail in special sections below.

Integument

Integument in Liposcelidae consists of cuticle, hypoderm, and basal membrane (Fig. 3.7a). Cuticle thickness in *Liposcelis* is 1.3–10.1 μm (average M = 2.5; n = 80). Cuticle consists of epicuticle, exocuticle, and endocuticle. Hypoderm is represented by strongly flattened cells up to 3 μm thick.

Digestive and excretory systems

Structure of digestive system in *Liposcelis* weakly different from those of other psocopterans. Digestive canal divided into fore-, mid-, and hindgut. Fore- and hindgut with thin cuticular lining is shown in Fig. 3.8a, b. Midgut without cuticular lining. Intestinal canal somewhat longer than body, forming short loop in posterior one-third of abdomen. Labial glands well-developed, localized in metathorax and anterior one-third of abdomen, and connected with salivarium via narrow ducts.

Foregut divided into pharynx and oesophagus. Oesophagus straight, running through entire thorax, in posterior one-third slightly dilated, with layer of circular musculature.

Midgut long and wide, occupying considerable part of volume of abdomen, in anterior part forming short processes directed anteriad. Walls of midgut formed by strongly flattened cells (9–13 in cross-section). Muscles of midgut not found. Peritrophic membrane not found.

Hindgut divided into small intestine and rectum, small intestine forming short loop. Rectum with several rectal ampullae.

Boundary between mid- and hindgut bearing four Malpighian tubules, shaped as long slender tubes that form several loops in posterior half of abdomen.

Circulatory system and fat body

Circulatory system represented by heart and long slender aorta, other vessels not found. Two Malpighian tubules attached to heart via slender cords on apices of Malpighian tubules. The role of this structure is unknown, and it has been described only in other psocopterans (Finlayson 1949) and some orthopterans (Wigglesworth 1953). Fat body occupying almost all cavities between organs in abdomen and to a smaller degree in thorax (Fig. 3.7b, d).

Tracheal system

Tracheal system strongly simplified, with one pair of thoracic and six pairs of abdominal spiracles. Two longitudinal stems connect spiracles and reach into head,

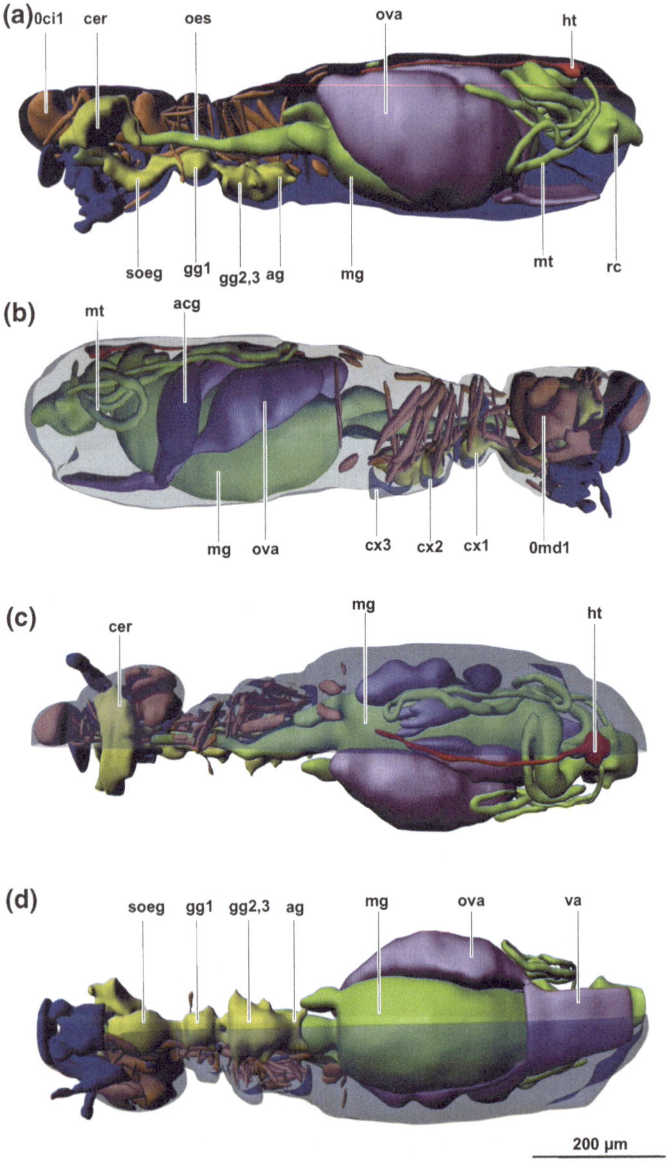

Fig. 3.6 Internal morphology of adult *Liposcelis bostrychophila*, 3D: **a** lateral internal view; **b** lateral external view; **c** dorsal view; **d** ventral view; *ag* abdominal ganglion, *cer* brain, *cx1.2.3* pro-, meso-, and metacoxae, *gg1.2.3* pro-, meso-, and metathoracic ganglia, *ht* heart, *mg* midgut, *mt* Malpighian tubules, *oes* oesophagus, *ova* ovary, *rc* rectum, *soeg* suboesophageal ganglion. Colors: *blue* cuticle, *green* digestive system, *yellow* central nervous system, *brown* musculature, *purple* reproductive system. Musculature see text

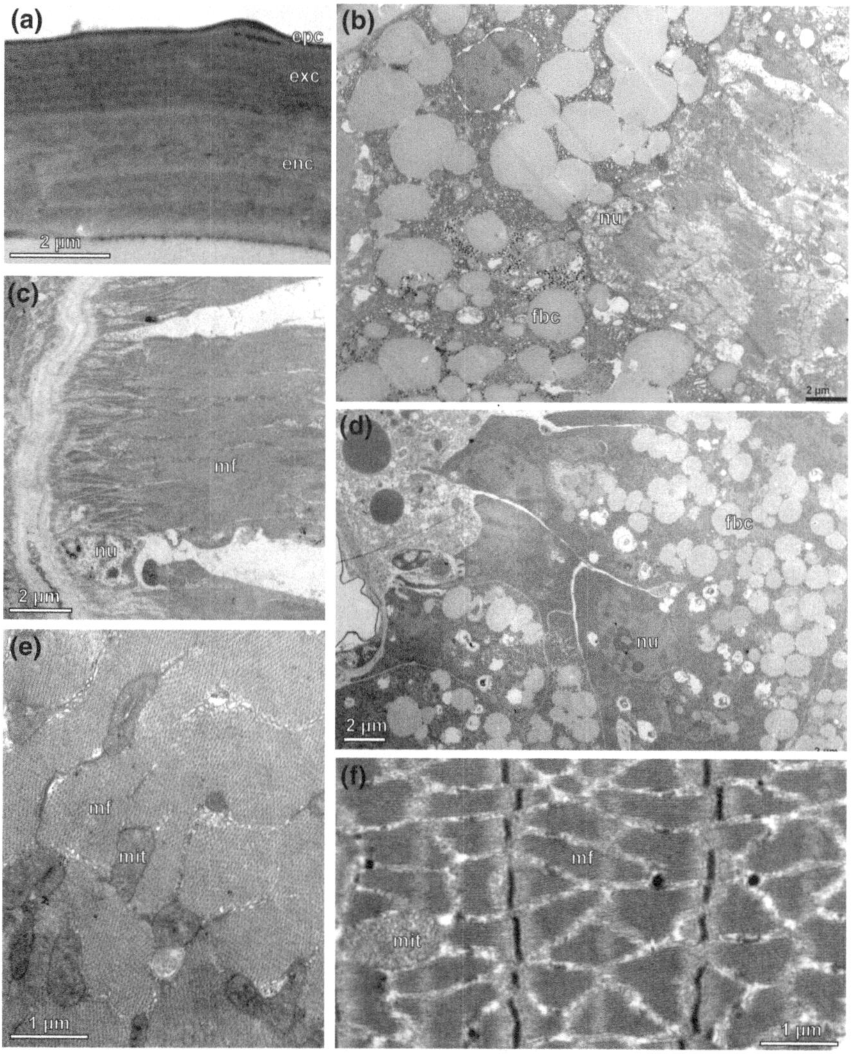

Fig. 3.7 Ultrastructure of adult *Liposcelis bostrychophila*, TEM: **a** integument, head; **b**, **d** fat body; **c** attachment of musculature to cuticle; **e** musculature, cross-section, mesothorax; **f** musculature, diagonal section, head; *enc* endocuticle, *epc* epicuticle, *exc* exocuticle, *fbc* cell of fat body, *mf* muscle fiber, *mit* mitochondrion, *nu* nucleus

forming branched system of tracheae. Numerous tracheae also diverge from stems in abdomen and thorax. Tracheae have a structure of walls typical of insects.

Nervous system

Central nervous system in Psocoptera is displaying minor oligomerization and ganglion condensation (Fig. 3.9a, b). Cerebrum and suboesophageal ganglion

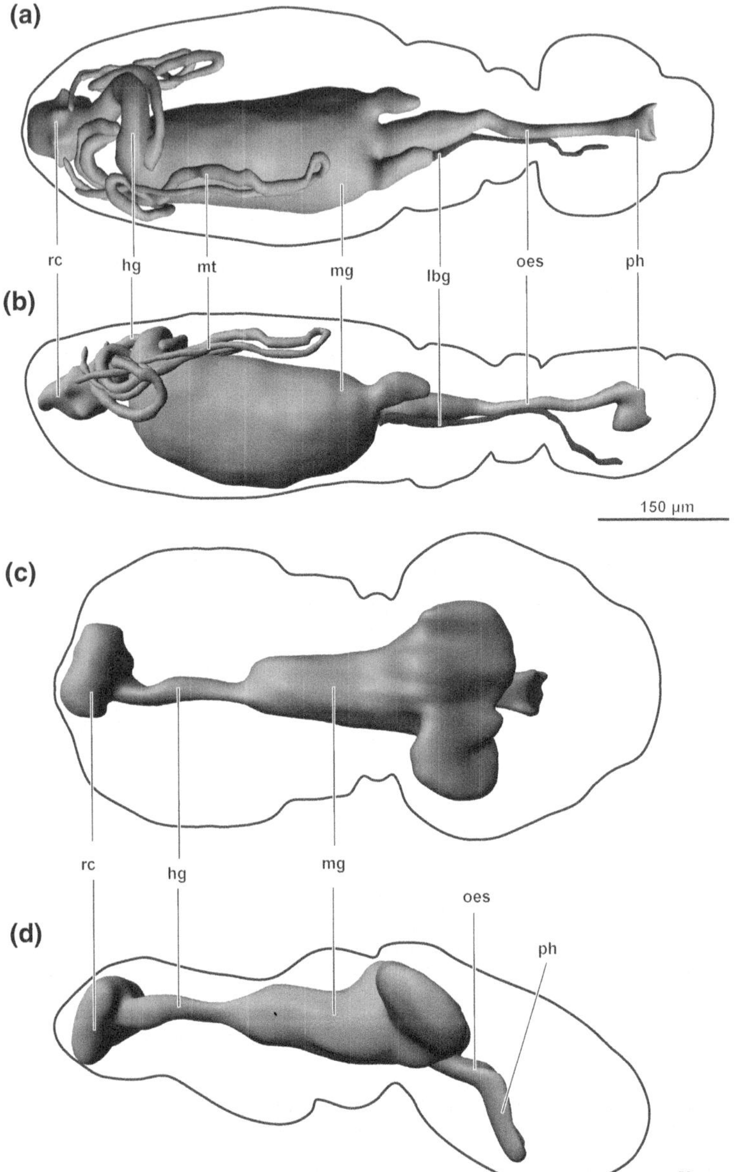

Fig. 3.8 Structure of digestive system in *Liposcelis bostrychophila*, 3D: **a**, **b** adult; **c**, **d** first instar nymph; **a**, **c** dorsal view; **b**, **d** lateral view; *hg* hindgut, *lbg* labial gland, *mg* midgut, *mt* Malpighian tubules, *oes* oesophagus, *ph* pharynx, *rc* rectum

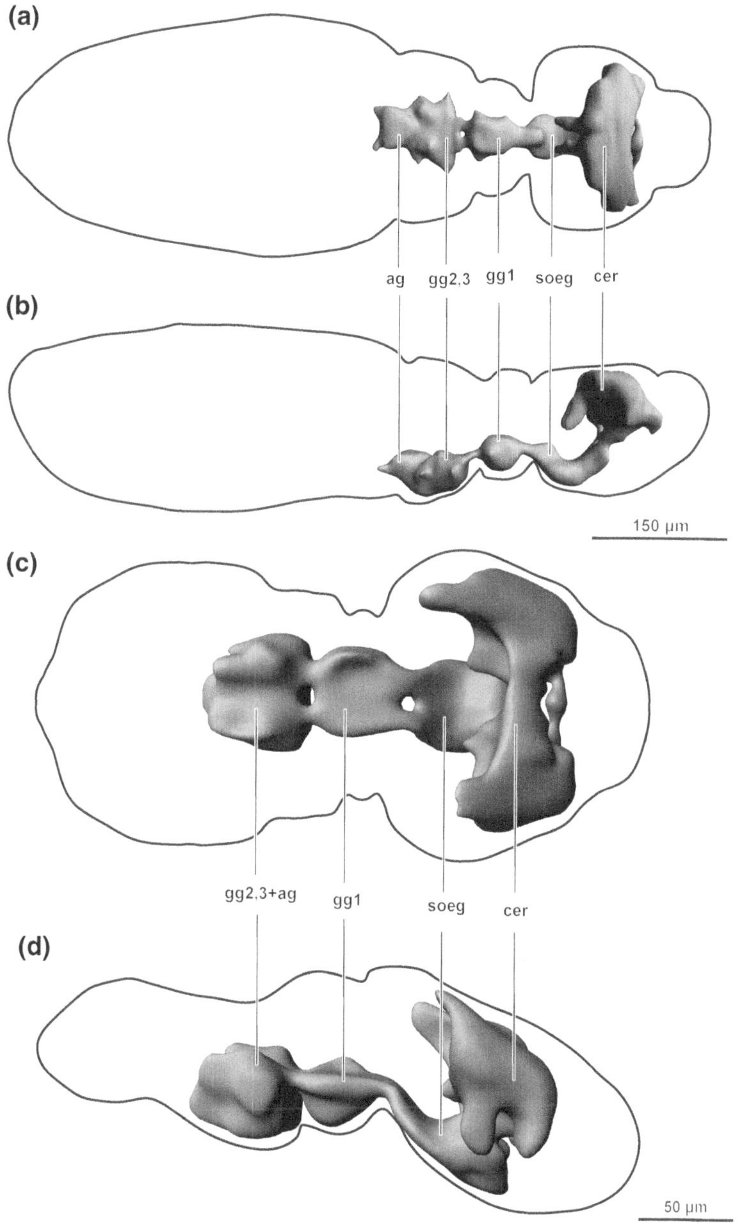

Fig. 3.9 Central nervous system in *Liposcelis bostrychophila*, 3D: **a**, **b** adult; **c**, **d** first instar nymph; **a**, **c** dorsal view; **b**, **d** lateral view; *ag* abdominal ganglion, *cer* brain, *gg1.2.3* pro-, meso-, and metathoracic ganglia, *soeg* suboesophageal ganglion

separated and situated in cranium. Prothoracic ganglion distinguished. Mesothoracic and metathoracic ganglia fused with abdominal synganglion.

Ensheathment of standard structure, typical of ganglion sheaths of insects. Represented by extraneural sheath and double ganglionic sheath. Extraneural sheath consists of thick layer of cells and strongly pronounced over entire perimeter of ganglia; their thickness is 0.6–2.5 μm. Content of fat droplets is high. Ganglionic sheath consists of neural lamella and perineurium. Thickness of neural lamella 0.08–0.12 μm. Thickness of perineurium 0.1–0.3 μm. Central nervous system with distinguished cortical glia. Neuropil glia, enveloping neurons and forming perineural layer of ganglionic sheath, strongly pronounced. All main organelles discernible in cytoplasm of cells bodies of neurons at cellular level (Fig. 3.10): mitochondria, granular, and agranular endoplasmic reticulum, Golgi complex, lysosomes. Neuropil with clearly discernible axons of neurons, their axolemmas, mitochondria, and components of cytoskeleton (microtubules, neurofibrils). Nuclei of cells in nervous system rounded. Much of nucleus occupied by evenly distributed euchromatin; heterochromatin forming isolated point areas in center of nucleus. Polarized synaptic contacts recorded. Vast majority of synaptic contacts represented by chemical synapses of asymmetrical type. Minimum diameter of projection about 0.072 μm.

Cell body diameter in nervous system of adult *Liposcelis bostrychophila* 2.59–4.29 μm (M = 2.96; n = 374). Cerebrum of adult *Liposcelis bostrychophila* containing only about 9000 cells.

Most of cerebrum, as in other insects, occupied by protocerebrum. Central body complex consisting of central body (differentiated into dorsal region of central body, ventral region of central body, also known as ellipsoid body, and two pairs of noduli) and protocerebral bridge (Fig. 3.11a–c).

Mushroom bodies situated on both sides of central body complex. Penduncle of mushroom bodies twisted, without lobes. Calyx of mushroom bodies rather small, microglobular, consisting of homogeneous neuropil core surrounded by ring of glomerular structures.

As a result of ecological specialization and consequent simplification of organs of vision (Wundrig 1936) and replacement of complex eyes with prosommatidia (Mazokhin-Porshnyakov and Kazyakina 1981), optic neuropils of *Liposcelis* strongly reduced. One optical neuropil completely lost and two others weakly developed. By analogy with holometabolous larvae with weakly developed stemmata, medulla probably also lost. Remaining optic plates shaped as small spherical neuropils.

The lateral protocerebrum zone is situated between the mushroom bodies and optic lobes. In contrast to above-listed regions, the lateral protocerebrum has no distinct structure or boundaries, and therefore it will not be included in comparative

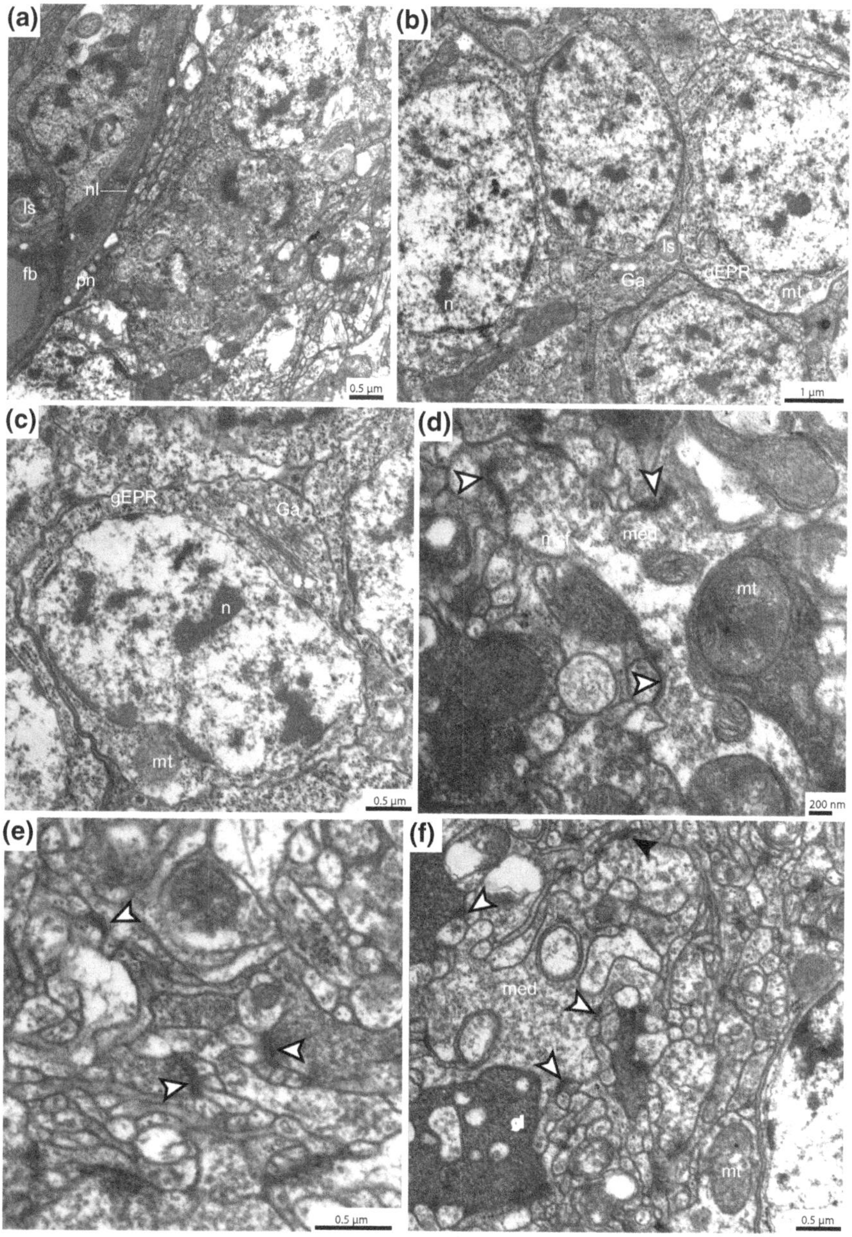

Fig. 3.10 Ultrastructure of brain in adult *Liposcelis bostrychophila*, TEM: *fb* extraneural layer of ensheathment, *Ga* Golgi apparatus, *gEPR* granular endoplasmic reticulum, *gl* cortical glia, *ls* lisosomes, *med* granules of mediator, *mt* mitochondrion, *n* nucleus, *nl* neural lamella, *pn* perineurium, *syn* synaptic contact. **d**, **e** *arrows* indicate synaptic contacts

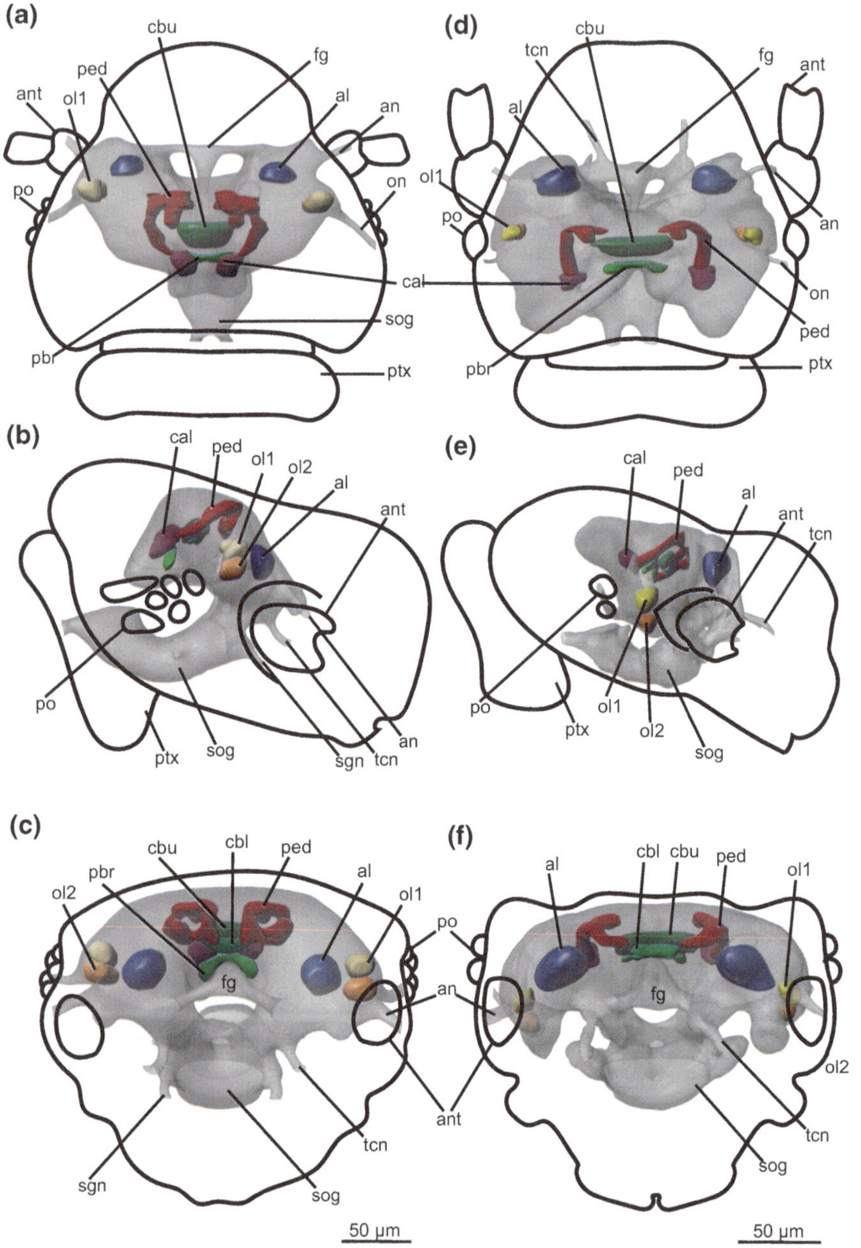

Fig. 3.11 Head ganglia in *Liposcelis bostrychophila,* 3D. **a–c** adult, **d–f** late first instar nymph. *al* antennal lobe; *an* antennal nerve; *ant* antenna; *cbu* upper region of central body; *cbl* lower region of central body; *cmb* calyx of mushroom body; *fg* frontal ganglion; *ol1. 2* optic ganglia; *on* optic nerve; *pcb* protocerebral bridge; *pmb* peduncle of mushroom body; *po* prosommatidium; *ptx* prothorax; *sgn* nerves of suboesophageal ganglion innervating mouthparts; *sog* suboesophageal ganglion; *tcn* nerves of tritocerebrum innervating labrum

analysis as a separate zone of the cerebrum. As a result of strong condensation and convergence of all structures of the cerebrum, it is difficult to discern any other, smaller structures, such as commissures, tracts, chiasms, etc.

The rather small antennal lobes of the deutocerebrum are situated in the ventral proximal part of the cerebrum, near pharynx. Because of the small size of the organism, it is difficult to exclude the presence of glomerular structure definitely, but compared to other studied organisms, the antennal lobes of *Liposcelis bostrychophila* appear rather aglomerular.

The tritocerebrum is well pronounced, slightly shifted anteriad relative to other regions of the cerebrum. The tritocerebral commissure is well developed.

Frontal ganglion of somatogastric nervous system clearly pronounced. Unpaired nerve disernible up to single formation probably resulting from fusion of occipital ganglion and postcerebral endocrinic complex.

Muscular system

Musculature of head (Fig. 3.12, Table 3.1). 0an1 (M. tentorioscapalis anterior): O, anterior arms of tentorium; I, anterior margin of base of scape. 0an2 (M. tentorioscapalis posterior): O, anterior arms of tentorium; I, posterior margin of base of scape. 0an3 (M. tentorioscapalis lateralis): O, anterior arms of tentorium; I, lateral margin of base of scape. 0lb1 (M. frontolabralis): O, medial part of frons; I, lateral margin of labrum. 0lb2 (M. frontoepipharyngalis): O, medial part of frons; I, central part of base of labrum. 0lb5 (M. labroepipharyngealis): O, wall of labrum; I, epipharynx. 0md1 (M. craniomandibularis internus), largest muscle of head: O, posterior and lateral part of cranium; I, medial part of base of mandible. 0md3 (M. craniomandibularis externus): O, posterolateral part of cranium; I, lateral part of base of mandible. 0md4 (M. hypopharyngomandibularis): O, lateral part of cranium; I, lateral margin of mandible. 0md6 (M. tentoriomandibularis lateralis): O, anterior arms of tentorium; I, lateral margin of mandible. 0md8 (M. tentoriomandibularis medialis inferior): O, anterior arms of tentorium; I, medial part of base of mandible. 0mx1 (M. craniocardinalis): O, ventral surface of cranium; I, base of cardo. 0mx2 (M. craniolacinialis): O, posterior part of gena; I, base of lacinia. 0mx4 (M. tentoriostipitalis anterior): O, tentorium; I, base of stipes. 0mx6 (M. stipitolacinialis): O, stipes; I, base of lacinia. 0la5 (M. tentoriopraementalis): O, base of tentorium and cranium next to it; I, base of prementum. 0la8 (M. submentopraementalis): O, gular area; I, base of prementum. 0hy1 (M. frontooralis): O, frons; I, lateral sclerites of hypopharynx. 0hy2 (M. tentoriooralis): O, sites of attachment of anterior arms to cranium; I, lateral sclerites of hypopharynx. 0hy3 (M. craniohypopharyngealis): O, tentorium; I, lateral wall of hypopharynx. 0hy7 (M. praementosalivaris anterior): O, lateral part of prementum; I, lateral wall of salivarium. 0hy8 (M. praementosalivaris posterior): O, base of prementum; I, lateral wall of salivarium. 0hy9 (M. oralis transversalis): transverse musculature between lateral sclerites of hypopharynx on dorsal surface. 0ci1 (M. clypeopalatalis), one of largest muscles of head: O, clypeus; I, dorsal wall of cibarium. 0bu1 (M. clypeobuccalis): O, lateral part of clypeus; I, upper wall of anatomical mouth. 0bu2

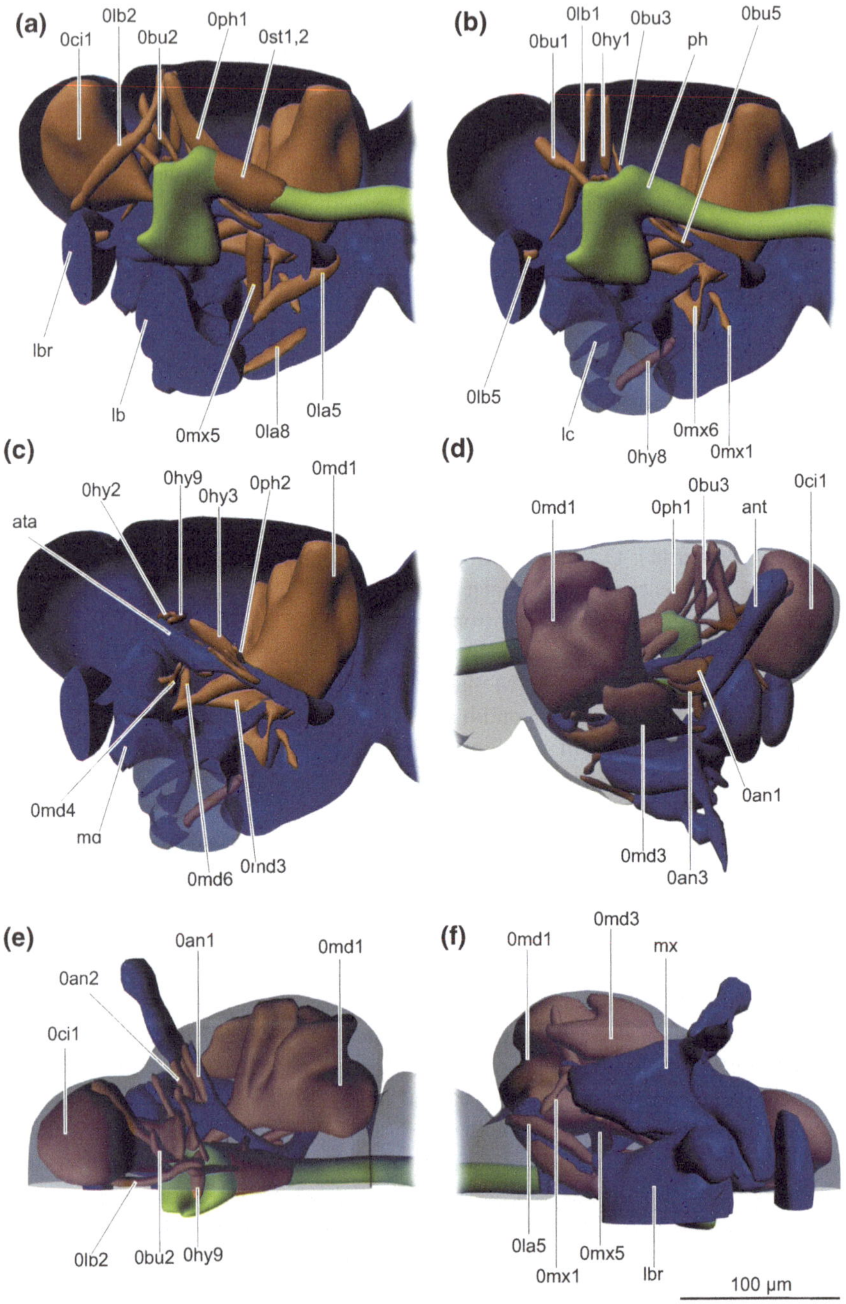

Fig. 3.12 Musculature of head in adult *Liposcelis bostrychophila*, 3D: **a–c** lateral internal view; **d** lateral external viewexternal view; **d** dorsal view; **f** ventral view; *ant* antenna, *ata* anterior arms of tentorium, *lb* labium, *lbr* labrum, *lc* lacinia, *md* mandible, *mx* maxilla, *ph* pharynx. Musculature see text

Table 3.1 Musculature of Psocoptera (partly from published sources: Badonnel 1934; Maki 1938; Matsuda 1970)

	Head muscles														
	0an1	0an2	0an3	0lb1	0lb2	0lb5	0md1	0md3	0md4	0md6	0md8	0mx1	0mx2	0mx4	0mx6
Liposcelis adult	+	+	+	+	+	+	+	+	+	+	+	+	+	+	+
Liposcelis nymph 1 ins.	+	+	+	+	+	+	+	+	+	+	+	–	+	+	+
Stenopsocus adult	+	+	+	+	+	+	+	+	+	+	+	+	+	+	+
	0ph1	0ph2	0st1	0st2											
Liposcelis adult	+	+	+	+											
Liposcelis nymph 1 ins.	+	+	+	+											
Stenopsocus adult	+	+	+	+											

	Head muscles														
	0la5	0la8	0hy1	0hy2	0hy3	Ohy5	0hy7	0hy8	0hy9	0ci1	0bu1	0bu2	0bu3	0bu5	0bu6
Liposcelis adult	+	+	+	+	+	–	+	+	+	+	+	+	+	+	+
Liposcelis nymph 1 ins.	+	+	+	+	+	–	–	–	+	+	+	+	+	a	a
Stenopsocus adult	+	+	+	+	+	+	+	+	+	+	+	+	+	+	+

	Thoracic muscles											
	Idlm1	Idlm2	Idlm5	Idvm1	Idvm2	Ivlm3	Idvm4	Idvm5	Idvm6	Idvm10	Idvm15	Idvm16
Liposcelis adult	+	+	+	+	+	+	+	–	+	+	+	+
Liposcelis nymph 1 ins.	+	+	+	+	+	+	+	–	+	+	+	+
Stenopsocus adult	+	+	+	+	+	+	+	–	+	+	+	+
Psocus adult	–	+	+	–	+	+	b	+	b	+	+	+

	Thoracic muscles														
	Idvm18	Idvm19	Itpm1	Itpm3	Ispm2	Ipcm2	Ipcm4	Ipcm5	Ipcm8	Ivlm4	Iscm1	Iscm4	Iscm5	Iscm6	Iscm7
Liposcelis adult	+	+	+	+	+	+	+	+	+	+	+	+	–	+	+
Liposcelis nymph 1 ins.	+	+	+	–	+	+	+	+	+	+	+	+	–	+	+

(continued)

Table 3.1 (continued)

	Thoracic muscles														
	Idvm18	Idvm19	Itpm1	Itpm3	Ispm2	Ipcm2	Ipcm4	Ipcm5	Ipcm8	Ivlm4	Iscm1	Iscm4	Iscm5	Iscm6	Iscm7
Stenopsocus adult	+	+	+	+	+	+	+	+	+	+	+	+	+	+	+
Psocus adult	+	+	+	+	–	+	+	+	+	+	+?	+?	+?	+	+?

	IIdlm1	IIdlm2	IIdvm1	IIdvm2	IIdvm4	IIdvm5	IIdvm6	IIdvm7	IIdvm8	IItpm2	IItpm3	IItpm4
Liposcelis adult	+	–	–	+	–	+	+	+	+	–	–	–
Liposcelis nymph 1 ins.	+	–	–	+	–	+	+	+	+	–	–	–
Stenopsocus adult	+	+	+	+	+	+	+	+	+	+	+	+
Psocus adult	+	+	–	+	+	+	+	+	+	+	+	+

	IItpm7	IIspm6	IIpcm1	IIpcm3	IIpcm5	IIvlm3	Ivlm7	Ivlm9	IIscm1	IIscm2	IIscm4	IIscm6	IIscm7
Liposcelis adult	–	+	+	+	+	+	+	+	+	+	+	+	+
Liposcelis nymph 1 ins.	–	+	–	+	+	+	+	+	+	+	+	+	+
Stenopsocus adult	+	+	+	+	+	+	+	+	+	+	+	+	+
Psocus adult	+	+?	+	+	+	+	+	+	?	?	?	+	?

	IIIdlm1	IIIdlm2	IIIdvm1	IIIdvm5	IIIdvm6	IIIdvm7	IIIdvm8	IIItpm1	IIItpm2	IIItpm3	IIItpm4
Liposcelis adult	–	+	+	+	+	+	–	–	–	–	–
Liposcelis nymph 1 ins.	–	+	+	+	+	+	–	–	–	–	–
Stenopsocus adult	+	+	+	+	+	+	+	+	+	+	+
Psocus adult	+	+	–	+	+	+	+	+	+	+	+

	IIItpm9	IIIspm2	IIIpcm1	IIIpcm3	IIIpcm4	IIIpcm5	IIIpcm7	IIIvlm2	IIIscm1	IIIscm2	IIIscm4	IIIscm6
Liposcelis adult	–	+	+	+	–	+	–	+	–	+	+	+
Liposcelis nymph 1 ins.	–	+	–	+	–	+	–	+	–	+	+	+
Stenopsocus adult	+	–	+	+	+	+	+	+	+	+	+	+
Psocus adult	+	+	+	+	+	+	+?	+	+?	+?	+?	+

+ present; − absent; *a*, *b* fused muscles, ? unknown

(M. frontobuccalis anterior): O, anterior part of frons; I, upper wall of anatomical mouth. 0bu3 (M. frontobuccalis posterior): O, lateral part of frons; I, upper wall of anatomical mouth. 0bu5 (M. tentoriobuccalis anterior): O, tentorium; I, ventral wall of anatomical mouth. 0bu6 (M. tentoriobuccalis posterior): O, tentorium; I, ventral wall of anatomical mouth. 0ph1 (M. verticopharyngealis): O, central part of frons; I, dorsal wall of pharynx. 0ph2 (M. tentoriopharyngealis): O, tentorium; I, ventral wall of pharynx. 0st1 (M. annularis stomodaei): circular musculature of pharynx. 0st2 (M. longitudinalis stomodaei): longitudinal musculature of pharynx. Internal muscles of mouthparts not studied because of extremely small size.

Musculature of thorax (Fig. 3.13). **Prothorax**. Idlm1 (M. prophragma-occipitalis): O, central part of prophragma (phragma of prothorax); I, posterior margin of occipital area of cranium. Idlm2 (M. pronoto-occipitalis): O, central part of pronotum; I, posterior margin of occipital area of cranium. Idlm5 (M. pronoto-phragmalis anterior): O, anterior part of pronotum; I, prophragma. Idvm1 (M. cervico-occipitalis anterior): O, cervical sclerite; I, posterolateral margin of occipital area of cranium. Idvm2 (M. cervico-occipitalis medialis): O, cervical sclerite; I, posterior margin of occipital area of cranium. Idvm4 (M. pronoto-cervicalis lateralis): O, anterolateral part of pronotum; I, cervical region. Idvm6 (M. pronoto-cervicalis medialis): O, anterior part of pronotum; I, cervical sclerite. Idvm10 (M. profurca-phragmalis): O, apex of profurca; I, prophragma. Idvm15 (M. pronoto-trochantinocoxalis): O, central part of pronotum; I, anterior margin of base of coxa. Idvm17 (M. pronoto-coxalis posterior): O, central part of pronotum; I, posterior margin of base of coxa. Idvm18 (M. pronoto-coxalis lateralis): O, posterior part of pronotum; I, posterolateral margin of base of coxa. Idvm19 (M. pronoto-trochanteralis): O, central part of pronotum; I, trochanter. Itpm1 (M. pleurocrista-occipitalis): O, posterolateral margins of occipital area of cranium; I, furca of pleurite of prothorax. Itpm3 (M. pronoto-pleuralis anterior): O, anterior part of pronotum; I, dorsal part of pleurite of prothorax. Ispm2 (M. prospina-mesopleuralis): O, prodorsum; I, pleural apodeme of mesothorax. Ipcm2 (M. procoxa-cervicalis transversalis): O, anterior margin of base of coxa; I, cervical sclerite of other half of the body. Ipcm4 (M. propleuro-coxalis superior): O, pleural apodeme of prothorax; I, anterior margin of base of coxa. Ipcm5 (M. propleuro-coxalis inferior): O, pleural apodeme of prothorax; I, anterior margin of base of coxa. Ipcm8 (M. propleuro-trochanteralis): O, dorsal part of pleurite; I, trochanter. Ivlm3 (M. profurca-tentorialis): O, apex of profurca; I, base of tentorium and posterior margin of cranium. Ivlm4 (M. profurca-spinalis): O, profurca; I, prodorsum. Ivlm7 (M. profurca-mesofurcalis): O, profurca; I, mesofurca (furca of mesothorax). Ivlm9 (M. prospina-mesofurcalis): O, prodorsum; I, mesofurca. Iscm1 (M. profurca-coxalis anterior): O, profurca; I, anterior margin of base of coxa. Iscm4 (M. profurca-coxalis lateralis): O, profurca; I, lateral margin of base of coxa. Iscm6 (M. profurca-trochanteralis): O, profurca; I, trochanter. Iscm7 (M. prospina-mesocoxalis): O, prodorsum; I, anterior margin of base of mesocoxa. Internal muscles of legs not studied because of extremely small size.

Mesothorax. IIdlm1 (M. prophragma-mesophragmalis): O, lateral margins of prophragma; I, mesonotum. IIdvm2 (M. mesonoto-trochantinalis anterior): O,

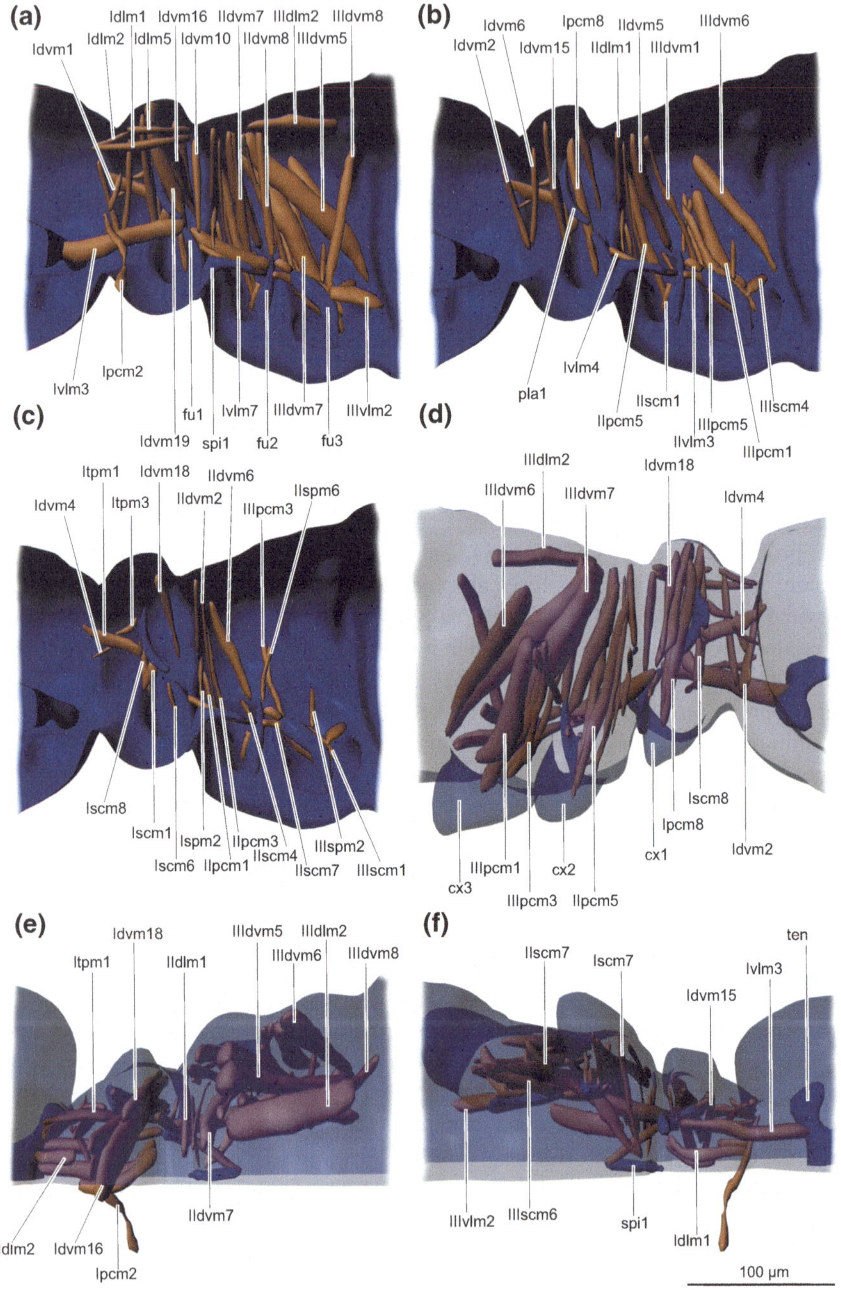

Fig. 3.13 Musculature of thorax in adult *Liposcelis bostrychophila*, 3D: **a**–**c** lateral internal view; **d** lateral external view; **e** dorsal view; **f** ventral view; *pla1* pleural apodeme of prothorax, *fu1.2.3* pro-, meso-, and metafurca, *cx1.2.3* pro-, meso-, and metacoxae, *spi1* prodorsum, *ten* tentorium. Musculature see text

anterior part of mesonotum (not distinguished anterior part of dorsal sclerite of meso- and metathorax); I, trochantin. IIdvm5 (M. mesonoto-coxalis posterior): O, mesonotum; I, posterior margin of base of coxae. IIdvm7 (M. mesonoto-trochanteralis): O, mesonotum; I, trochanter. IIdvm6 (M. mesocoxa-subalaris): O, lateral part of mesonotum; I, posterolateral margin of base of coxa. IIdvm8 (M. mesofurca-phragmalis): O, mesonotum; I, apex of mesofurca. IIspm6 (M. mesofurca-metanepisternalis): O, mesofurca; I, anterior margin of pleurite of metathorax. IIpcm1 (M. mesanepisterno-trochantinalis): O, pleural apodeme of mesothorax; I, trochantin. IIpcm3 (M. mesanepisterno-coxalis anterior): O, dorsal part of episternum of mesothorax; I, anterior margin of base of coxa. IIpcm5 (M. mesanepisterno-trochanteralis): O, dorsal part of episternum of mesothorax; I, trochanter. IIvlm3 (M. mesofurca-metafurcalis): O, mesofurca; I, metafurca (furca of metathorax). IIscm1 (M. mesofurca-coxalis anterior): O, mesofurca; I, anterior margin of base of coxae. IIscm2 (M. mesofurca-coxalis posterior): O, mesofurca; I, posterior margin of base of coxae. IIscm4 (M. mesofurca-coxalis lateralis): O, mesofurca; I, lateral margin of base of coxae. IIscm6 (M. mesofurca-trochanteralis): O, mesofurca; I, trochanter. IIscm7 (M. mesospina-metacoxalis): O, base of mesofurca; I, anterior margin of base of metacoxae. Internal muscles of legs not studied because of extremely small size.

Metathorax. IIIdlm2 (M. metanoto-phragmalis): O, central part of metanotum (not distinguished posterior part of dorsal sclerite of meso- and metathorax); I, metaphragma. IIIdvm1 (M. metanoto-sternalis): O, anterior part of metanotum; I, uncertain, lateral part of sternite or margin of base of coxa. IIIdvm5 (M. metanoto-coxalis posterior): O, lateral part of metanotum; I, posterior margin of base of coxae. IIIdvm6 (M. metacoxa-subalaris): O, posterior margin of base of coxae; I, lateral part of metanotum. IIIdvm7 (M. metanoto-trochanteralis): O, central part of metanotum; I, trochanter. IIIspm2 (M. metafurca-pleuralis): O, metafurca; I, pleurite of metathorax. IIIpcm1 (M. metanepisterno-trochantinalis): O, anterior margin of pleurite of metathorax; I, trochantin or anterior margin of base of coxa. IIIpcm3 (M. metanepisterno-coxalis anterior): O, dorsal part of pleurite of metathorax; I, anterior margin of base of coxa. IIIpcm5 (M. metanepisterno-trochanteralis): O, lateral margin of metanotum; I, trochanter. IIIvlm2 (M. metafurca-abdominosternalis): O, metafurca; I, anterior margin of sclerite 2 of abdomen. IIIscm1 (M. metafurca-coxalis anterior): O, metafurca; I, anterior margin of base of coxa. IIIscm2 (M. metafurca-coxalis posterior): O, metafurca; I, posterior margin of base of coxa. IIIscm4 (M. metafurca-coxalis lateralis): O, metafurca; I, lateral margin of base of coxa. IIIscm6 (M. metafurca-trochanteralis): O, metafurca; I, trochanter. Internal muscles of legs not studied because of extremely small size.

Musculature of abdomen. Dorsal longitudinal muscles (Mm. dorsales), several parallel tufts in each segment: O, anterior phragma, I, posterior phragma. Ventral longitudinal muscles (Mm. ventrales), one tuft in each segment: O, anterior margin of segment, I, posterior margin of segment. Dorsoventral muscles, two pairs of tergosternal muscles in each segment: O, lateral part of tergite, I, lateral part of sternite.

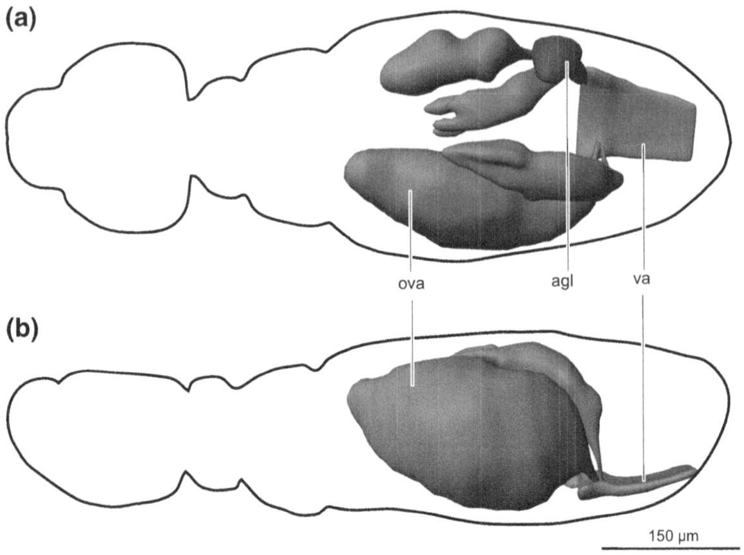

Fig. 3.14 Reproductive system of *Liposcelis bostrychophila*, 3D: **a** dorsal view; **b** lateral view; *agl* accessory glands, *ova* ovary, *va* vagina

Reproductive system

Male reproductive system consists of paired testes, seminal vesicles, spermiducts, accessory glands, and copulatory apparatus.

Female reproductive system consists of paired ovaries, paired oviducts with transition to vagina, and unpaired accessory gland (Fig. 3.14). Each ovary consisting of ovarioles. Ovipositor described in section on structure of abdomen.

3.4 Morphology of First Instar Nymph of *Liposcelis bostrychophila*

The morphology of first instar nymphs is almost identical to that of adults, except for proportions of body and particular organs, measurements, and some peculiar features, described below (Fig. 3.15 and 3.16).

3.4.1 External Morphology

Body size of first instar nymph of *Liposcelis bostrychophila* only 300–350 μm (M = 329, n = 10). Head longer than one-third of body and wider than other parts

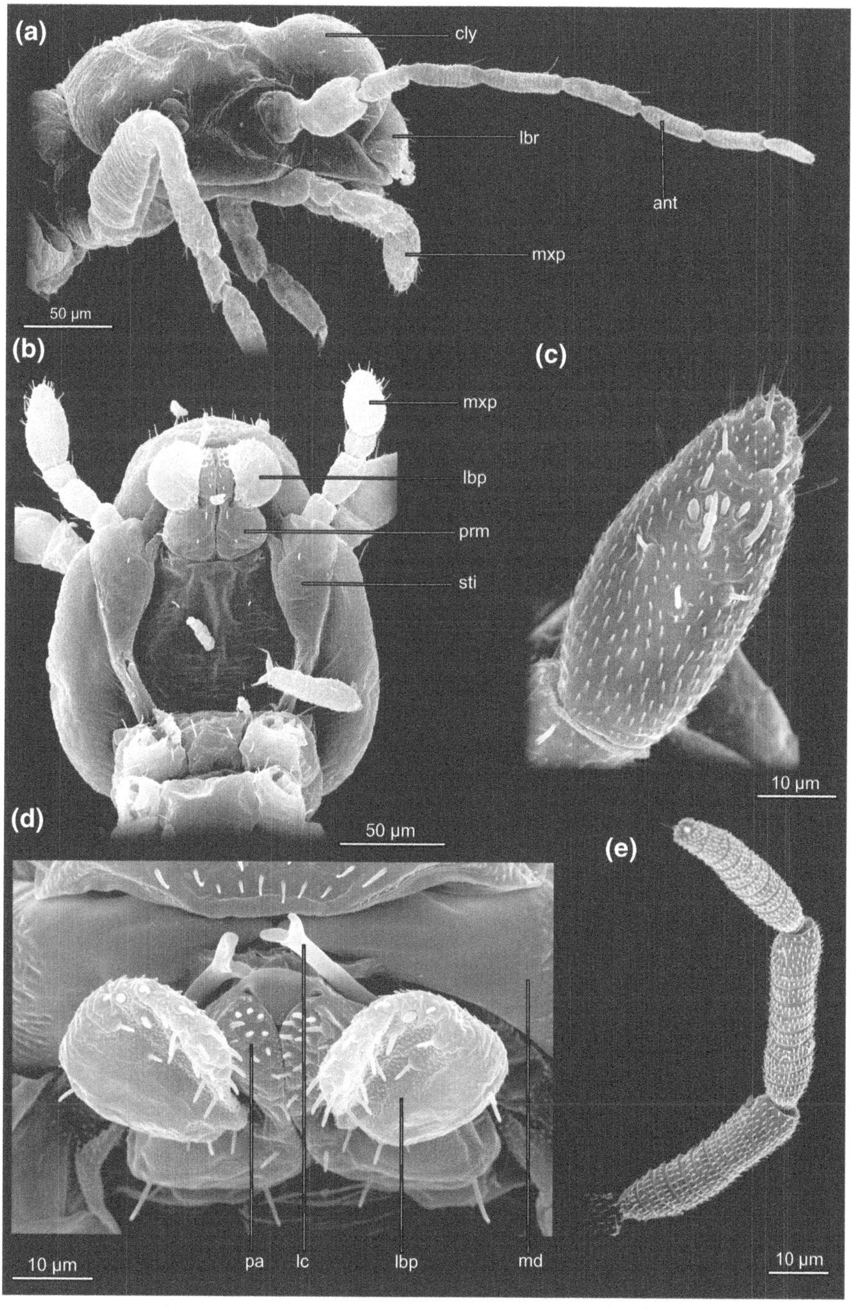

Fig. 3.15 Head of first instar nymph in *Liposcelis bostrychophila*, SEM: **a** head, lateral view; **b** head, ventral view; **c** apical maxillary palpomere; **d** mouthparts, frontal view; **e** apical antennomeres; *cly* clypeus; *lbp* labial palp, *lbr* labrum, *lc* lacinia, *md* mandible; *pa* paraglossae, *prm* prementum, *sti* stipes

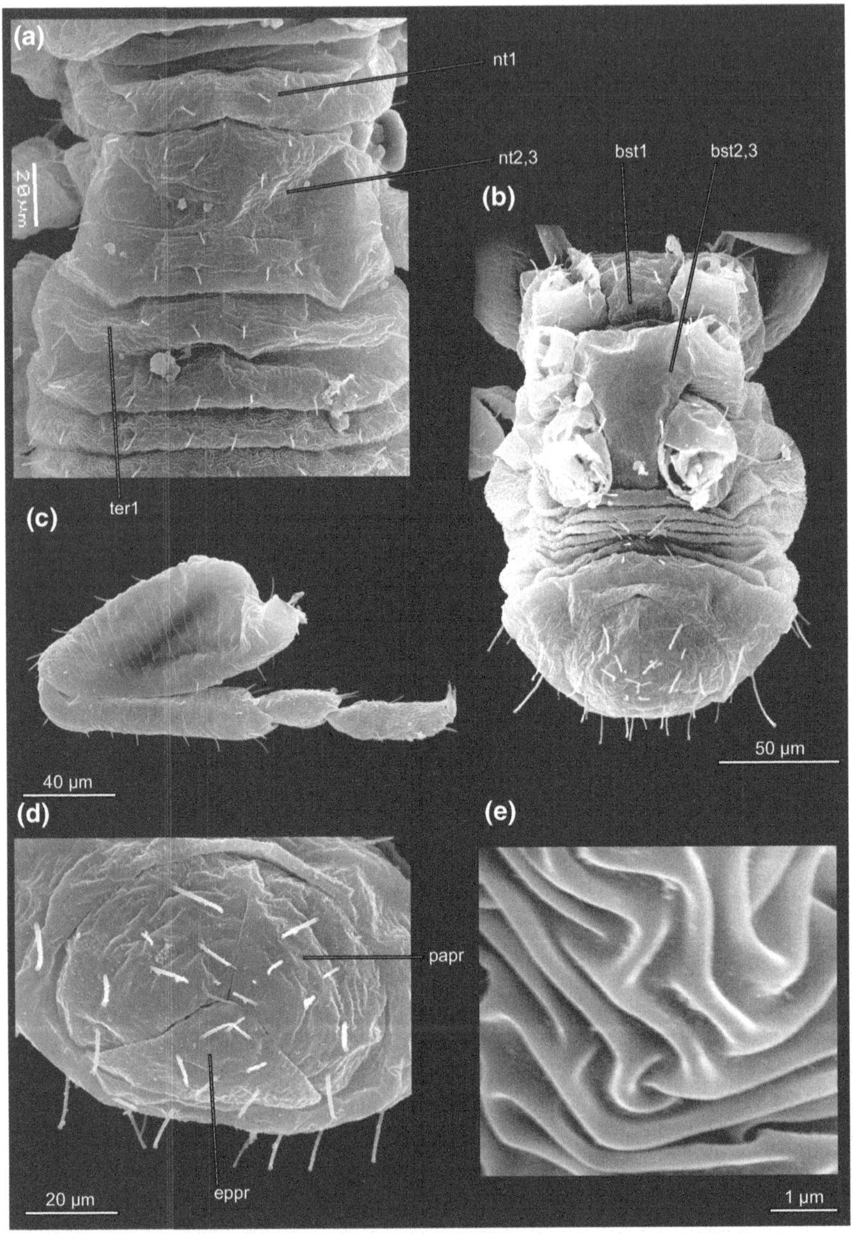

Fig. 3.16 External morphology of first instar nymph *Liposcelis bostrychophila*, SEM: **a** thorax and base of abdomen, dorsal view; **b** thorax and abdomen, ventral view; **c** hindleg; **d** anal valves; **e** cuticular folds on surface of abdomen; *bst1.2.3* pro-, meso-, and metabasisternite, *eppr* epiproct, *nt1.2.3* pro-, meso-, and metanotum, *papr* paraproct, *ter* tergite

of body. Integument weakly pigmented and almost not sclerotized, in most areas wrinkled and folded. Coloration of integument light yellow or almost white. Frontoclypeal suture not distinct. Dorsal arms of tentorium absent. Eyes consisting of two ommatidia each. Antennae 9-segmented. Pleurite of mesothorax not divided into episternum and epimeron. Tarsi 2-segmented. Number of abdominal sternites and tergites difficult to count because of very strongly wrinkled abdomen. Visible division into sternites and tergites absent. Number spiracles could not be determined with certainty because of strongly wrinkled abdomen. Copulatory apparatus and ovipositor absent.

3.4.2 Internal Morphology

Abdomen strongly flattened and wrinkled (Fig. 3.16b, e), therefore internal organs shifted anteriad into thorax and abdomen. All organs situated very closely to each other, and musculature often deeply sunk into many organs (Fig. 3.17). Cuticle thickness 0.8–4.5 μm (M = 1.6; n = 80). Digestive canal without loops or strong curves. Midgut short, situated in thorax and head, its anterior part forming process reaching anterior half of head. Number of Malpighian tubules could not be determined. Heart and blood vessels absent or indiscernible, cavities of body filled with hemolymph almost absent because of very closely packed organs and parenchyma-like fat body between them.

Nervous system

Central nervous system of first instar nymphs similar to that of adults, but different in following features (Fig. 3.9c, d). Degree of oligomerization and concentration of ganglia considerably higher than in adults. Protocerebrum with asymmetrical distal processes reaching into prothorax larger on right side. In contrast to larvae of smallest coleopterans, in nymphs of *Liposcelis bostrychophila* discernible connectives and circum-oesophageal connective clearly discernible. Prothoracic ganglion distinguished. Mesothoracic, metathoracic, and abdominal ganglia fused into one synganglion.

Nymphal cerebrum distinguished by considerable degree of neuropil compartmentalization and generally similar to adult brain. Nymphal central body distinguished only by absence of formed noduli. Optic lobes in first instar nymphs developed even less than in adults, but with rather small spherical neuropilar formations.

Cell size in nervous system of first instar nymphs about 1.98–3.61 μm (M = 2.30; n = 95). Brain of first instar nymphs of *Liposcelis bostrychophila* contains about 8500 cells.

Muscular system

Musculature of first instar nymphs completely identical to musculature of adults except for several features, listed below.

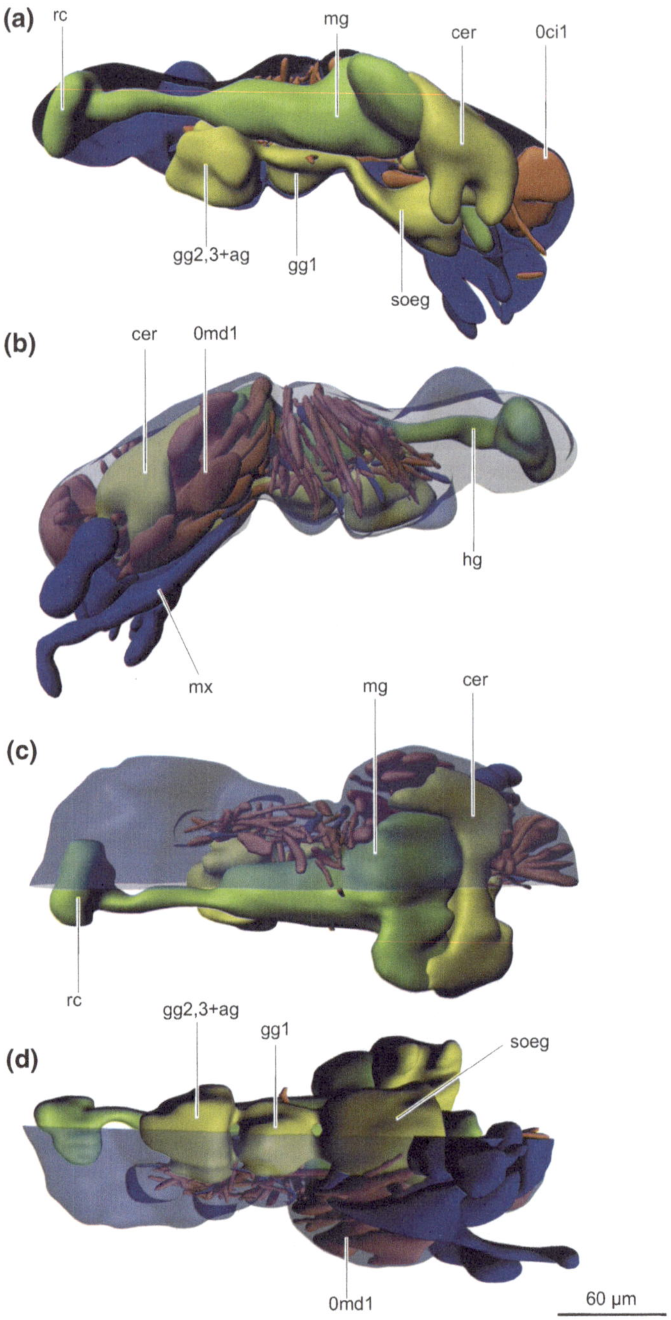

Fig. 3.17 Internal morphology of first instar nymph *Liposcelis bostrychophila*, 3D: **a** lateral internal view; **b** lateral external view; **c** dorsal view; **d** ventral view; *ag* abdominal ganglion, *cer* cerebrum, *hg* hindgut, *gg1.2.3* pro-, meso-, and metathoracic ganglia, *mg* midgut, *mx* maxilla; *rc* rectum, *soeg* suboesophageal ganglion. Colors: *blue* cuticle, *green* digestive system, *yellow* central nervous system, *brown* musculature, *purple* reproductive system. Musculature see text

Musculature of head. 0mx1. 0hy7 and 0hy8 absent. 0bu5 and 0bu6 fused or one of them absent.

Musculature of thorax. Itpm3, IIpcm1 absent.

Musculature of abdomen. The musculature of the abdomen could not be examined in detail, because in first instar nymphs abdomen is strongly flattened and all internal organs area extremely closely set, hindering interpretation. Muscles include dorsal longitudinal (Mm. dorsales), ventral longitudinal (Mm. ventrales), and dorsoventral muscles.

Reproductive system absent.

References

Badonnel, A. (1934). Recherches sur l'anatomie des Psoques. *Bulletin biologique de la France et de la Belgique, 18*, 1–241.

Badonnel, A. (1951). Ordre des Psocoptères. In P.-P. Grassé (Ed.) *Traité de zoologie, anatomie, systématique, biologie* (Vol. 10(2), pp. 1301–1340). Masson, Paris.

Biining, J., & Sohst, S. (1990). Ultrastructure and cluster formation in ovaries of bark lice, Peripsocus phaecopterus Stephen and Stenopsocus stigmaticus Imhov & Labram (Insecta: Psocoptera). *International Journal of Insect Morphology and Embryology, 19*, 227–241.

Chapman, G. B. (2003). Pharynx, esophagus, and associated structures in the booklouse, *Liposcelis divinatorius*. *Invertebrate Biology, 122*(1), 52–60.

Emeljanov, A. F., Golub, N. V., & Kuznetzova, V. G. (2001). Эволюционные преобразования структуры семенников и яичников у сеноедов, пухоедов и вшей (Psocoptera, Phthiraptera: Mallophaga, Anoplura) (Evolutionary transformations of testes and ovaries in booklice, birdlice, and sucking lice (Psocoptera, Phthiraptera: Mallophaga, Anoplura)). *Entomologicheskoe Obozrenie, 80*(3), 557–577.

Finlayson, L. H. (1949). The life-history and anatomy of *Lepinotus patruelis* Pearman (Psocoptera–Atropidae). *Proceedings of the Zoological Society of London, 119*(2), 301–323.

Golub, N. V. (2003). Репродуктивная система сеноедов (Psocoptera) (Reproductive System of Psocoptera). *Entomologiceskoe Obozrenie, 82*(1), 90–101.

Goss, R. J. (1954). Ovarien development and oogenesis in the booklouse, *Liposcelis divergens* Badonnel (Psocoptera: Liposcelidae). *Annals of the Entomological Society of America, 47*, 190–207.

Hu, F., Zhang, G.-N., & Wang, J.-J. (2009). Antennal sensillae of five stored-product psocids pests (Psocoptera: Liposcelididae). *Micron 5*, pp. 628–634.

Jentsch, S. (1940). Zur Morphologie des Gehirns und der Lichtsinnesorgane der Psocopteren. *Zoologische Jahrbücher. Abteilung für Anatomie und Ontogenie der Tiere, 66*, 403–436.

Kucerova, Z., Li, Zhihong, & Hromadkova, J. (2009). Morphology of nymphs of common storedproduct psocids (Psocoptera, Liposcelididae). *Journal of Stored Products Research, 45*, 54–60.

Maki, T. (1938). Studies of the thoracic musculature of insects. *Memoirs of the Faculty of Science and Agriculture; Taihoku Imperial University, 24*, 1343.

Matsuda, R. (1970). Morphology and evolution of the insect thorax. *Memoirs of the Entomological Society of Canada, 76*, 1–431.

Mazokhin-Porshnyakov, G. A., & Kazyakina, V. I. (1981). К строению органов зрения взрослых сеноедов (Psocoptera) (On the structure of the organs of vision in adult Psocoptera). *Entomologiceskoe Obozrenie, 60*(2), 252–259.

Noland, R. C. (1924). The anatomy of *Troctes divinatorius* Muell. *Transactions of the Wisconsin Academy of Sciences, Arts, and Letters, 21*, 195–211.

Paulus, T., & Honomichl, K. (1989). Feinstruktur des ventralen mandibularen Muskelrezeptors bei *Liposcelis bostrichophilus* (Insecta: Psocoptera). *Zoomorphology, 108*, 353–358.

Polilov, A. A. (2016). Anatomy of adult and first instar nymph of book lice *Liposcelis bostrychophila* (Psocoptera: Liposcelididae). *Entomological Review* (in press).

Seeger, W. (1975). Funktionsmorphologie an spezialbildungen der fühlergeißel von Psocoptera und anderen Paraneoptera. *Zeitschrift für Morphologie der Tiere, 81*(2), 137–159.

Vyashnikova, V. N. (1970). Особенности строения половой системы сеноедов и ее систематическое значение (Peculiar features of the reproductive system of Psocoptera and its taxonomic importance). *Entomologiceskoe Obozrenie, 69*(2), 317–346.

Wigglesworth, V. B. (1953). *The principles of insect physiology*. London: E. P. Dutton.

Wong, S. K., & Thornton, I. W. B. (1968). The internal morphology of the reproductive systems of some psocid species. *Proceedings of the Royal Entomological Society of London Series A General Entomology, 43*, 1–12.

Wundrig, G. (1936). Die sehorgane der Mallophagen, nebst vergleichenden untersuchungen an Liposceliden und Anopluren. *Zoologische Jahrbuecher Abteilung fuer Anatomie und Ontogenie der Tiere, 62*(45), 1–10.

Yoshizawa, K. (2005). Morphology of Psocomorpha (Psocodea: 'Psocoptera'). *Insecta matsumurana. Series Entomology, 62*, 1–44.

Yoshizawa, K., & Johnson, K. P. (2006). Morphology of male genitalia in lice and their relatives and phylogenetic implications. *Systematic Entomology, 31*(2), 350–361.

Chapter 4
Structure of the Principal Groups of Microinsects. II. Thrips (Thysanoptera: Thripidae)

4.1 Introduction

The order Thysanoptera (thrips) includes about 6000 species of small insects (0.5–14 mm) of very diverse biology. Most of them are phytophagous, usually feeding on plant sap (some are anthophilous). Others are predaceous or mycophagous (feeding on fungal spores or hyphae), or ectoparasitic (Pinent et al. 2003). The smallest thrips belong to the family Thripidae, the largest and one of the best studied families of Thysanoptera. The anatomy and morphology of all stages of the life cycle are described in detail for *Hercinothrips* (Moritz 1988a,b, c, 1989a, b, c, d). The external and internal morphology of the head (Mickoleit 1963) and thorax (Mickoleit 1961) have been described in detail for the genus *Thrips*. The anatomy of adults and nymphs of *Heliothrips haemorrhoidalis* was recently studied (Polilov and Shmakov 2016).

Many researchers paid special attention to the structure of thrips mouthparts, unique in their asymmetry, and to date mouthparts have been studied in many species, including representatives of the family Thripidae (Garman 1890; Peterson 1915; Reyne 1927; Davies 1958; Mound 1971; Heming 1978; Moritz 1982d; Hunter and Ullman 1992, 1994).

In other groups of thrips, external and internal morphology have been studied in detail in *Aeolothrips* (Aeolothripidae; Mickoleit 1963; Moritz 1978, 1979, 1982a, b, c): morphology of the head and of thorax has been studied in detail in *Haplothrips* and *Phlaeothrips* (Phlaeothripidae; Mickoleit 1961, 1963).

4.2 Brief Description of the Structure of Thripidae

Small insects 0.5–5 mm long (on average about 2 mm). Body elongate, flattened. Coloration of integument monotonous, in different representatives of this family varying from yellow to black. Usually integument covered with sparse hairs. Structure of integument varying from simple sparse punctation to complex cellulate

A.A. Polilov, *At the Size Limit—Effects of Miniaturization in Insects*,
DOI 10.1007/978-3-319-39499-2_4

microsculpture. **Head** opisthognathous, directed obliquely posteriad. Cranium solid, except for weakly pronounced occipital suture. Compound eyes consisting of 55–70 ommatidia each. Three ocelli present. Antennae flagelliform or moniliform, 6–9-segmented; 2–3 apical antennomeres forming weakly articulate style. Mouthparts of unique structure: asymmetrical and represented by mouth cone formed by anteclypeus, labrum, maxillae, and labium. Mouth cone inside with unpaired (left) mandible and paired stylets of maxillae. **Thorax** consisting of detached prothorax and fused meso- and metathorax. **Wings:** two pairs, forewing slightly longer than hindwing, consisting of narrow wing blade with depleted venation and wide fringe of setae on the perimeter. At rest folded along body. **Legs** with normal division into coxa, trochanter, femur, tibia, and tarsus. Tarsi 1–2-segmented, on apex of apical antennomere with sucker, claws reduced. **Abdomen** consisting of 10 segments. Number of visible sternites 6; number of visible tergites 10. Segments 9–10 in females divided by longitudinal cleft. **External male genitalia** represented by elongate aedeagus, parameres, and endotheca with short pseudovirga. Females with short **ovipositor,** formed by two pairs of dentate valves.

4.3 Morphology of Adult *Heliothrips haemorrhoidalis* (Thripidae)

4.3.1 External Morphology

Body size of adult *H. haemorrhoidalis* 1380–1470 μm (M = 1427, n = 10) (Fig. 4.1).

Structure of head

Head opisthognathous, directed obliquely posteriad (Fig. 4.2). Cranium solid, except for weakly pronounced occipital suture. Head connected with prothorax via articular membrane and pair of cervical sclerites. Tentorium strongly simplified. Represented by short dorsal arms, anterior arms, fused with cranium, and posterior arms, over much of their length fused with cranium. Tentorial bridge absent.

Eyes consisting of 55–70 ommatidia. Three ocelli present.

Antennae flagelliform or moniliform, 6–9-segmented, 2–3 apical antennomeres forming weakly articulate style. All segments of flagellum bearing trichoid sensilla.

Mouthparts of peculiar structure typical of thrips: asymmetrical and represented by moth cone formed by anteclypeus, labrum, maxillary plates, and labium (Fig. 4.2). Mouth cone inside with paired stylets of maxillae and unpaired (left) mandible. Anteclypeus asymmetrical, triangular or trapezoid, membranous. Labrum also asymmetrical, triangular, movably joined with cranium. Only left **mandible** developed, shaped as long stylet. Right mandible represented by small primordium. **Maxillae** consisting of plates (fused stipes and galea), on external surface bearing 2–3-segmented palps and on posterior margin bearing lacinia represented by clever

Fig. 4.1 External morphology of *Heliothrips haemorrhoidalis*, SEM: **a–c** adult; **d–f** nymph; **a**, **d** dorsal view; **b**, **e** lateral view; **c**, **f** ventral view

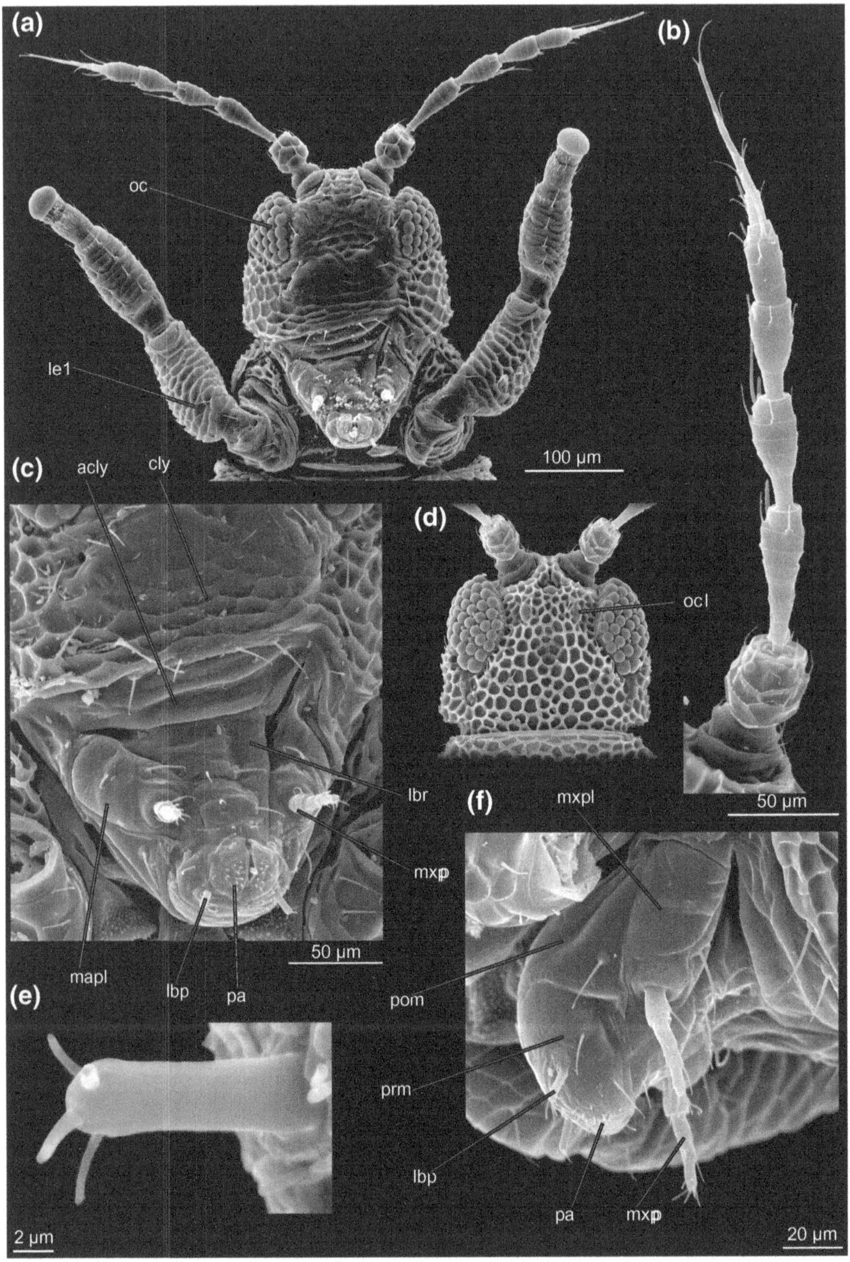

Fig. 4.2 External morphology of head in adult *Heliothrips haemorrhoidalis*, SEM: **a**, **d** head; **b** antenna; **c**, **f** mouthparts; **e** labial palp; **a**, **c** ventral view; **b**, **e**, **f** lateral view; **d** dorsal view; *acly* anteclypeus, *ant* antenna, *cly* clypeus, *lbp* labial palp, *lbr* labrum, *le1* foreleg, *mxp* maxillary palp, *mxpl* maxillary plates, *oc* eye, *pa* paraglossae, *pom* postmentum, *prm* prementum

and stylet. According to some authors, the clever is homologous to cardo, but most believe that it is part of lacinia, while cardo is absent in Thripidae. **Labium** consisting of well developed postmentum and prementum, bearing 2-segmented palps, paraglossae, and fused glossae. Apical palpomere of labial palps and apex of paraglossa bearing groups of trichoid sensilla. Hypopharynx with well developed salivary pump. The mouthparts work as follows: the mandible is used for piercing through the food item, then maxillae (together forming the alimentary canal) are inserted into the pierced hole, and the contents of the item are sucked out with the strong pharyngeal pump (Heming 1978).

Structure of thorax

Prothorax in Thripidae formed by pronotum, narrow pleurite, weakly divided into episternum and epimeron, weakly sclerotized basisternite, narrow furcosternite (Matsuda 1970; Mickoleit 1961: eusternum), and spinosternite (Fig. 4.3). Rather small trochantin present in front of base of coxa. Pleurite with well developed pleural apodeme. Furcosternite bearing profurca shaped as two straight rods. Prodorsum slender, directed posteriad.

Mesothorax. Notal part represented by wide scutum, narrow postnotum sunk into fold forming mesophragma, narrow prescutum (Mickoleit 1961: intertergite). Tegulae well developed. Pleural part consisting of episternum and epimeron, divided by well pronounced suture, in front of episternum with narrow interpleurite, fused with prealar sclerite (Matsuda 1970: prealar arm). Ventral sclerite in *Thrips* (Mickoleit 1961) distinguished from episternum and formed by fused pre-episternum, katepisternum, and basisternite. In *Heliothrips* and *Hercinothrips* (Moritz 1989c), episternum fused with ventral sclerite. In some thrips posterior part of ventral sclerite separated by transverse suture (Mickoleit 1961: sternellum). In *Heliothrips* ventral sclerites of meso- and metathorax fused. Spiracle present in front of episternum. Mesofurca well developed, U-shaped. Mesodorsum (dorsum of mesothorax) shaped as short protrusion of cuticle.

Metathorax. Metanotum consisting of two sclerites (Mickoleit 1961: metanotum cranialis and caudalis). Pleural and ventral part formed by rather small epimeron, separated by well pronounced suture and large sclerite, formed by fused episternum, pre-episternum, katepisternum, and basisternite (Matsuda 1970: metapleurosternum). A unique feature of thrips is the shift of the basilare of metathorax anteriad, into the mesothorax. Spiracle present above basalare. Metafurca appearing as short rods, curved anteriad.

Wings of peculiar structure, represented by narrow wing blade with depleted venation and wide fringe of setae on perimeter of wing (Fig. 4.3f, 4.4). In Thripidae, only one or two longitudinal veins present in forewing, and one longitudinal veins present in hindwing; homology of these veins unclear.

Legs ambulatorial, with normal division into coxa, trochanter, femur, tibia, and tarsus (Fig. 4.3d, e). Femora often dilated. Tarsi 1–2-segmented, on apex of apical tarsomere with sucker (arolium); claws reduced.

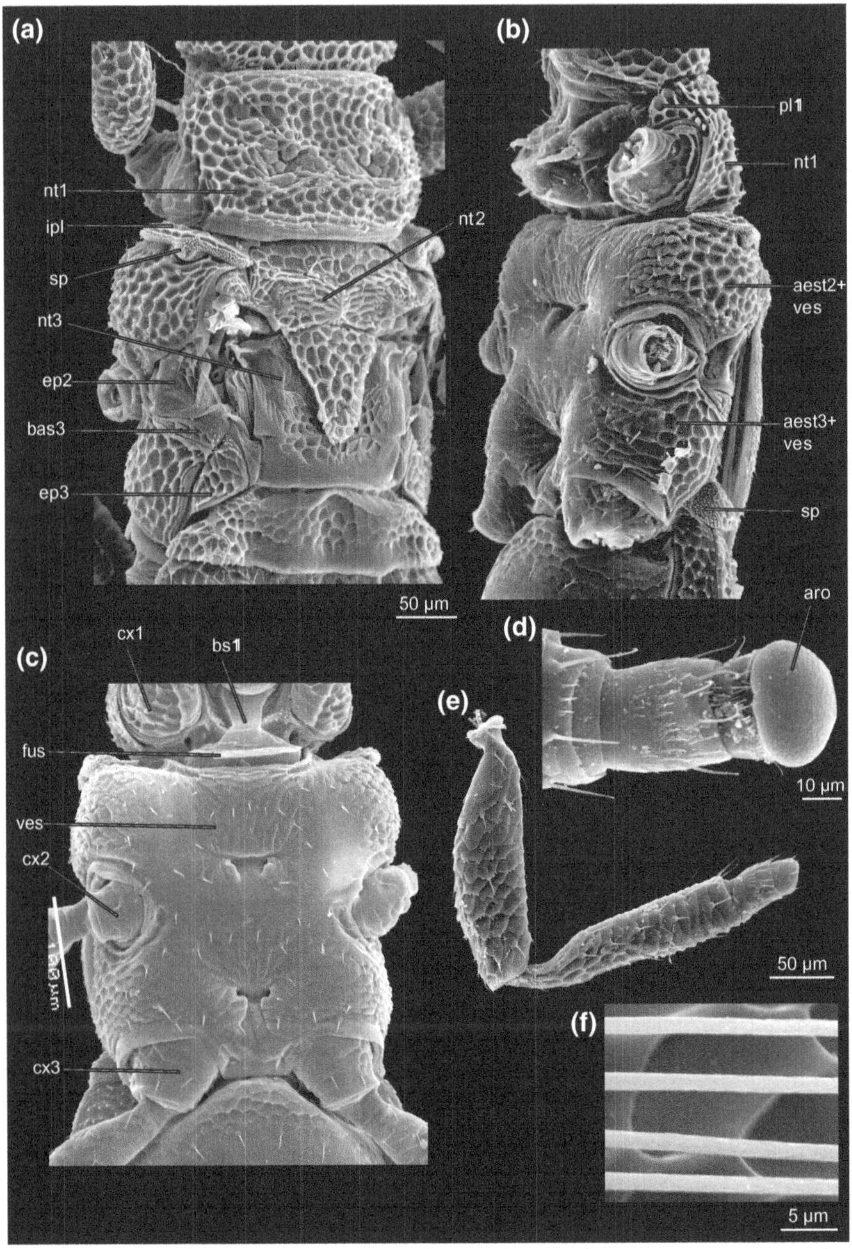

Fig. 4.3 External morphology of thorax in adult *Heliothrips haemorrhoidalis*, SEM: **a–c** thorax; **a** dorsal view; **b** lateral view; **c** ventral view; **d** tarsus and pretarsus; **e** midleg; **f** wing setae; *aest2* episternum of mesothorax, *aro* arolium, *bas3* basalare of metathorax, *bst1* basisternite of prothorax, *cx1.2.3* pro-, meso-, and metacoxae, *ep2.3* epimeron of meso- and metathorax, *fus* furcosternite, *ipl* interpleurite, *nt1.2.3* pro-, meso- and metanotum, *pl1* pleurite of prothorax, *sp* spiracle, *ves* ventral sclerite

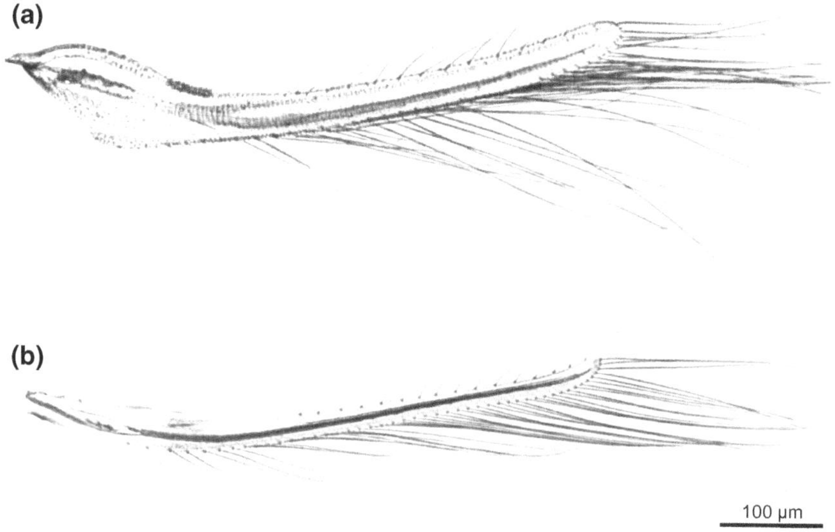

Fig. 4.4 Wings of *Heliothrips haemorrhoidalis*: **a** forewing; **b** hindwing

Structure of abdomen

Abdomen consisting of 10 segments (Fig. 4.5). Number of complete visible sternites 6. Number of visible tergites 10. Segments 9–10 in females formed by cylindrical tergites, divided ventrally by longitudinal cleft. Sternites of these segments strongly reduced, triangular. In males of manz species sternite with pore areas (Mound 2009). Spiracles usually present on tergites 1 and 8. **External male genitalia** represented by elongate aedeagus, parameres, and endotheca with short pseudovirga. Females with short **ovipositor,** formed by two pairs of dentate valves.

4.3.2 Internal Morphology

General plan of internal structure: medial part of body cavities occupied by intestine; most of thorax occupied by musculature, nervous system strongly oligomerized and condensed in thorax; reproductive system occupying considerable part of abdomen (Fig. 4.6).

Particular organ systems are described in more detail in special sections below:

Integument

Integument consisting of cuticle, hypoderm, and basal membrane (Fig. 4.7a, b). Cuticle thickness 2.7–13,2 µm (M = 7,5; n = 80). Cuticle consisting of epicuticle and exo- and endocuticle. Hypoderm consisting of flattened cells and has a thickness of up to 3 µm.

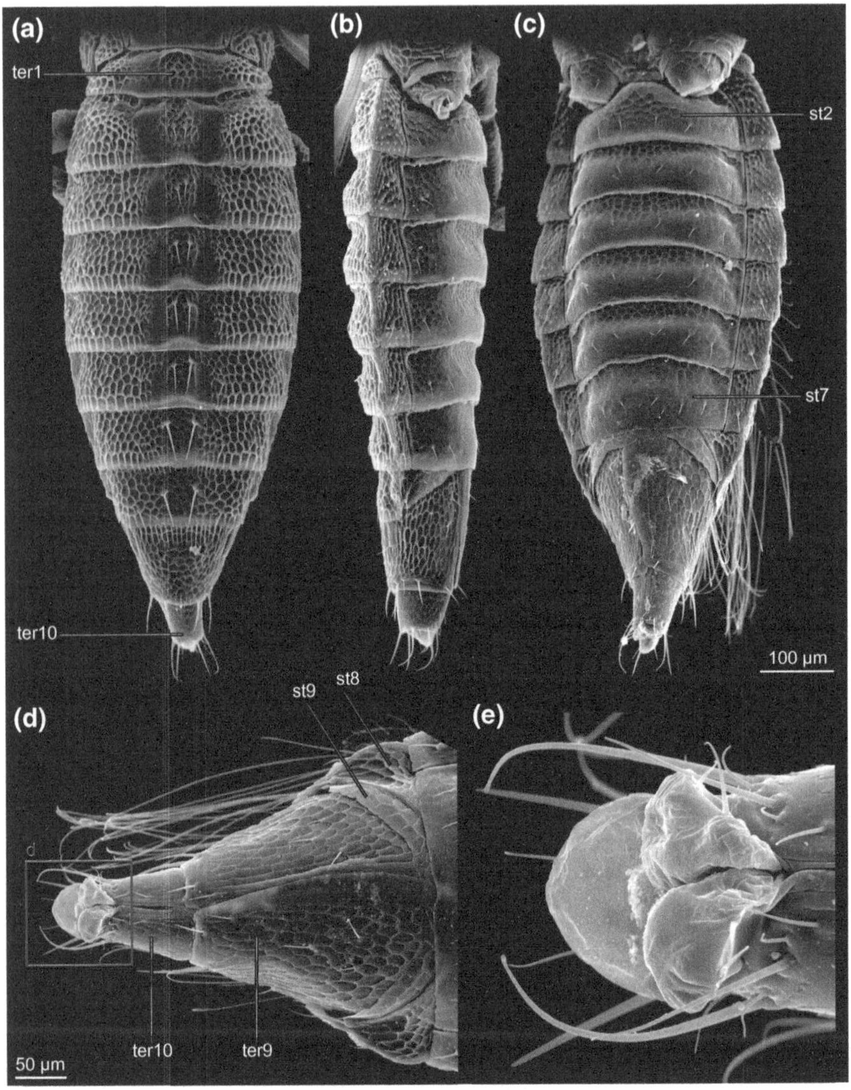

Fig. 4.5 External morphology of abdomen in adult *Heliothrips haemorrhoidalis*, SEM: **a** dorsal view; **b** lateral view; **c–e** ventral view; *st* sternite, *ter* tergite

Digestive and excretory systems

Digestive canal divided into fore-, mid-, and hindgut (Fig. 4.8a, b). Fore- and hindgut with thin cuticular lining. Midgut without cuticular lining. Intestinal canal somewhat longer than body, forming loop in metathorax. With two pairs of salivary glands: ovoid and tubular. Ovoid glands short, oval; tubular glands elongate, reaching middle of abdomen and attached by apex to wall of midgut.

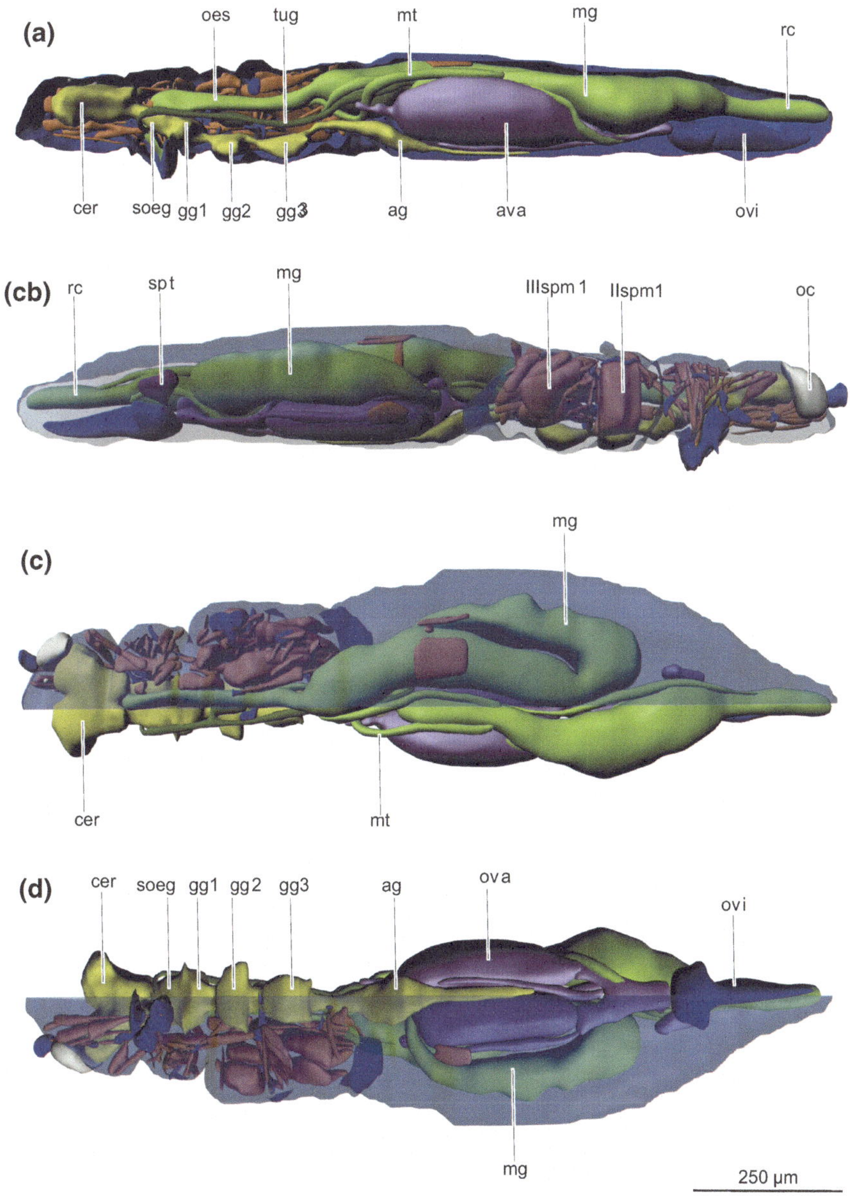

Fig. 4.6 Internal morphology of adult *Heliothrips haemorrhoidalis*, 3D: **a** lateral internal view; **b** lateral external view; **c** dorsal view; **d** ventral view; *ag* abdominal ganglion, *cer* cerebrum, *gg1.2.3* pro-, meso-, and metathoracic ganglia, *mg* midgut, *mt* Malpighian tubules, *oc* eye, *oes* oesophagus, *ova* ovary, *ovi* ovipositor, *rc* rectum, *spt* spermatheca, *soeg* suboesophageal ganglion, *tug* tubular gland. Colors: *blue* cuticle, *green* digestive system, *yellow* central nervous system, *brown* musculature, *purple* reproductive system. Musculature see text

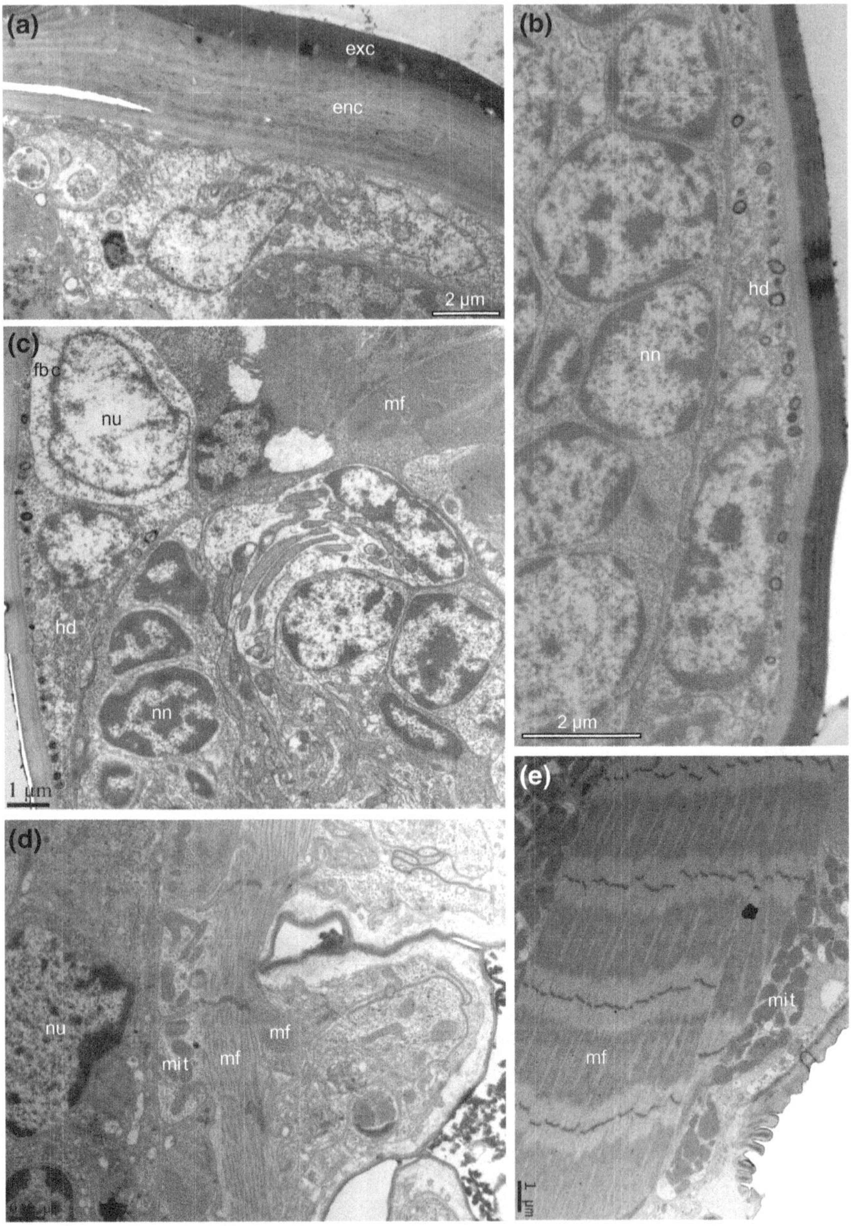

Fig. 4.7 Ultrastructure of adult *Heliothrips haemorrhoidalis*, TEM: **a** integument of mesothorax; **b** integument and cortical layer of cerebrum; **c** section of head (fragment); **d** cross section of walls of oesophagus; **e** longitudinal section of muscles, mesothorax; *enc* endocuticle, *exc* exocuticle, *fbc* cell of fat body, *hd* hypoderm, *mf* muscle fiber, *mit* mitochondrion, *nn* nucleus of neuron, *nu* nucleus

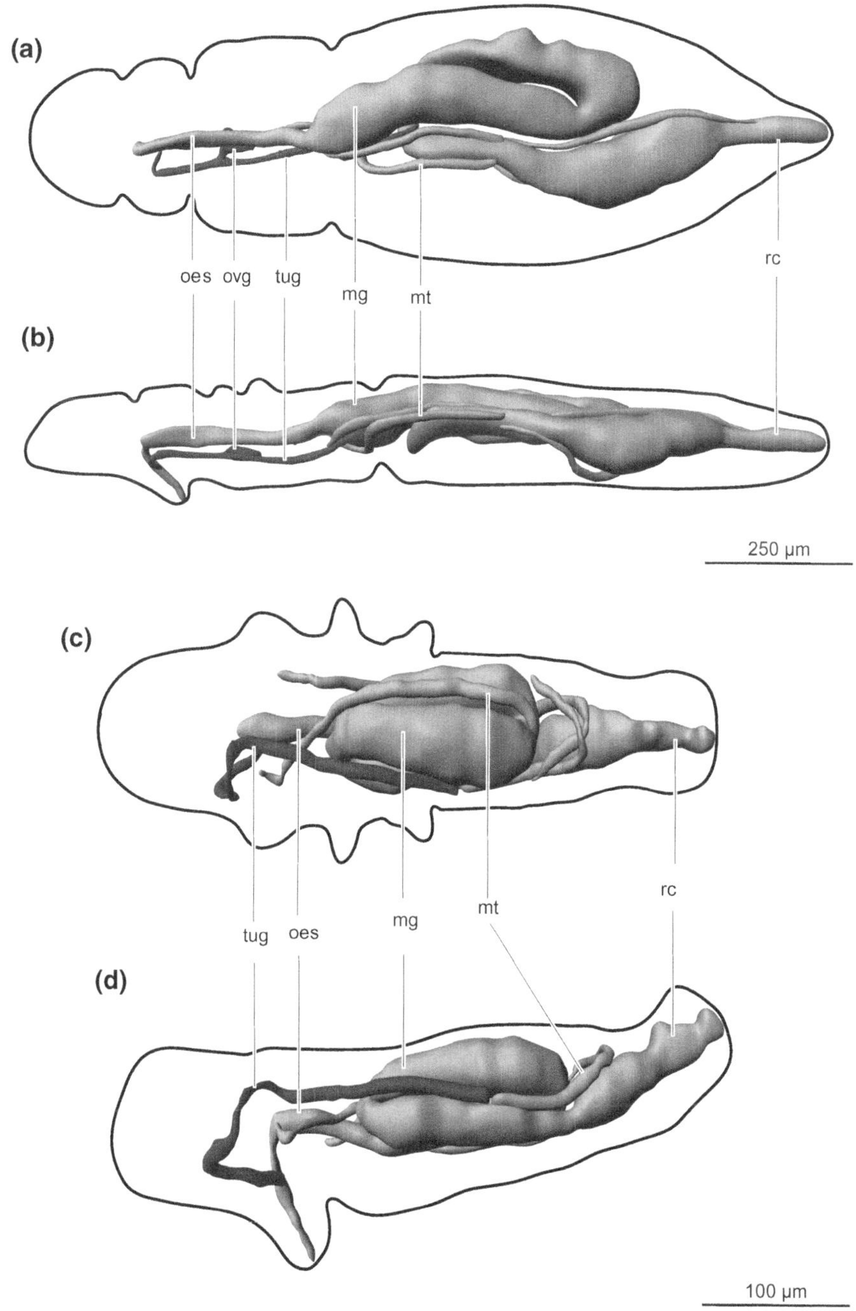

Fig. 4.8 Structure of digestive system in *Heliothrips haemorrhoidalis*, 3D: **a**, **b** adult; **c**, **d** first instar nymph; **a**, **c** dorsal view; **b**, **d** lateral view; *mg* midgut, *mt* Malpighian tubules, *oes* oesophagus, *ovg* ovoid gland, *rc* rectum, *tug* tubular gland

Foregut divided into pharynx and oesophagus. Pharynx of rather great diameter and strong musculature. Oesophagus straight, with layer of circular and longitudinal muscles.

Midgut appearing as tubes of variable diameter, forming loop in abdomen. Wall consisting of unilayered epithelium, in cross section with 3–7 homogeneous cells visible, much fewer than in other insects. Rather sparse layer of circular musculature present.

Hindgut indistinctly divided into small intestine and rectum. Small intestine short, rounded in section; rectum long, oval in section and bearing several rectal papillae.

Boundary between mid- and hindgut bearing four **Malpighian tubules**, two directed anteriad, two posteriad.

Circulatory system and fat body

The circulatory system of insects includes the hemolymph and fat body. Circulatory system developed weakly, with only short heart and nondividing aorta. **Fat body** distributed irregularly, consisting of cells of various shape, containing electron transparent inclusions.

Tracheal system

Tracheal system represented by pair of longitudinal stems with weakly branching tracheae diverging from them. Two pairs of thoracic (on meso- and metathorax) and two pairs of abdominal spiracles present (on tergites 1 and 8). Tracheae with morphology typical of insects, consisting of hypoderm and intima; intima with spiral thickenings (taenidia).

Nervous system

Degree of oligomerization and condensation of ganglia in adult *Heliothrips* lowest among studied microinsects. Cerebrum situated in posterior part of cranium. Suboesophageal ganglion situated in prothorax and fused with prothoracic ganglion. Meso- and metathoracic ganglia and abdominal synganglion separated (Fig. 4.9a, b).

Ensheathment of standard structure, typical of ganglion sheaths of insects. Thickness of neural lamella 0.04–0.05 μm. Thickness of perineurium 0.1–0.4 μm. Cerebral and cortical layers with strongly pronounced cortical glia and neuroglia of sheaths. Most of volume of cells in nervous system occupied by nucleus (70–80 %). In *H. haemorrhoidalis* chromatin forming rather small accumulations in center of nucleus and on periphery (dark ring of heterochromatin). Cytoplasm of bodies and projections of cells in nervous system containing all principal organelles (Fig. 4.10), as in *Liposcelis*. Neuropil of *H. haemorrhoidalis* containing polarized and non-polarized synaptic contacts.

Diameter of body in cells of central nervous system 2.23–4.73 μm (M = 3.07; n = 360). Cerebrum of *H. haemorrhoidalis* containing about 10,000 cells.

Central body complex in *H. haemorrhoidalis* formed by two distinguished structures: central body and protocerebral bridge (Fig. 4.11). Central body

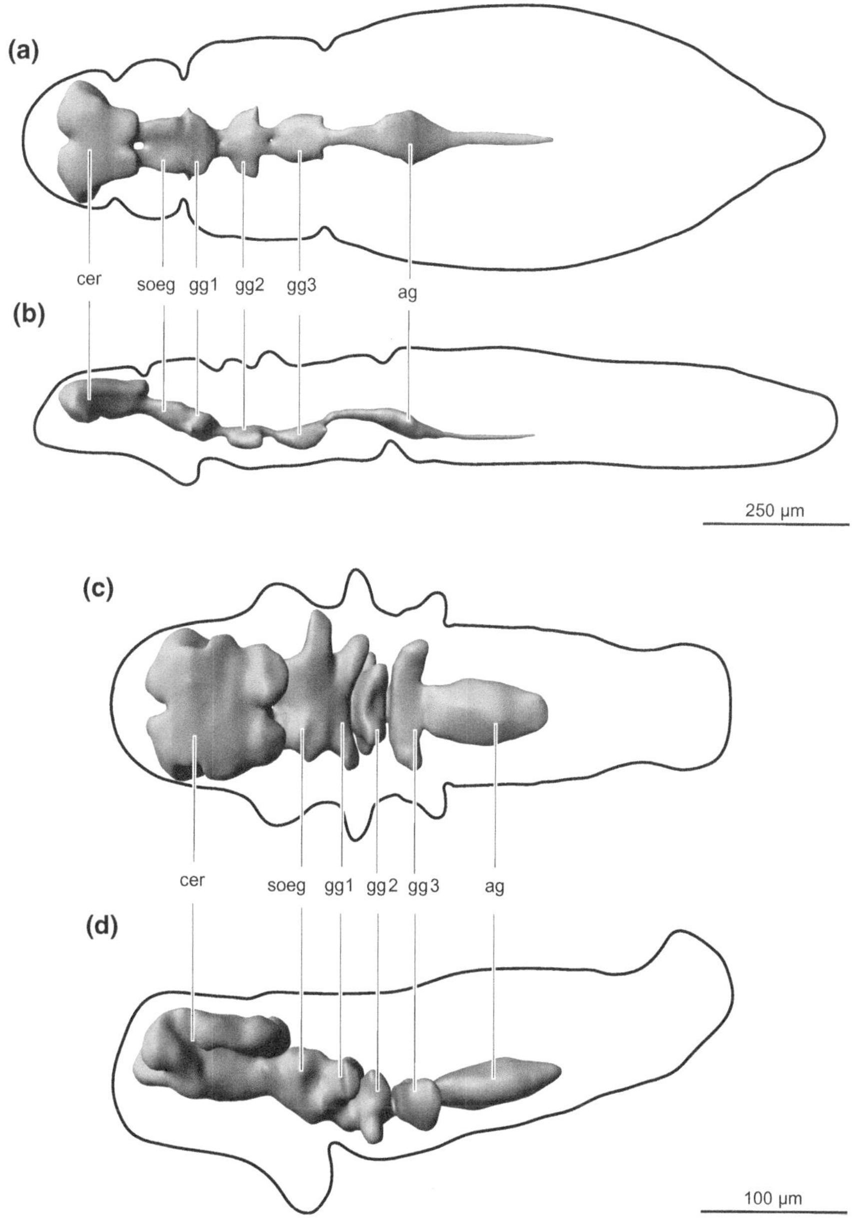

Fig. 4.9 Structure central nervous system *Heliothrips haemorrhoidalis*, 3D: **a**, **b** adult; **c**, **d** first instar nymph; **a**, **c** dorsal view; **b**, **d** lateral view; *ag* abdominal ganglion, *cer* cerebrum, *gg1.2.3* pro-, meso-, and metathoracic ganglia, *soeg* suboesophageal ganglion

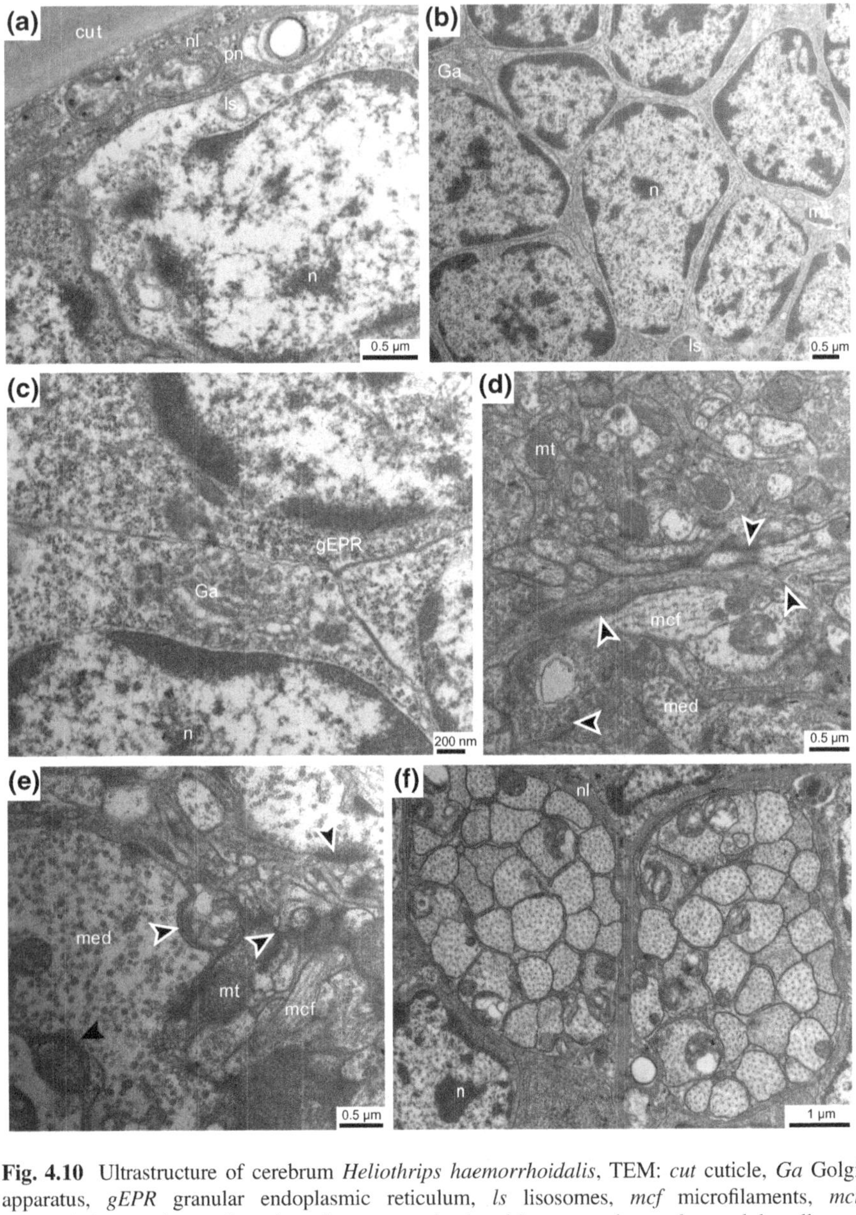

Fig. 4.10 Ultrastructure of cerebrum *Heliothrips haemorrhoidalis*, TEM: *cut* cuticle, *Ga* Golgi apparatus, *gEPR* granular endoplasmic reticulum, *ls* lisosomes, *mcf* microfilaments, *mct* microtubules, *med* granules of mediator, *mt* mitochondrion, *n* nucleus, *nl* neural lamella, *pn* perineurium, *syn* synaptic contact. **d**–**e** *Arrows* indicate synaptic contacts

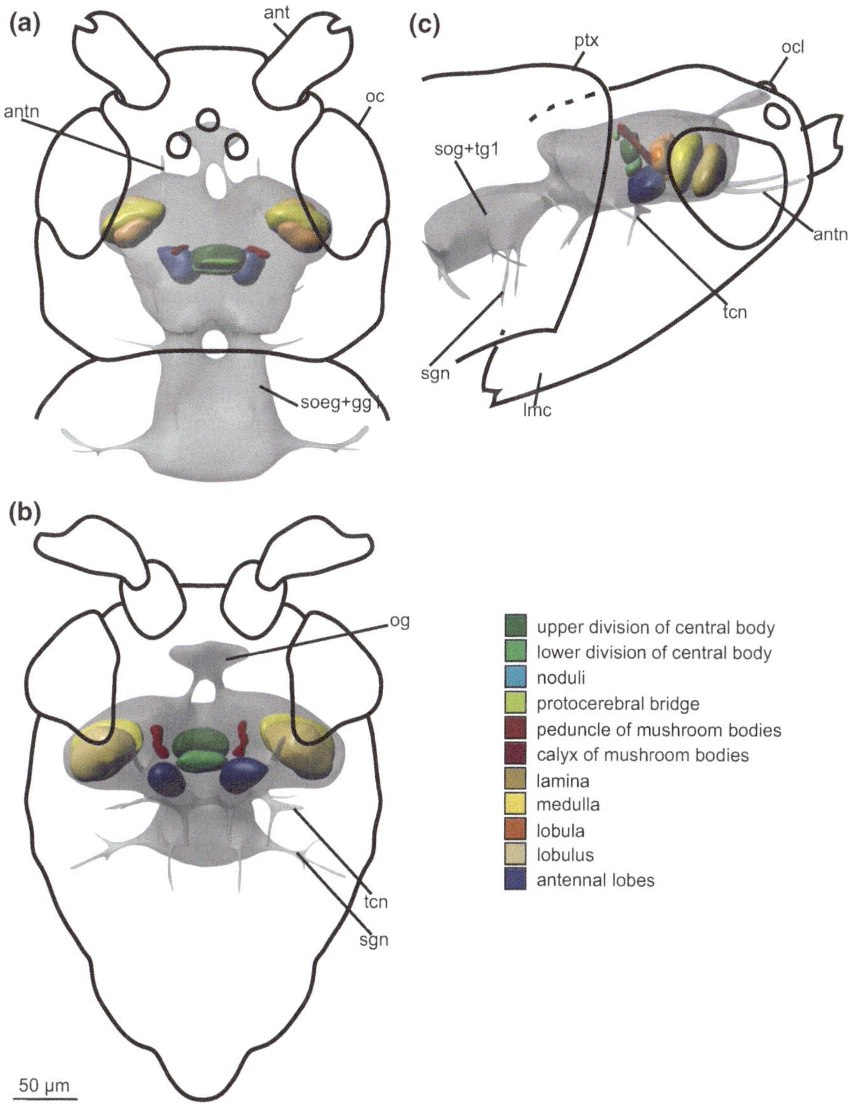

Fig. 4.11 3D reconstructions of cerebrum adult *Heliothrips haemorrhoidalis*: **a** dorsal view; **b** lateral view; **c** frontal view; *antn* antennal nerves, *ant* antennae, *oc* compound eyes, *ocl* ocelli, *og* ocellar ganglion, *lmc* mouthparts, *sgn* nerves of suboesophageal ganglion, innervating mouthparts, *soeg+gg1* synganglion of suboesophageal and prothoracic ganglia, *tcn* nerves of tritocerebrum

differentiated into dorsal region and ventral region. Noduli absent. Mushroom bodies with simplified structure and represented only by peduncular apparatus (peduncle). Calyx reduced. Division of peduncle into α- and β-lobes also not found. Optic lobes shifted anteriad and represented by three optic neuropils. All three optic neuropils elongate, situated in one row, close to each other. Three ocellar ganglia fused into one formation.

Deutocerebrum shifted toward medial plane of cerebrum. Rather small antennal lobes spherical in shape, without glomerular structure.

Tritocerebrum very small; tritocerebral commissure barely visible. Frontal ganglion fused with cerebrum. Occipital ganglion fused with postcerebral endocrine complex.

Muscular system

Musculature of head (Fig. 4.12, Table 4.1). 0an2 (M. tentorioscapalis posterior): O, anterior tentorial arms; I, posterior margin of base of scape. 0an3 (M. tentorioscapalis lateralis): O, anterior tentorial arms; I, anterolateral margin of base of scape. 0lb1 (M. frontolabralis): O, central part of frons; I, medial part of base of labrum. 0lb2 (M. frontoepipharyngalis): O, medial and posterior part of frons; I, lateral part of base of labrum. 0lb5 (M. labroepipharyngealis): O, external wall of labrum; I, epipharynx. 0md1 (M. craniomandibularis internus): O, lateral part of cranium; I, medial part of base of mandible. 0md6 (M. tentoriomandibularis lateralis): O, anterior tentorial arms; I, base of mandibles. tm, muscle of unclear homology (Mickoleit 1963: 10. m. lacinio-mandibularis): O, base of lacinia; I, stylet of mandibles. 0mx2 (M. craniolacinialis): O, gena; I, base of stylet of lacinia. 0mx6 (M. stipitolacinialis): O, stipes; I, base of stylet of lacinia. 0la8 (M. submentopraementalis): O, base of tentorium; I, lateral margin prementum. 0la12 (M. praementoglossalis): O, prementum; I, base of paraglossa. 0hy3 (M. craniohypopharyngealis): O, base of tentorium; I, posterior margin of salivarium. 0hy6 (M. postmentoloralis): O, postmentum; I, hypopharynx. 0hy7 (M. praementosalivaris anterior): O, prementum; I, lateral margin of salivarium. 0hy8 (M. praementosalivaris posterior): O, base of prementum; I, lateral margin of salivarium. 0hy12 (M. hypopharyngosalivaris): O, base of tentorium; I, salivarium. 0ci1 (M. clypeopalatalis), largest muscle of head consisting of several subunits: O, clypeus and frons; I, anterior wall of cibarium. 0st1 (M. annularis stomodaei): circular musculature of pharynx. 0st2 (M. longitudinalis stomodaei): longitudinal musculature of pharynx. Internal musculature of antennae and mouthparts not studied because of extremely small size.

Musculature of thorax (Figs. 4.13 and 4.14). **Prothorax.** Idlm1 (M. prophragma-occipitalis): O, prophragma; I, posterior margin of occipital area. Idlm2 (M. pronoto-occipitalis): O, central part of pronotum; I, posterior margin of occipital area. Idlm5 (M. pronoto-phragmalis anterior): O, central part of pronotum; I, lateral part of prophragma. Idvm2 (M. cervico-occipitalis medialis): O, cervical sclerite; I, lateral part of posterior margin of occipital area. Idvm5 (M. pronoto-cervicalis anterior): O, anterior part of pronotum; I, cervical sclerite. Idvm6 (M. pronoto-cervicalis medialis): O, anterior part of pronotum; I, cervical sclerite. Idvm10 (M. profurca-phragmalis): O, apical part of profurca; I, lateral part of prophragma. Idvm14 (M. pronoto-trochantinalis posterior): O, central part of pronotum; I, trochantin. Idvm15 (M. pronoto-trochantinocoxalis): O, central part of pronotum; I, anterior margin of base of coxa. Idvm17 (M. pronoto-coxalis

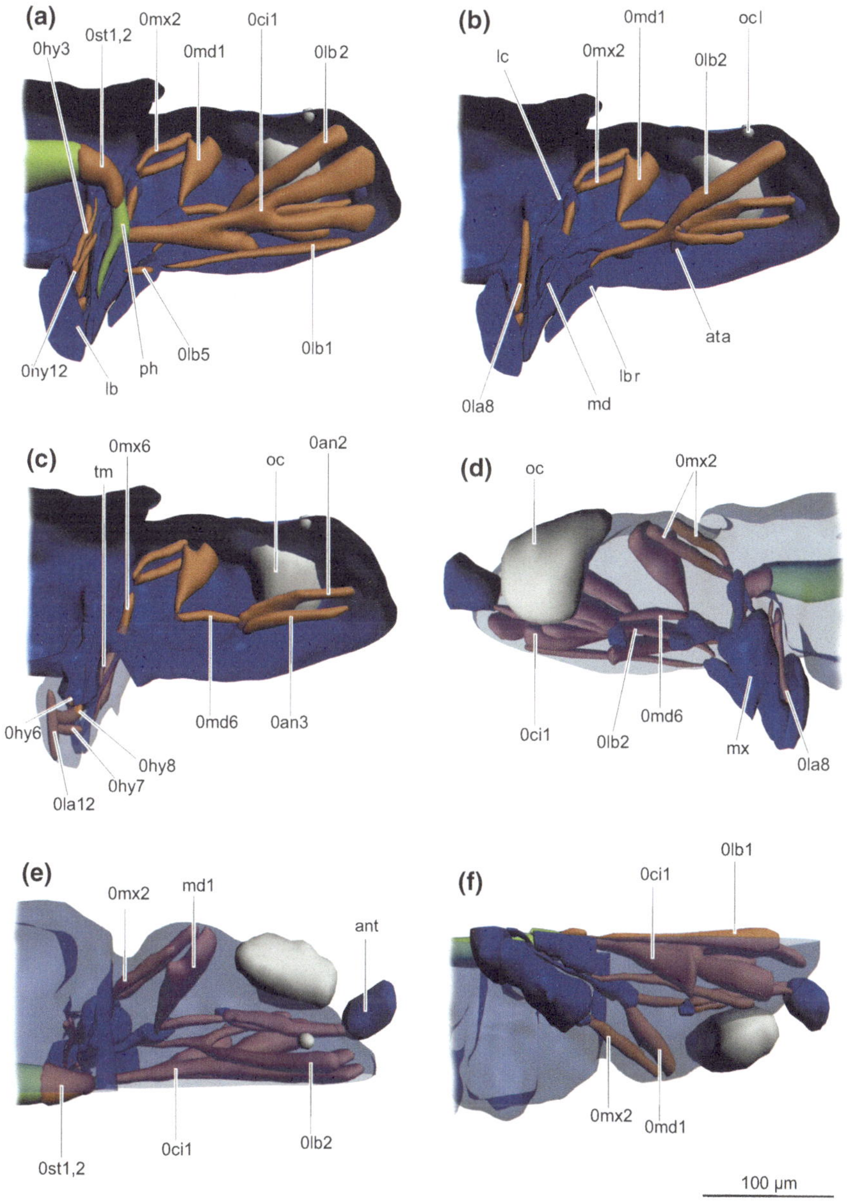

Fig. 4.12 Musculature of head adult *Heliothrips haemorrhoidalis*, 3D: **a–c** lateral internal view; **d** lateral external view; **e** dorsal view; **f** ventral view; *ant* antenna, *ata* anterior tentorial arms, *lb* labium, *lbr* labrum, *lc* lacinia, *md* mandible, *mx* maxilla, *oc* eye, *ocl* ocellus, *ph* pharynx. Musculature see text

Table 4.1 Musculature of Thysanoptera (partly from published sources: Mickoleit 1961, 1963; Moritz 1989c; Matsuda 1970)

	Head muscles												
	0an2	0an3	0lb1	0lb2	0lb5	0md1	0md6	tm	tm2	0mx2	0mx3	0mx6	0la8
Heliothrips adult	+	+	+	+	+	+	+	+	–	+	–	+	+
Heliothrips nymph 1 ins.	+	+	+	+	+	+	+	–	–	+	–	+	+
Thrips adult	+	+	+	+	+	+	?	–	–	+	–	+	+
Hercinothrips adult	+	+	+	+	+	+	+	–	–	+	–	+	+
Hercinothrips nymph	+	+	+	+	+	+	+	–	–	+	–	+	+
Aeolothrips adult	+	+	+	+	+	+	+	+	–	+	+	+	+
Phlaeothrips adult	+	+	+	+	+	+	?	–	+	+	–	+	+
Haplothrips adult	+	+	+	+	+	+	?	–	+	+	–	+	+

	Head muscles													
	0la11	0la12	0hy3	0hy6	0hy7	0hy8	0hy9	0hy12	0hy13	0ci1	0bu2	0bu5	0st1	0st2
Heliothrips adult	–	+	+	+	+	+	–	+	–	+	–	–	+	+
Heliothrips nymph 1 ins.	–	+	+	+	+	+	–	+	–	+	–	–	+	+
Thrips adult	–	+	+	+	+	+	–	+	–	+	+	–	+	+
Hercinothrips adult	–	+	+	+	+	+	+	+	–	+	+	–	+	+
Hercinothrips nymph	–	+	+	+	+	+	–	+	–	+	+	–	+	+
Aeolothrips adult	+	+	+	+	+	+	–	+	+	+	+	–	+	+
Phlaeothrips adult	–	+	+	+	+	–	+	+	+	+	+	+	+	+
Haplothrips adult	–	+	+	+	+	–	+	+	+	+	+	+	+	+

	Thoracic muscles											
	Idlm1	Idlm2	Idlm5	Idvm1	Idvm2	Idvm5	Idvm6	Idvm10	Idvm14	Idvm15	Idvm16	Idvm17
Heliothrips adult	+	+	+	–	+	+	+	+	+	+	+	+
Heliothrips nymph 1 ins.	+	+	+	–	–	+	+	+	+	+	+	+
Thrips adult	+	+	+	–	+	a	a	+	+	+	+	+

(continued)

Table 4.1 (continued)

	Thoracic muscles											
	Idlm1	Idlm2	Idlm5	Idvm1	Idvm2	Idvm5	Idvm6	Idvm10	Idvm14	Idvm15	Idvm16	Idvm17
Hercinothrips adult	+	+	+	–	+	a	a	+	+	+	+	+
Hercinothrips nymph	+	+	–	–	+	a	a	+	+	–	+	–
Aeolothrips adult	+	+	+	–	+	a	a	+	+	+	+	+
Phlaeothrips adult	+	+	+	+	+	a	a	+	+	+	+	+

	Thoracic muscles													
	Idvm19	Itpm2	Itpm3	Itpm4	Ispm1	Ipcm5	Ipcm8	Ivlm3	Ivlm4	Ivlm7	Ivlm9	Iscm3	Iscm4	Iscm6
Heliothrips adult	+	+	–	–	+	+	+	+	+	+	–	+	+	–
Heliothrips nymph 1 ins.	+	–	–	–	+	+	+	+	+	+	–	+	+	–
Thrips adult	+	–	–	+	+	+	+	+	+	+	+	+	+	+
Hercinothrips adult	+	+	+	–	–	+	+	+	+	+	+	–	+	+
Hercinothrips nymph	–	–	–	+	–	+	+	+	+	+	+	–	–	–
Aeolothrips adult	+	–	+	+	–	+	+	+	+	+	+	+	+	+
Phlaeothrips adult	+	–	–	–	–	+	+	+	+	+	+	+	+	–

	IIdlm1	IIdvm1	IIdvm5	IIdvm8	IItpm1	IItpm2	IItpm4	IItpm7	IItpm9	IItpm11	IItpm12	IIppm1	IIspm1
Heliothrips adult	+	+	+	+	+	+	+	+	–	+	+	–	+
Heliothrips nymph 1 ins.	+	–	+	+	–	–	–	–	–	+	–	–	–
Thrips adult	+	+	+	+	+	+	–	+	+	+	+	+	+
Hercinothrips adult	+	+	+	+	+	+	+	+	+	+	+	–	+
Hercinothrips nymph	–	–	+	+	–	–	–	–	–	+	–	–	–
Aeolothrips adult	+	+	+	+	+	+	+	+	+	+	+	+	+
Phlaeothrips adult	+	+	+	+	+	–	–	+	+	+	+	–	+

	IIspm2	IIspm6	IIpcm1	IIpcm4	IIvlm3	IIvlm7	IIscm1	IIscm2	IIscm4	IIscm6	IIscm7
Heliothrips adult	+	–	–	+	+	+	+	+	+	+	–
Heliothrips nymph 1 ins.	+	–	–	+	+	+	+	+	+	+	–

(continued)

Table 4.1 (continued)

	IIspm2	IIspm6	IIpcm1	IIpcm4	IIvlm3	IIvlm7	IIscm1	IIscm2	IIscm4	IIscm6	IIscm7
Thrips adult	+	–	+	+	+	+	+	+	+	+	+
Hercinothrips adult	+	–	+	+	+	+	+	+	+	+	+
Hercinothrips nymph	+	–	–	+	+	+	+	+	+	+	–
Aeolothrips adult	+	+	+	+	+	+	+	+	+	+	+
Phlaeothrips adult	+	–	+	+	+	+	+	+	+	+	+

	IIIdlm1	IIIdlm2	IIIdvm1	IIIdvm5	IIIdvm6	IIIdvm7	IIIdvm8	IIItpm2	IIItpm4	IIItpm9	IIppm2
Heliothrips adult	+	+	+	+	+	+	+	+	+	+	+
Heliothrips nymph 1 ins.	+	+	–	+	–	–	+	–	–	–	–
Thrips adult	+	+	+	+	+	+	+	+	+	+	+
Hercinothrips adult	+	+	+	+	+	–	+	+	+	+	–
Hercinothrips nymph	+	–	–	+	–	–	+	–	–	–	–
Aeolothrips adult	+	+	+	+	–	–	+	+	+	+	–
Phlaeothrips adult	+	+	+	+	+	–	+	+	+	+	+

	IIIspm1	IIIspm2	IIIpcm4	IIIpcm5	IIIvlm2	IIIscm1	IIIscm2	IIIscm4	IIIscm6
Heliothrips adult	+	+	+	+	+	+	+	+	+
Heliothrips nymph 1 ins.	–	–	+	–	+	+	+	+	–
Thrips adult	+	+	+	+	+	+	+	+	+
Hercinothrips adult	+	–	+	+	+	+	+	–	+
Hercinothrips nymph	–	–	+	–	+	+	+	–	–
Aeolothrips adult	+	+	+	–	+	+	+	–	+
Phlaeothrips adult	+	+	+	–	+	+	+	+	+

+ present; – absent; *a*, *b* fused muscles, ? unknown

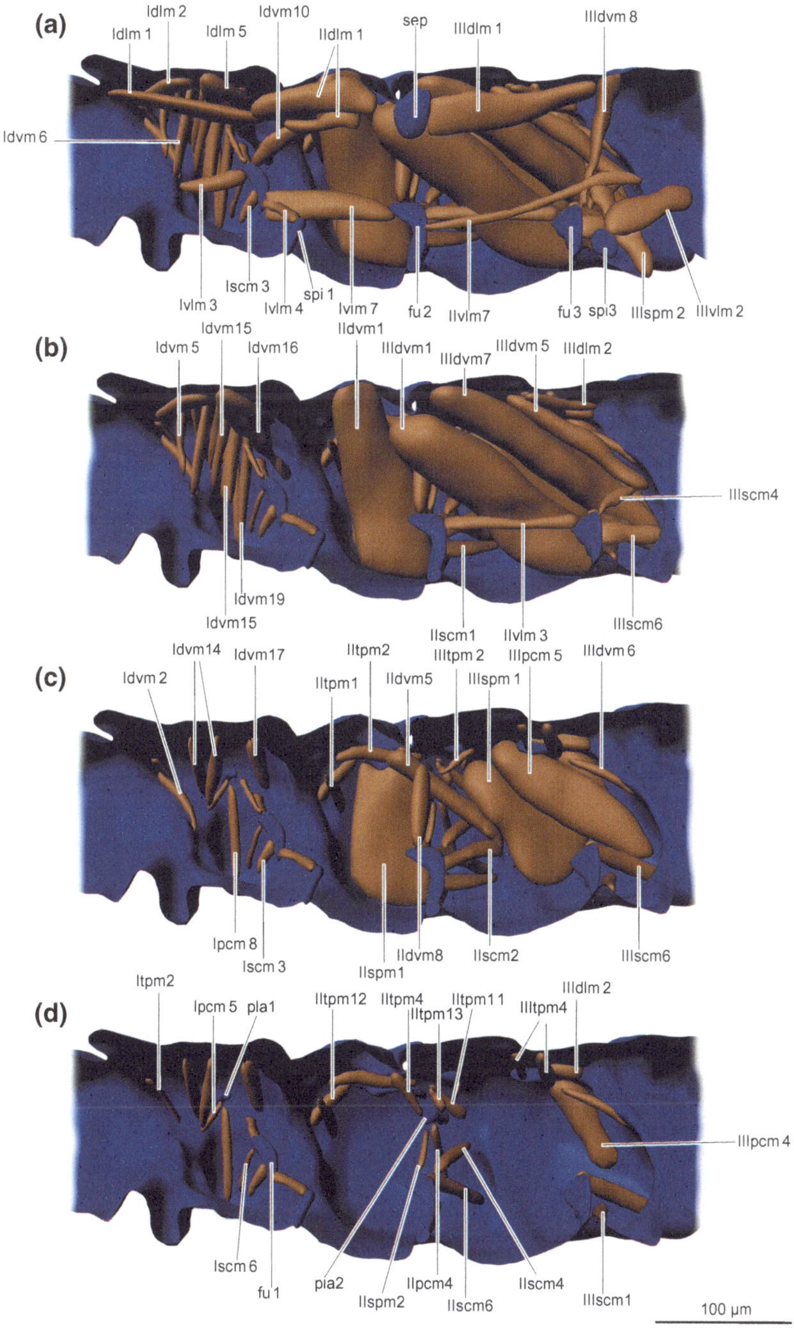

Fig. 4.13 Musculature of thorax adult *Heliothrips haemorrhoidalis*, 3D, lateral internal view: *fu1.2.3* pro-, meso-, and metafurca, *pla1.2* pleural apodeme pro- and of mesothorax, *sep* septa, *spi1.3* pro- and metaspina. Musculature see text

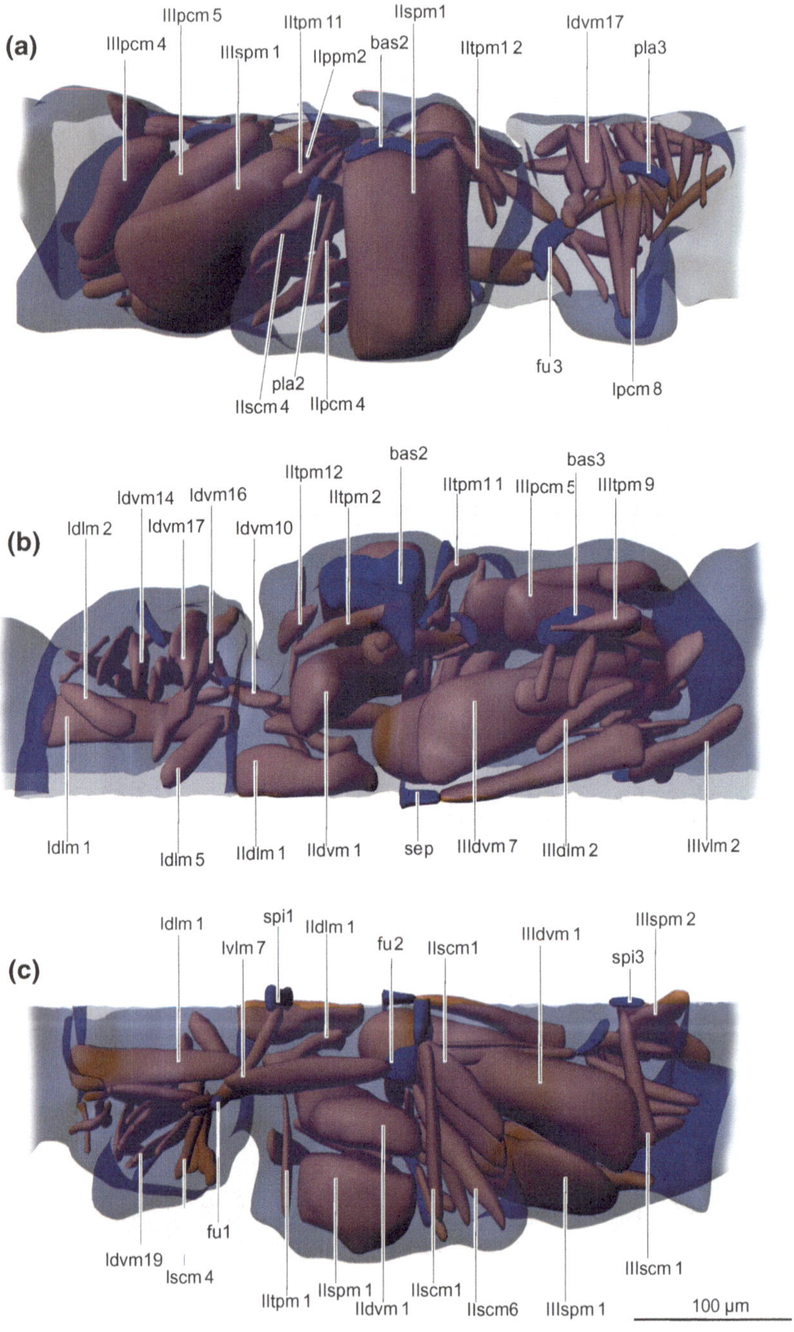

Fig. 4.14 Musculature of thorax adult *Heliothrips haemorrhoidalis*, 3D: **a** lateral view; **b** dorsal view; **c** ventral view; *bas2.3* basalare of meso- and metathorax, *fu1.2.3* pro-, meso-, and metafurca, *pla1.2* pleural apodeme of pro- and mesothorax, *sep* septa, *spi1.3* pro- and metaspina. Musculature see text

posterior): O, central part of pronotum; I, posterior margin of base of coxa. Idvm17 (M. pronoto-coxalis posterior): O, posterior part of pronotum; I, lateral margin of base of coxa. Idvm19 (M. pronoto-trochanteralis): O, lateral part of pronotum; I, trochanter. Itpm2 (M. propleuro-occipitalis): O, lateral part of posterior margin of occipital area; I, episternum prothorax. Ispm1 (M. profurca-apodemalis): O, apical part of profurca; I, pleural apodeme. IIpcm5 (M. mesanepisterno-trochanteralis): O, pleural apodeme; I, anterior margin of base of coxa. Ipcm8 (M. propleuro-trochanteralis): O, pleural apodeme; I, trochanter. Ivlm3 (M. profurca-tentorialis): O, profurca; I, ventrolateral part of posterior margin of cranium. Ivlm4 (M. profurca-spinalis): O, profurca; I, prodorsum. Ivlm7 (M. profurca-mesofurcalis): O, profurca; I, mesofurca. Iscm3 (M. profurca-coxalis medialis): O, profurca; I, medial margin of base of coxa. Iscm4 (M. profurca-coxalis lateralis): O, profurca; I, lateral margin of base of coxa. Internal musculature of legs not studied because of extremely small size.

Mesothorax. IIdlm1 (M. prophragma-mesophragmalis): O, central part of prophragma; I, central part of mesophragma (phragma of mesothorax). IIdvm1 (M. mesonoto-sternalis): O, lateral part of mesonotum; I, basisternite. IIdvm5 (M. mesonoto-coxalis posterior): O, lateral part of mesonotum; I, posterior margin of base of coxa. IIdvm8 (M. mesofurca-phragmalis): O, mesofurca; I, mesophragma. IItpm1 (M. prophragma-mesanepisternalis): O, anterior margin of mesonotum; I, anterior part of episternum. IItpm2 (M. mesopleura-praealaris): O, dorsal part of pleural suture; I, anterior margin of mesonotum. IItpm4 (M. mesonoto-pleuralis anterior): O, pleural apodeme; I, anterolateral margin of mesonotum. IItpm7 (M. mesanepisterno-axillaris): O, dorsal part of episternum; I, base of wing. IItpm11 (M. mesopleura-subalaris): O, epimeron of mesothorax; I, lateral margin of mesonotum. IItpm12 (M. prophragma-intersegmentalis): O, anterior margin of episternum; I, connecting membrane between pleurite and mesonotum. IIspm1 (M. mesopleura-sternalis): O, basalare; I, basisternite. IIspm2 (M. mesofurca-pleuralis): O, apex of mesofurca; I, apex of pleural apodeme. IIpcm4 (M. mesanepisterno-coxalis posterior): O, pleural apodeme; I, anterior margin of base of coxa. IIvlm3 (M. mesofurca-metafurcalis): O, mesofurca; I, metafurca. IIvlm7 (M. mesofurca-abdominosternalis): O, mesofurca; I, anterior margin of abdominal sternite 2. IIscm1 (M. mesofurca-coxalis anterior): O, mesofurca; I, anterior margin of base of coxa. IIscm2 (M. mesofurca-coxalis posterior): O, mesofurca; I, posterior margin of base of coxa. IIscm4 (M. mesofurca-coxalis lateralis): O, mesofurca; I, lateral margin of base of coxa. IIscm6 (M. mesofurca-trochanteralis): O, mesofurca; I, trochanter. Internal musculature of legs not studied because of extremely small size.

Metathorax. IIIdlm1 (M. mesophragma-metaphragmalis): O, septa of mesophragma; I, central part of metaphragma (phragma of metathorax). IIIdlm2 (M. metanoto-phragmalis): O, metanotum; I, metaphragma. IIIdvm1 (M. metanoto-sternalis): O, anterior part of metanotum; I, basisternite. IIIdvm5 (M. metanoto-coxalis posterior): O, central part of metanotum; I, posterior margin of base of coxae. IIIdvm6 (M. metacoxa-subalaris): O, lateral part of metanotum; I, posterior margin of base of coxae. IIIdvm7 (M. metanoto-trochanteralis): O, anterior part of metanotum; I, trochanter. IIIdvm8 (M. metafurca-phragmalis): O, apical part of metafurca; I, lateral part of metaphragma. IIItpm2 (M. metapleura-praealaris): O, dorsal margin of pleural ridge; I, lateral margin of metanotum. IIItpm4 (M. mesonoto-pleuralis anterior): O, dorsal margin of epimeron; I, lateral margin of pronotum. IIItpm9 (M. metepimero-axillaris tertius): O, dorsal margin of epimeron; I, lateral margin of pronotum. IIppm2 (M. metabasalare-intersegmentalis): O, basalare; I, membranous area between pleurite and metanotum. IIIspm1 (M. metapleura-sternalis): O, basalare; I, basisternite. IIIspm2 (M. metafurca-pleuralis): O, metafurca; I, pleural apodeme. IIIpcm4 (M. metanepisterno-coxalis posterior): O, dorsal part of episternum; I, anterior margin of base of coxa. IIIpcm5 (M. metanepisterno-trochanteralis): O, dorsal part of episternum; I, trochanter. IIIvlm2 (M. metafurca-abdominosternalis): O, metafurca; I, anterior margin of abdominal segment 2. IIIscm1 (M. metafurca-coxalis anterior): O, metafurca; I, anterior margin of base of coxa. IIIscm2 (M. metafurca-coxalis posterior): O, metafurca; I, posterior margin of base of coxa. IIIscm4 (M. metafurca-coxalis lateralis): O, metafurca; I, lateral margin of base of coxa. IIIscm6 (M. metafurca-trochanteralis): O, metafurca; I, trochanter. Internal musculature of legs not studied because of extremely small size.

Musculature of abdomen. Dorsal longitudinal musculature (Mm. longitudinalis dorsales), two subunits (interior, exterior) of several parallel fibers: O, anterior phragma, I, posterior phragma. Ventral longitudinal musculature (Mm. longitudinalis ventrales), several parallel fibers: O, anterior margin of

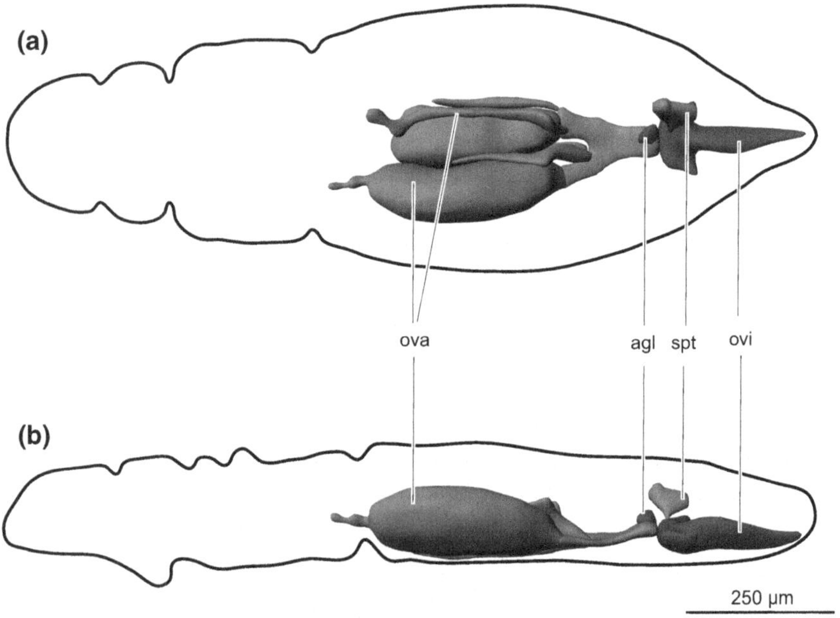

Fig. 4.15 Reproductive system of *Heliothrips haemorrhoidalis*, 3D: **a** dorsal view; **b** lateral view; *agl* accessory glands, *ova* ovary, *ovi* ovipositor, *spt* spermatheca

segment, I, posterior margin of segment. Dorsoventral musculature (Mm. dorsoventral is abdominalis): O, lateral part of tergite; I, lateral part of sternite.

Reproductive system

Male. Many Thripidae, including *Heliothrips haemorrhoidalis*, consist of populations that consist exclusively of females, and only some of them sporadically include males (Crawford 1940). In other species, the reproductive system consists of paired testes, spermiducts, accessory glands, and copulatory apparatus (Moritz 1989d).

Female reproductive system consists of paired ovaries, oviducts, spermatheca, and accessory gland (Fig. 4.15). Each ovary consists of three panoistic ovarioles.

4.4 Morphology of First Instar Nymph of *Heliothrips haemorrhoidalis* (Thripidae)

First instar nymphs (larvae) are almost identical morphologically to adults, except for proportions of body and particular organs, measurements, and some peculiar features described below.

4.4.1 External Morphology

Body size of first instar nymph in *H. haemorrhoidalis* 430–480 μm (M = 452; n = 7). Integument weakly pigmented and sclerotized, almost all body with wrinkled microsculpture. Eyes consisting of three ommatidia; ocelli absent. Antennal flagellum indistinctly divided into antennomeres (Fig. 4.16d). Labial palps one-segmented. Boundaries between all sclerites of thorax indiscernible (Fig. 4.16a, b, e). Tarsi one-segmented, with rather small claws (Fig. 4.17b, d). Abdomen consisting of ten ring-shaped segments (except for segment 1, with reduced sternal part). Boundaries between tergites and sternites indiscernible.

4.4.2 Internal Morphology

Cuticle 1.1–2.5 μm thick (M = 1.6; n = 80). Tracheal system strongly simplified, represented by two pairs spiracles (on mesothorax and abdominal segment 8), longitudinal stems and weakly branching tracheae.

Nervous system

The central nervous system of first instar nymphs is similar in structure to that of adults and differs from it in the following features (Figs. 4.8c, d and 4.18). Degree

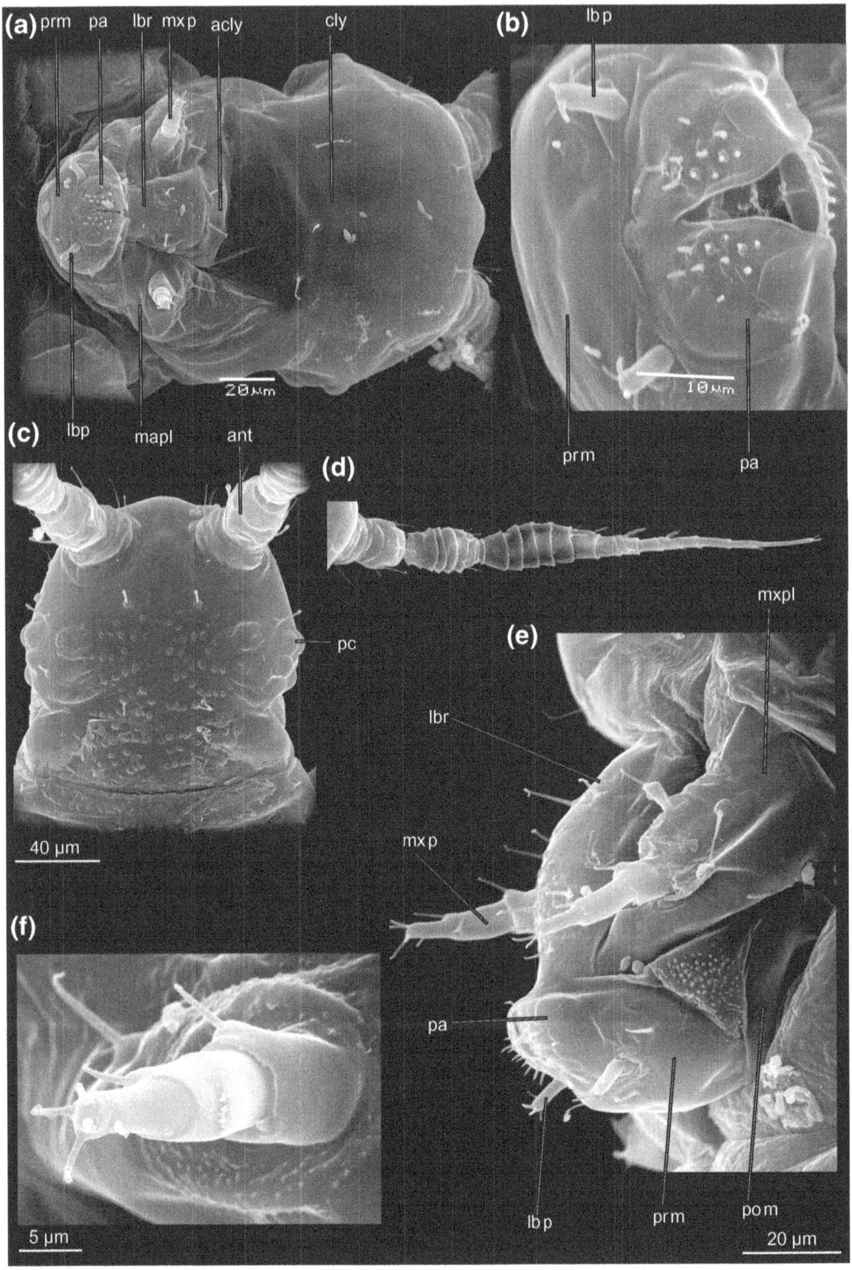

Fig. 4.16 External morphology of first instar nymph *Heliothrips haemorrhoidalis*, SEM: **a**, **c** head; **b**, **e** mouthparts; **d** antenna; **f** maxillary palp; *acly* anteclypeus, *ant* antenna, *cly* clypeus, *lbp* labial palp, *lbr* labrum, *mxp* maxillary palp, *mxpl* maxillary plates, *oc* eye, *pa* paraglossae, *pom* postmentum, *prm* prementum

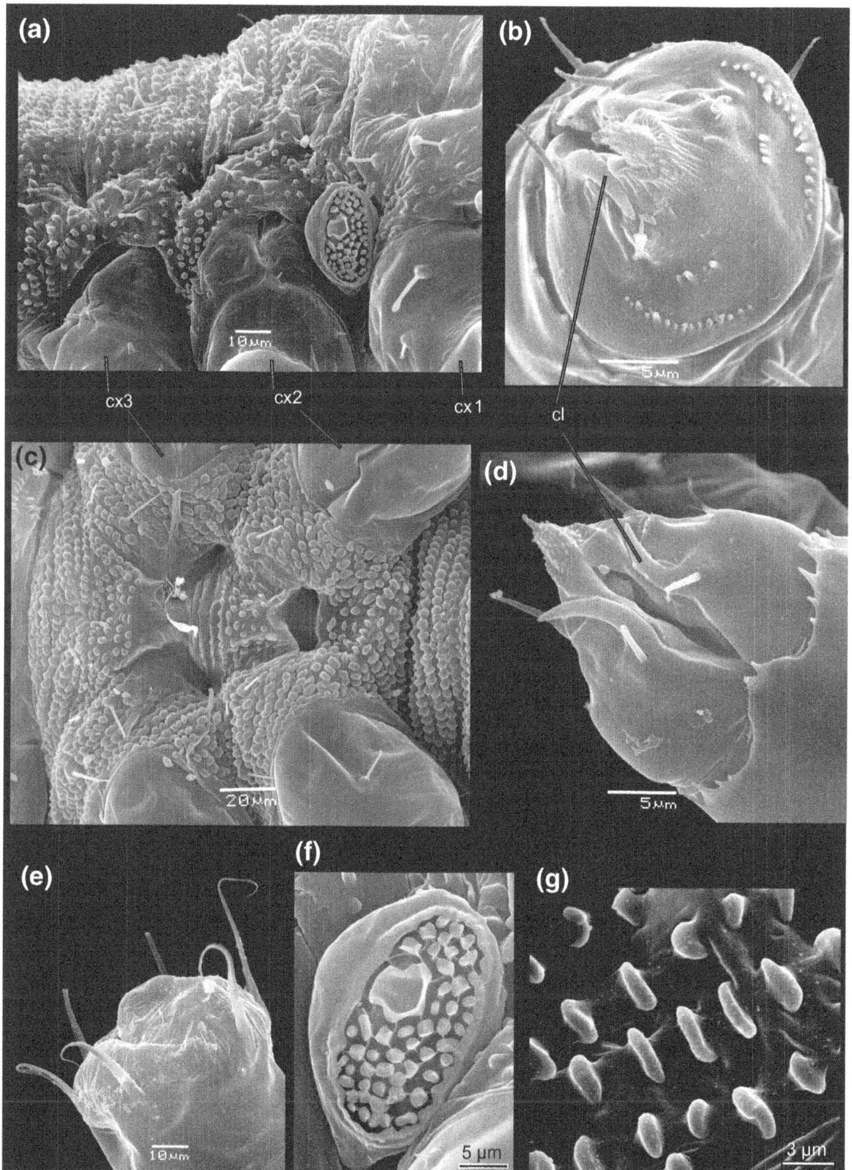

Fig. 4.17 External morphology of first instar nymph *Heliothrips haemorrhoidalis*, SEM: **a**, **c** thorax; **b**, **d** pretarsus; **e** apical part of abdomen; **f** mesthoracic spiracle; cuticular folds on abdominal segment; **a**, **b**, **e**, **f** lateral view; **c** ventral view; **d**, **g** dorsal view; *cl* claw, *cx1.2.3* pro-, meso-, and metacoxae

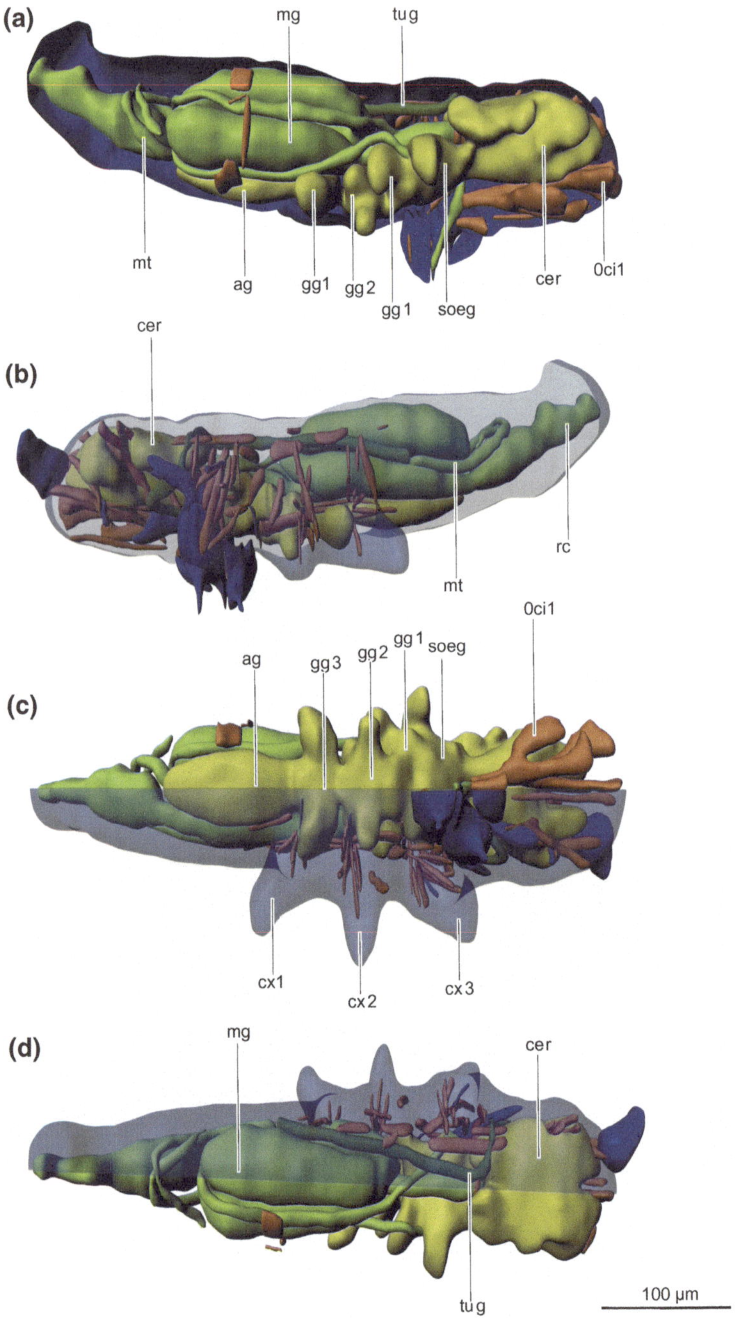
(a)
mg
tug
mt
ag
gg1
gg2
gg1
soeg
cer
Oci1
(b)
cer
mt
rc
(c)
Oci1
ag
gg3
gg2
gg1
soeg
cx1
cx2
cx3
(d)
mg
cer
tug
100 μm

◀ **Fig. 4.18** Internal morphology of first instar nymph *Heliothrips haemorrhoidalis*, 3D: **a** lateral internal view; **b** lateral external view; **c** dorsal view; **d** ventral view; *ag* abdominal ganglion, *cer* cerebrum, *cx1.2.3* pro-, meso-, and metacoxae, *gg1.2.3* pro-, meso-, and metathoracic ganglia, *mg* midgut, *mt* Malpighian tubules, *oes* oesophagus, *rc* rectum, *soeg* suboesophageal ganglion, *tug* tubular gland. Colors: *blue* cuticle, *green* digestive system, *yellow* central nervous system, *brown* musculature, *purple* reproductive system. Musculature see text

of ganglion compactization considerably higher in first instar nymphs than in adults. Ganglia of thorax very closely set and with blade-like processes in coxal parts of legs. Prothoracic ganglion with unpaired asymmetrical process. Cerebrum of nymphs situated mostly in cranium, but rather small part of protocerebrum shifted into prothorax. All three ganglia of thorax and abdominal ganglion separated, but very closely set.

Cell size in nervous system of first instar nymphs of *H. haemorrhoidalis* 2.69–5,35 μm (M = 3,18; n = 395) (Fig. 4.19). Cerebrum of first instar nymphs containing about 6000 cells.

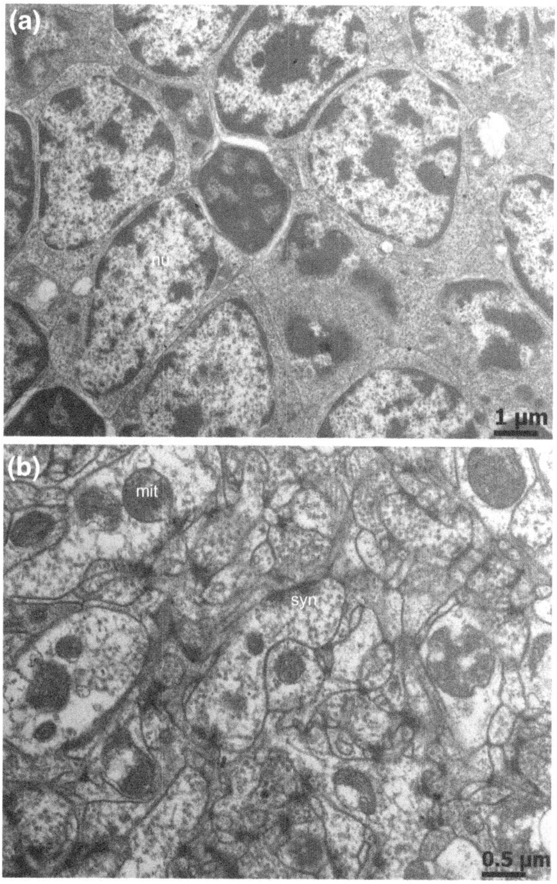

Fig. 4.19 Ultrastructure of cerebrum of first instar nymph *Heliothrips haemorrhoidalis*, TEM: **a** cortical слой; **b** neuropil; *mit* mitochondrion, *nu* nucleus, *syn* synapse

Budding central body complexle in neuropil of protocerebrum. All modular organization of central body still indiscernible in first instar nymphs, but its general double structure already discernible. Primordia of mushroom bodies present. Optic lobes developed weakly. Only primordia of two optic neuropils present in nymphs of *H. haemorrhoidalis*. Antennal lobes of first instar nymphs weakly developed. Absent glomerular organization absent. It is known that antennal lobes at larval stages lag behind in development compared to the main structures of the protocerebrum.

Muscular system

Musculature of head. tm1 absent.

Musculature of thorax. Idvm2. Itpm2. IIdvm1. IItpm1. IItpm4. IItpm7, IItpm12. IIspm1. IIIdvm1. IIIdvm6, IIIdvm7, IIItpm2. IIItpm4. IIItpm9, IIppm2. IIIspm1. IIIspm2. IIIpcm5, IIIscm6 absent.

Reproductive system absent.

References

Crawford, J. C. (1940). The male of *Heliothrips Haemorrhoidalis* (Bouche) (Thysanoptera). *Proceedings of the Entomological Society of Washington, 42*, 90–91.

Davies, R. G. (1958). Observations on the morphology of the head and mouthparts in the Thysanoptera. *Proceedings of the Royal Entomological Society of London Series A General Entomology, 33*(7–9), 97–106.

Garman, H. (1890). The mouth-parts of the Thysanoptera. *Bulletin of the Essex Institute, 22*, 1–30.

Heming, B. S. (1978). Structure and function of the mouthparts in larvae of *Haplothrips verbasci* (Osborn) (Thysanoptera, Tubulifera, Phlaeothripidae). *Journal of Morphology, 156*(1), 1–37.

Hunter, W. B., & Ullman, D. E. (1992). Anatomy and ultrastructure of the piercing-sucking mouthparts and paraglossal sensilla of *Frankliniella occidentalis* (Pergande) (Thysanoptera: Thripidae). *International Journal of Insect Morphology and Embryology, 21*(1), 17–35.

Hunter, W. B., & Ullman, D. E. (1994). Precibarial and cibarial chemosensilla in the western flower thrips, *Frankliniella occidentalis* (Pergande) (Thysanoptera: Thripidae). *International Journal of Insect Morphology and Embryology, 23*(2), 69–83.

Matsuda, R. (1970). Morphology and evolution of the insect thorax. *Memoirs of the Entomological Society of Canada, Suppl 76*, 1–431.

Mickoleit, E. (1961). Zur Thoraxmorphologie der Thysanoptera. *Zoologische Jahrbuecher Abteilung fuer Anatomie und Ontogenie der Tiere, 79*, 1–92.

Mickoleit, E. (1963). Untersuchungen zur Kopfmorphologie der Thysanoptera. *Zoologische Jahrbuecher Abteilung fuer Anatomie und Ontogenie der Tiere, 81*, 101–150.

Moritz, G. (1978). Das Abdomen von *Aeolothrips astutus* Priesner (Thysanoptera, Aeolothripidae). *Wissenschaftliche Hefte, 2*(1978), 143–144.

Moritz, G. (1979). Die Anatomie und Morphologie des Kopfinnenskeletts (Tentorium) der Aeolothripidae (Thysanoptera, Insecta). *Wissenschaftliche Hefte, 2*(1979), 97–108.

Moritz, G. (1982a). Zur Morphologie und Anatomie des Fransenflüglers *Aeolothrips intermedius* Bagnall, 1934 (Aeolothripidae, Thysanoptera, Insecta) – 1. Mitteilung: Der Kopf. *Zoologische Jahrbuecher Abteilung fuer Anatomie und Ontogenie der Tiere, 107*, 557–608.

Moritz, G. (1982b). Zur Morphologie und Anatomie des Fransenflüglers *Aeolothrips intermedius* Bagnall, 1934 (Aeolothripidae, Thysanoptera, Insecta) – 2. Mitteilung: Der

Thorax. *Zoologische Jahrbuecher Abteilung fuer Anatomie und Ontogenie der Tiere, 108*, 55–106.

Moritz, G. (1982c). Zur Morphologie und Anatomie des Fransenflüglers *Aeolothrips intermedius* Bagnall, 1934 (Aeolothripidae, Thysanoptera, Insecta) – 3. Mitteilung: Das Abdomen. *Zoologische Jahrbuecher Abteilung fuer Anatomie und Ontogenie der Tiere, 108*, 293–340.

Moritz G (1982d) Zur Morphologie des Kopfinnenskeletts (Tentorium) bei den Thysanoptera. *Deutsche Entomologische Zeitschrift 29*(1–3), 17–26.

Moritz G (1988a) On the genesis of the wings and the wing musculature of *Hercinothrips femoralis* (O.M. Reuter) (Thysanoptera, Insecta). *Acta Phytopathologica et Entomologica Hungarica 23*(3–4), 313–319.

Moritz, G. (1988b). Die Ontogenese der Thysanoptera unter besonderer Berücksichtigung des Fransenflüglers *Hercinothrips femoralis* (O.M. Reuter 1891) 1. Embryonalentwicklung. *Zoologische Jahrbuecher Abteilung fuer Anatomie und Ontogenie der Tiere, 117*, 1–64.

Moritz, G. (1988c). Die Ontogenese der Thysanoptera unter besonderer Berücksichtigung des Fransenflüglers *Hercinothrips femoralis* (O.M. Reuter 1891) 2. Erst- und Zweitlarve. *Zoologische Jahrbuecher Abteilung fuer Anatomie und Ontogenie der Tiere, 117*, 299–351.

Moritz, G. (1989a). Die Ontogenese der Thysanoptera unter besonderer Berücksichtigung des Fransenflüglers *Hercinothrips femoralis* (O.M. Reuter 1891) 3. Praepupa und Pupa. *Zoologische Jahrbuecher Abteilung fuer Anatomie und Ontogenie der Tiere, 118*, 15–54.

Moritz, G. (1989b). Die Ontogenese der Thysanoptera unter besonderer Berücksichtigung des Fransenflüglers *Hercinothrips femoralis* (O.M. Reuter 1891) 4 Imago-Kopf. *Zoologische Jahrbuecher Abteilung fuer Anatomie und Ontogenie der Tiere, 118*, 273–307.

Moritz, G. (1989c). Die Ontogenese der Thysanoptera unter besonderer Berücksichtigung des Fransenflüglers *Hercinothrips femoralis* (O.M. Reuter 1891) 5. Imago-Thorax. *Zoologische Jahrbuecher Abteilung fuer Anatomie und Ontogenie der Tiere, 118*, 391–427.

Moritz, G. (1989d). Die Ontogenese der Thysanoptera unter besonderer Berücksichtigung des Fransenflüglers *Hercinothrips femoralis* (O.M. Reuter 1891) 6. Imago-Abdomen. *Zoologische Jahrbuecher Abteilung fuer Anatomie und Ontogenie der Tiere, 119*, 157–217.

Mound, L. A. (1971). The feeding apparatus of thrips. *Bulletin of Entomological Research, 60*, 547–548.

Mound, L. A. (2009). Sternal pore plates (glandular areas) of male Thripidae (Thysanoptera). *Zootaxa, 2129*, 29–46.

Peterson, A. (1915). Morphological studies of the head and mouth parts of the Thysanoptera. *Annals of the Entomological Society of America, 8*, 20–66.

Polilov, A. A., & Shmakov, A. S. (2016). The anatomy of the thrips *Heliothrips haemorrhoidalis* (Thysanoptera, Thripidae) and its specific features caused by miniaturization. *Arthropod Structure & Development* (in press).

Reyne, A. (1927). Untersuchungen über die Mundteile der Thysanopteren. *Zoologische Jahrbuecher Abteilung fuer Anatomie und Ontogenie der Tiere, 49*, 391–500.

Chapter 5
Structure of the Principal Groups of Microinsects. III. Featherwing Beetles (Coleoptera: Ptiliidae)

5.1 Introduction

The family Ptiliidae includes the smallest Coleoptera and the smallest free-living insects. The smallest representative of this family is *Scydosella musawasensis,* in which adults have a body length of 325 µm. Featherwing beetles are distributed almost ubiquitously and occur in various decomposing plant debris, such as forest litter, decaying wood, etc., as well as in polypore fungi and feces of various animals.

Minute size complicated the study of the morphology of Ptiliidae, and until recently their external and internal morphology remained almost unknown. The external morphology of adults was treated only in outdated general studies (Gillmeister 1845; Matthews 1872, 1900) and several modern studies on the morphology of particular genera (Seevers and Dybas 1943; Sorensson 1997; Hall 1999). Data on the morphology of larvae are even scantier: larvae are known in 12 genera (of more than 70), many of them only briefly described (Boving and Craighead 1931; Paulian 1941) or simply figured in some illustrations (De Marzo 2000, 2001). There is only one general study (Dybas 1976), in which a general morphological portrait of ptiliid larvae is provided, based on representatives of nine genera. In recent years, the morphology of Ptiliidae came to be investigated intensively in the light of the study of miniaturization in insects (Grebennikov and Beutel 2002; Polilov and Beutel 2009).

The internal morphology of adult ptiliids remained unstudied until recently. There were two publications on the structure of the reproductive system (De Marzo 1992; De Coninck and Coessens 1981) and specialized studies on the structure of the spermatozoon (Dybas and Dybas 1987, 1981), which is longer than the body of adults in some ptiliid species. At present, ptiliids, as the smallest living insects, have become model organisms for studying the effects of miniaturization on structure in insects, and the internal morphology of many representatives of this family has been studied in rather much detail (Grebennikov and Beutel 2002; Polilov 2005, 2008; Polilov and Beutel 2009).

A.A. Polilov, *At the Size Limit—Effects of Miniaturization in Insects*,
DOI 10.1007/978-3-319-39499-2_5

The taxonomic position of Ptiliidae remains unchanged since the superfamily Staphylinoidea was recognized (Ganglbauer 1895); most experts believe that the family most closely related to Ptiliidae is Hydraenidae (Lawrence and Newton 1982, 1995; Newton and Thayer 1988, 1992; Hansen 1997; Hunt et al. 2007; Lawrence et al. 2011).

The superfamily Staphylinoidea is one of the largest coleopteran superfamilies; it includes over 50,000 species. Few studies summarize this large material. Several studies should be specially noted: one that described the external morphology of adults and larvae of all families of the superfamily Staphylinoidea (Hansen 1997), another on the morphology of larvae in some representatives of most subfamilies of this superfamily (Beutel and Molenda 1997), and three studies on the structure of mouthparts (Betz et al. 2003; Weide and Betz 2009; Weide et al. 2014).

The external morphology of adults and larvae of Staphylinidae is well studied. There are many studies on particular aspects of this subject, but there also many summarizing publications (Paulian 1941; Pototskaya 1967, 1971; Herman 1970; Tikhomirova 1973, 1974; Hammond 1976; Naomi 1985, 1987a, b, 1988a, b, c, d, 1989a, b, c, d, 1990; Leschen and Beutel 2001; Weide et al. 2010). Data on the internal morphology of Staphylinidae are much fewer; they are represented only by studies of particular organ systems (Carstens and Storch 1980; Araujo and Pasteels 1985; Panov 1986; Leschen and Beutel 2001; Weide et al. 2010, 2014).

The morphology Hydraenidae has been studied much less thoroughly: there are only a few studies on particular aspects of external morphology (Beutel 1994; Beutel and Jäch 1995; Delgado and Soler 1996, 1997a, b, c) and morphological sections of large taxonomic studies (Perkins 1980, 1997; Jach et al. 2000). The internal morphology has not been studied, except for brief descriptions of the reproductive system (Perkins 1980) and musculature of the head (Jach et al. 2000; Beutel et al. 2003).

5.2 Brief Description of the Structure of Ptiliidae

Small beetles, adults 0.3–3 mm long (on average about 0.8 mm). Body rounded, oval, or elongate (Figs. 5.1 and 5.2). Coloration of integument monotonous, in different representatives of this family varying from yellow (weakly pigmented wingless forms) to black. Usually, densely covered with hairs, sometimes with double pubescence (hairs of different length, structure, or orientation). Structure of integument diverse: from simple sparse punctation to complex cellulate microsculpture. Because of microsculpture, pubescence, and covering of hydrophobic compounds, body nonwettable. **Head** prognathous, not sunk into pronotum. Cranium solid, without sutures. Compound eyes lateral, consisting of 35–90 ommatidia each. Ocelli absent. Antennae with not compacted 2–3-segmented club, first two basal antennomeres considerably larger than others. Number of antennomeres usually 11, but in some representatives reduced, sometimes to eight. Mouthparts consisting of well developed mandibles, maxillae, labrum, and labium. **Thorax** in ptiliids, as in other beetles,

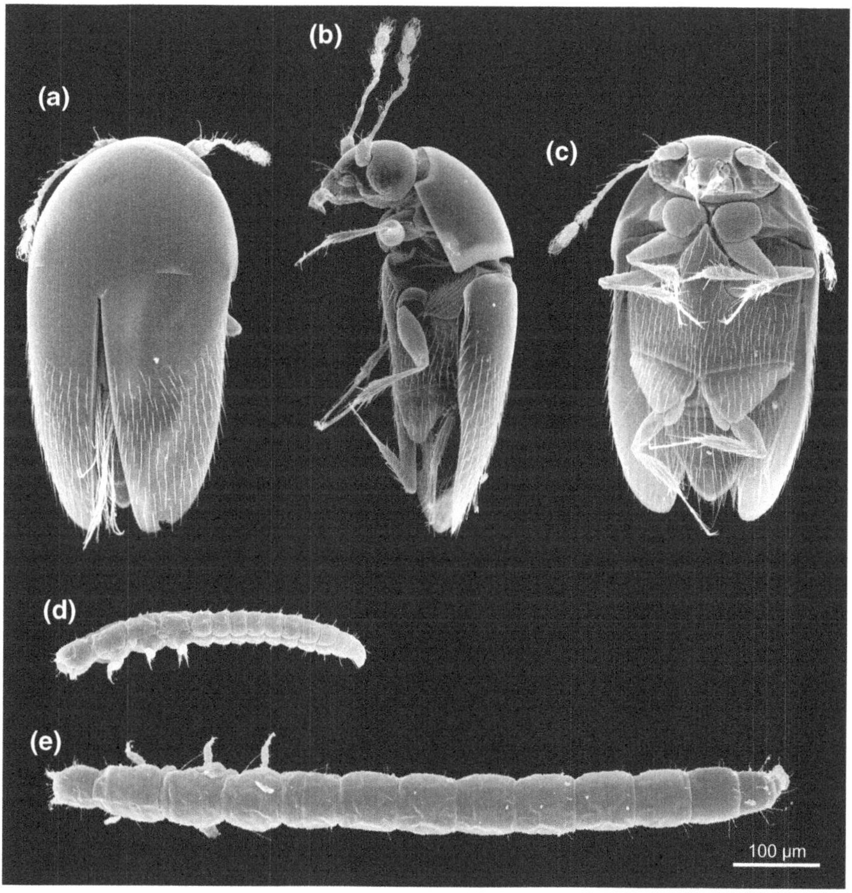

Fig. 5.1 *Mikado* sp., SEM: **a–c** adult; **d** first instar larva; **e** last instar larva; **a**, **e** dorsal view; **b**, **d** lateral view; **c** ventral view

consisting of detached prothorax and fused meso- and metathorax. Pronotum strongly sclerotized. Procoxal cavities usually open, but in some representatives completely closed. Mesothorax relatively small; metathorax larger than either pro- or metathorax. **Elytra** of various shape and microsculpture, without pronounced longitudinal striae. In most species elytra with narrow epipleura. Usually, elytra slightly shortened and not covering distal 2 or 3 abdominal segments, less often not shortened and covering all abdomen or strongly shortened and covering only basal abdominal segments. **Hindwings** typically with characteristic 'feathered' structure, sometimes without it, at rest folded transversely and situated under elytra. **Legs** with normal division into coxa, trochanter, femur, tibia, and tarsus, their relative size strongly varying among ptiliids. Tarsi 2–3-segmented, with two simple claws. **Abdomen** consisting of 10 segments. Number of visible sternites 6–7, because sternite 1 reduced, and several distal ones modified and sunk into abdomen. Number of visible tergites 9–10.

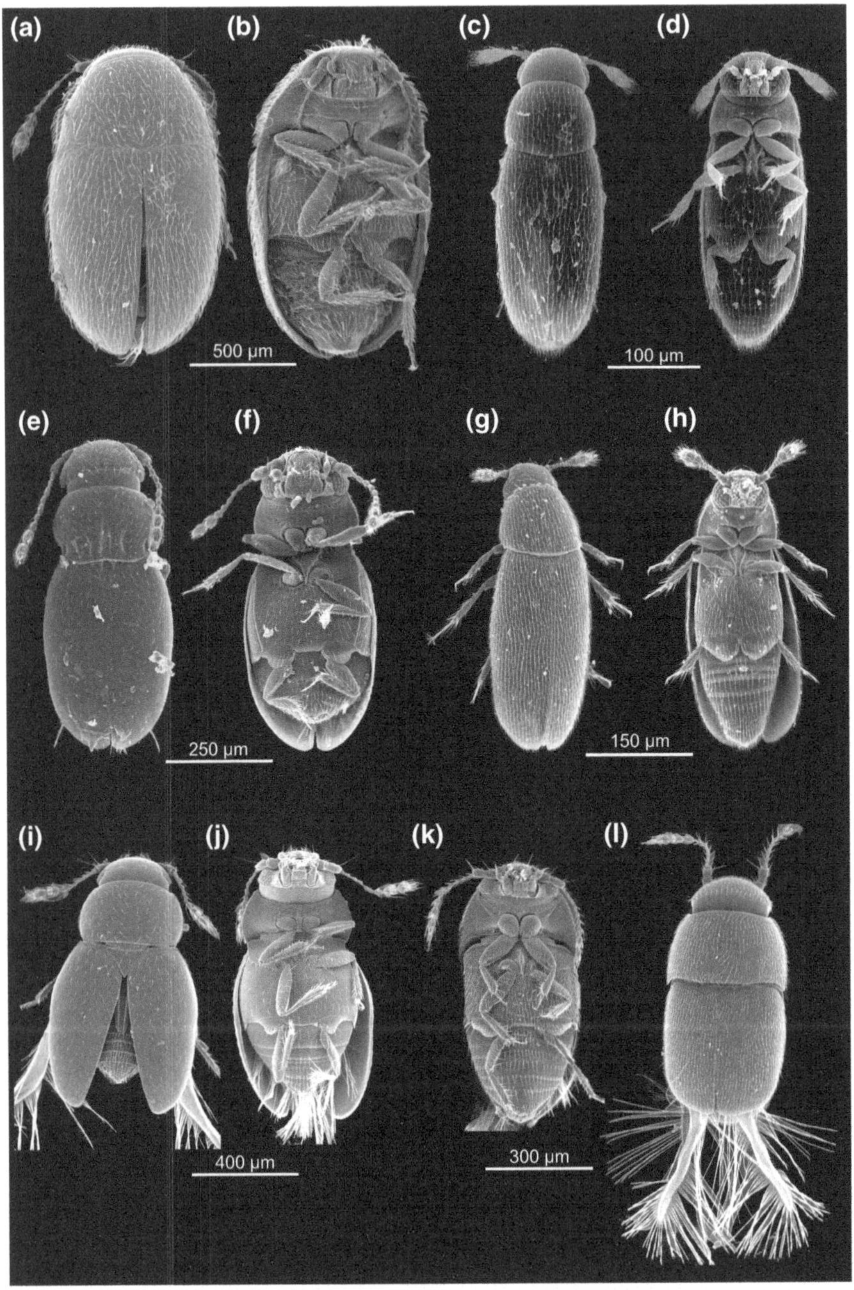

Fig. 5.2 Diversity of ptiliids, SEM: **a**, **b** *Nossidium pilosellum*; **c**, **d** *Nanosella* sp.; **e**, **f** *Millidium minutissimum*; **g**, **h** *Primorskella anodonta*; **i**, **j** *Ptenidium formicetorum*; **k**, **l** *Acrotrichis montandoni*; **a**, **c**, **e**, **g**, **i**, **k** dorsal view; **b**, **d**, **f**, **h**, **j**, **l** ventral view

External male genitalia in most species represented by simple penis, only in species of one genus (*Nossidium*) with parameres. On important diagnostic character is the structure of the **spermatheca**, very strongly varying among ptiliids. External sexual dimorphism weakly pronounced, sometimes males somewhat smaller than females and with stronger armature of protibiae and abdominal sternites. In the genera *Ptinella, Pteryx, Ptinelloides,* and *Astatopteryx* both sexes with winged and wingless forms (Dybas 1978). Wingless forms also distinguished by reduced eyes, less strongly sclerotized and less strongly pigmented integument and some other characters.

Larvae campodeiform, slightly curved in dorsoventral direction, length 0.8–2 mm (last instar). Body elongate, parallel-sided, almost cylindrical in cross-section. Integument weakly sclerotized and not pigmented. Body with scant setae of different length. **Head** prognathous. Cranium solid, without sutures. Ocelli absent, except for one species of the genus *Nossidium* (Dybas 1976). Antennae 3-segmented. Mouthparts consisting of well-developed mandibles, maxillae, labrum, and labium. **Thorax** consisting of three distinguished segments, subequal in size, slightly wider than abdomen. Thorax with only weakly sclerotized nonpigmented dorsal sclerites. Pair of spiracles present between pro- and mesothorax. **Legs** consisting of coxa, trochanter, femur, fused tibia and tarsus, and claw. **Abdomen** consisting of 10 segments. All segments without pronounced sclerites. In most species spiracles on abdominal segments absent. Segment 9 with 1-segmented urogomphi, in some representatives absent. Segment 10 with anal hooks and membranous anal vesicle.

5.3 Morphology of Adult Featherwing Beetles (Coleoptera: Ptiliidae)

Adult body length 980–1050 μm (M = 1015, n = 10) in *Acrotrichis grandicollis,* 390–455 μm (M = 430, n = 10) in *Mikado* sp., 325–352 μm (M = 338, n = 10) in *Scydosella musawasensis*

5.3.1 External Morphology

Structure of head

Cranium solid, without sutures (Fig. 5.3). Head connected with prothorax only via articular membrane, cervical sclerites absent. In contrast to many staphylinids and hydraenids, in ptiliids clypeus fused with frons (frontoclypeal suture absent), gular sutures also absent, and therefore with gular plate distinguished. Tentorium different from that of Staphylinidae and Hydraenidae in absence of laminitentorium, in many representatives of Nanosellini also in absence of dorsal tentorial arms (Fig. 5.4a, b).

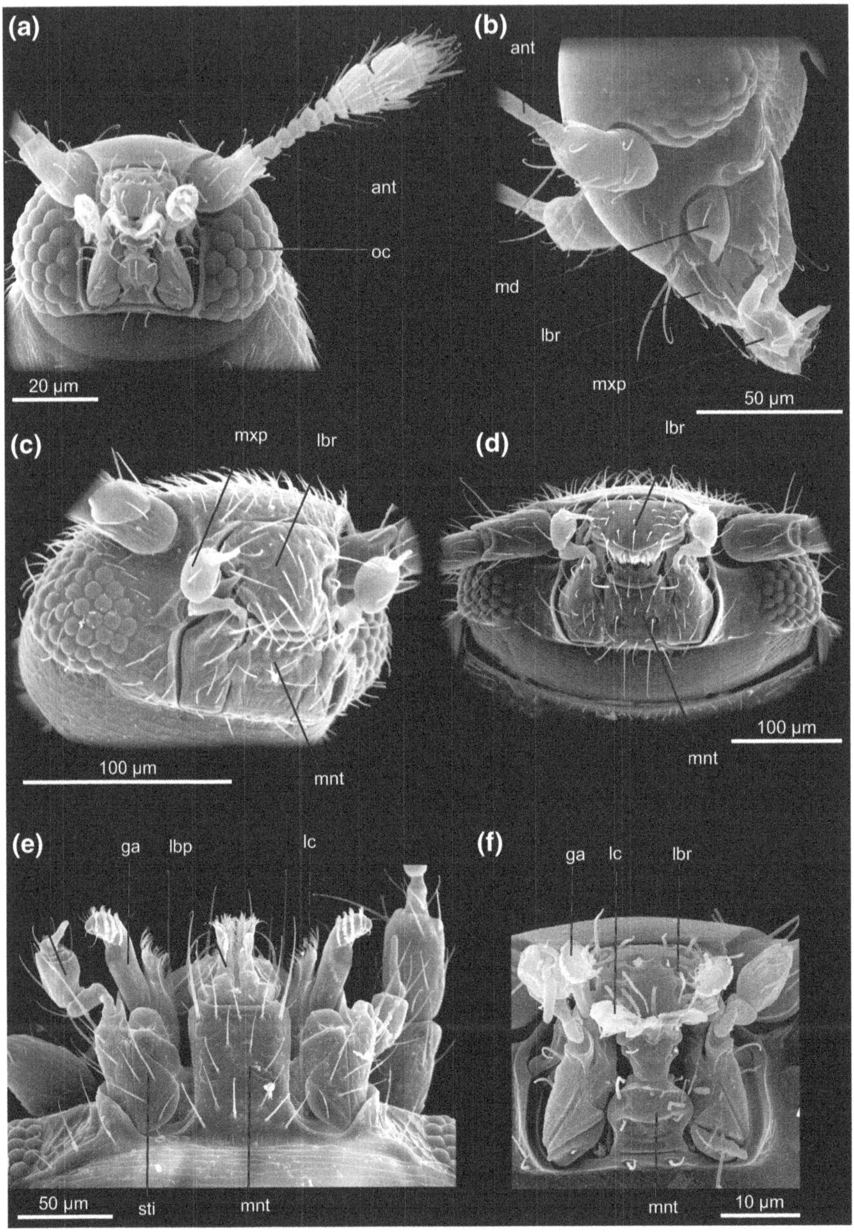

Fig. 5.3 Head and mouthparts of adult ptiliids, SEM: **a** *Nanosella* sp.; **b**, **f** *Mikado* sp.; **c** *Ptilium myrmecophilum*; **d** *Acrotrichis grandicollis*; **e** *Acrotrichis sericans*; **a**, **c**, **d** folded mouthparts at rest; **b**, **e**, **f** unfolded mouthparts at work; **a**, **c**, **d**, **f** frontal view; **b** lateral view; **e** ventral view; *ant* antenna, *ga* galea, *lbp* labial palp, *lbr* labrum, *lc* lacinia, *md* mandible, *mnt* mentum, *mxp* maxillary palp, *oc* eye, *sti* stipes

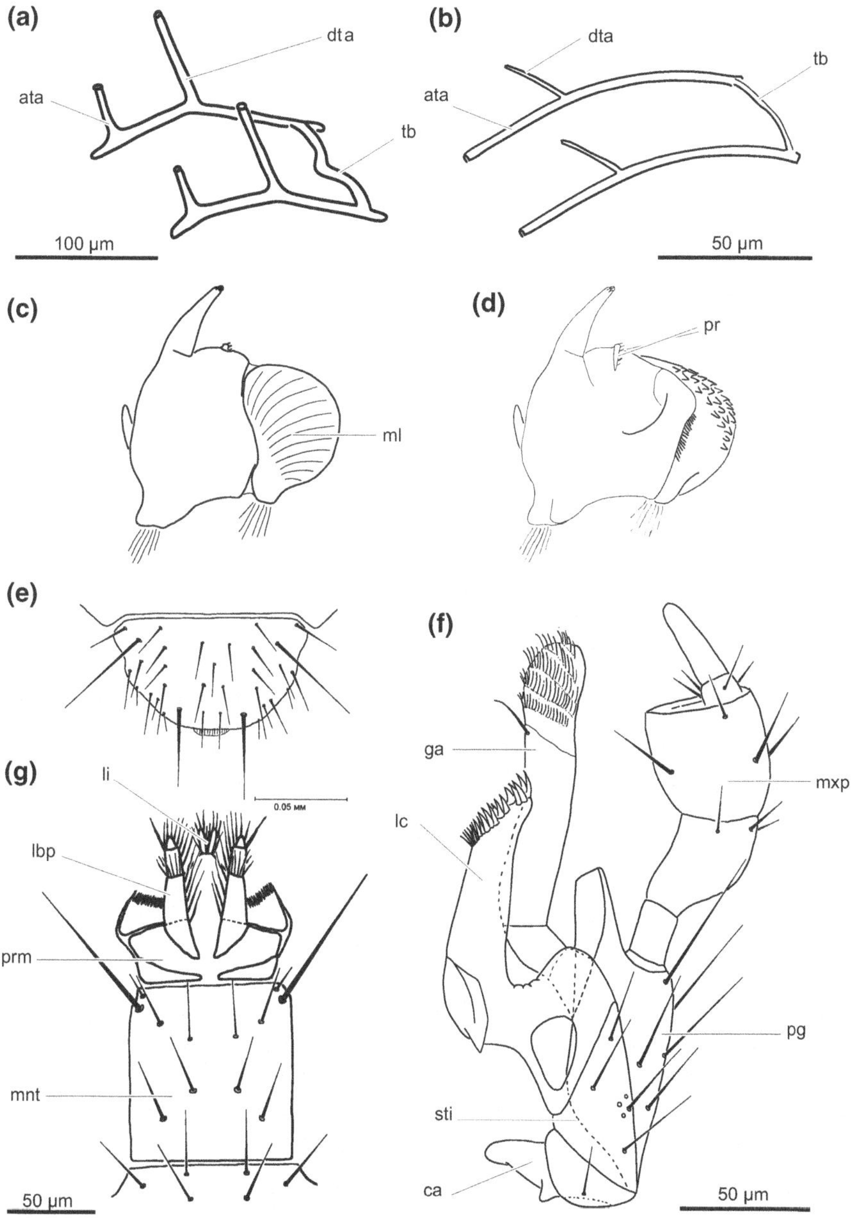

Fig. 5.4 Tentorium and mouthparts of ptiliids, schematic: **a**, **c–d** *Acrotrichis montandoni*; **b** *Nanosella* sp; **a**, **b** tentorium; **c**, **d** mandibles, **e** labrum, **f** maxilla, **g** labium; *ata* anterior tentorial arms, *dta* dorsal tentorial arms, *ca* cardo, *ga* galea, *lbp* labial palp, *lc* lacinia, *li* ligula, *ml* mobile lobe, *mnt* mentum, *mxp* maxillary palp, *pg* palpiger, *pr* prostheca, *prm* prementum, *sti* stipes, *tb* tentorial bridge

Compound eyes consisting of 35–90 ommatidia (in cotrast to tens of times as many in most staphylinids and other beetles). **Ocelli** absent.

Antennae in most representatives of Ptiliidae 11-segmented, as in most representatives of Staphylinidae and Hydraenidae, but in many species of Nanosellini and Cephaloplectinae 10-segmented, and in some species eight-segmented. Antennae attached laterally in front of eyes. Two types of sensilla recorded on ptiliid antennae: trichoid (probably mechanoreceptors, situated on all antennomeres, 3–40 on each antennomere) and digitiform (probably olfactory, situated on two apical antennomeres, 8–25 on each antennomere) (Fig. 5.5).

Mouthparts peculiarly structured (Figs. 5.3, 5.4, 5.6, 5.7, 5.8 and 5.9). **Labrum** movably joined with cranium and appearing as simple plate bearing 20–40 setae of different length. Internal surface of labrum and dorsal surface of oral cavity with complex epipharynx bearing tufts of setae and cuticular ridges. **Mandibles** well

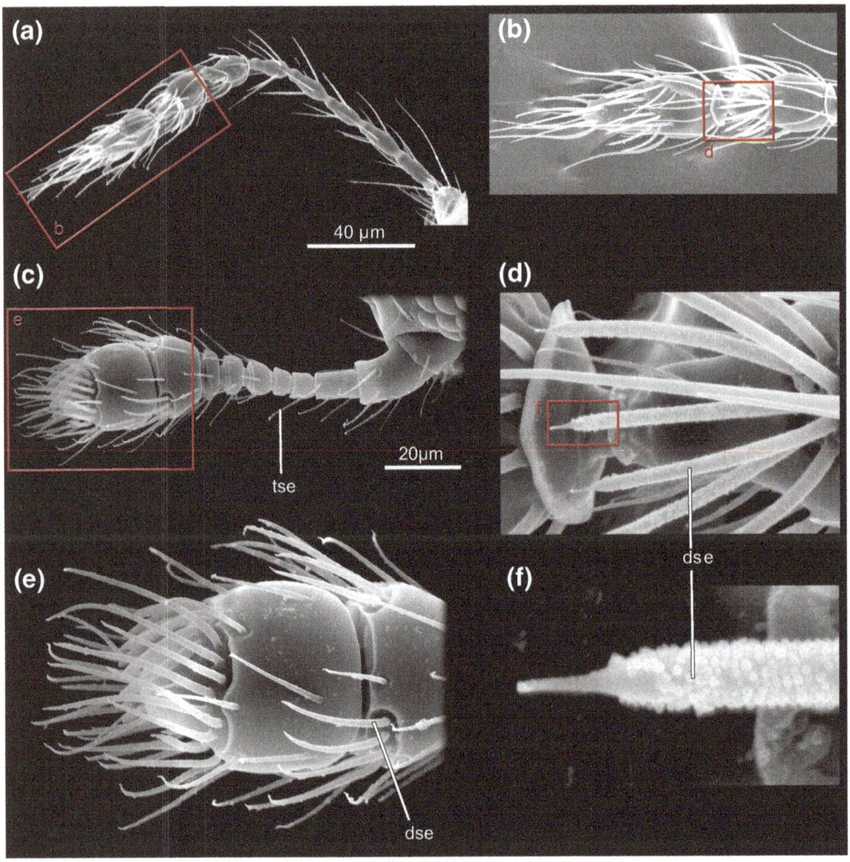

Fig. 5.5 Antennae of adult ptiliids, SEM: **a**, **b**, **d**, **f** *Acrotrichis montandoni*; **c**, **e** *Primorskella anodonta*; *dse* digitiform sensillum, *tse* trichoid sensillum

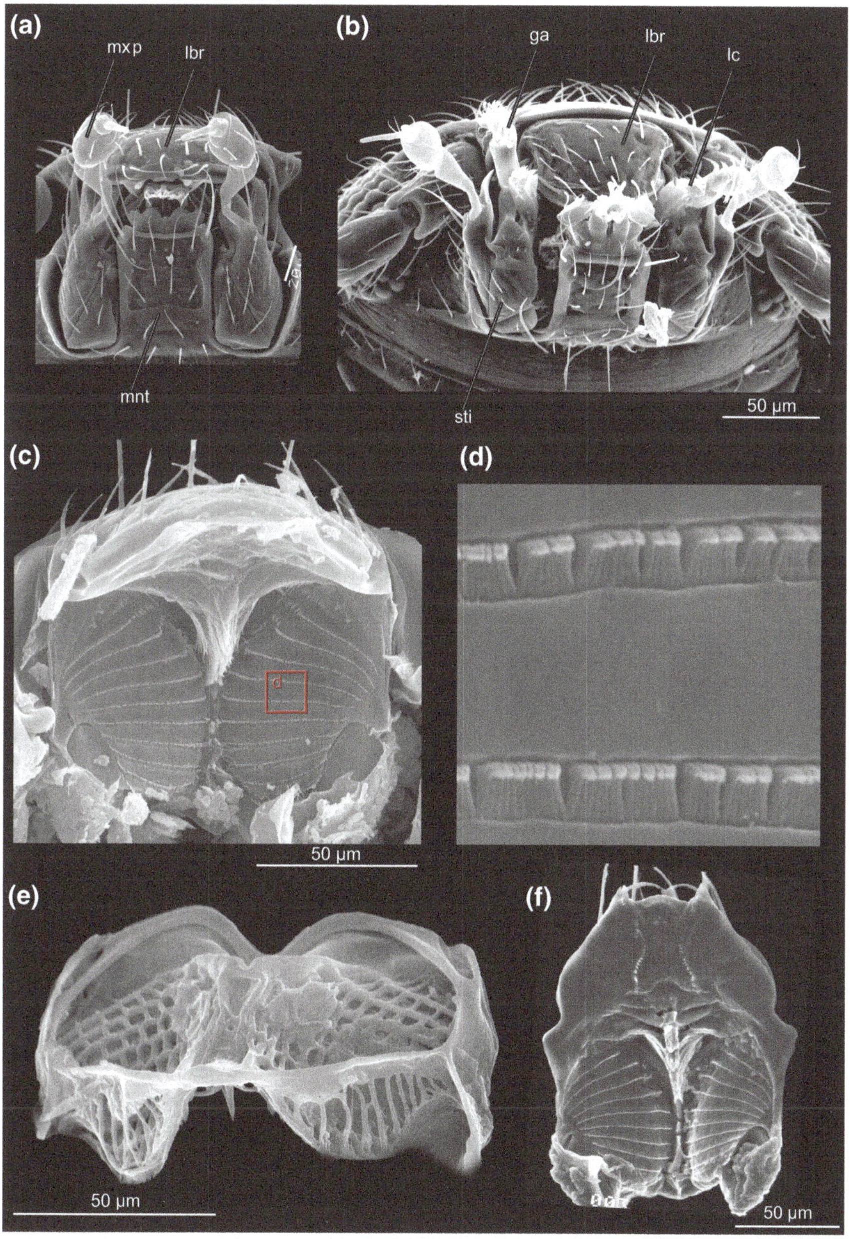

Fig. 5.6 Mouthparts of adult ptiliids, SEM: **a**, **b** mouthparts of *Baeocrara variolosa*; **c**–**e** epipharynx of *Acrotrichis sericans*; **f** epipharynx of *Acrotrichis montandoni*; **a**–**d**, **f** ventral view; **e** dorsal view; *ga* galea, *lbr* labrum, *lc* lacinia, *mnt* mentum, *mxp* maxillary palp, *sti* stipes

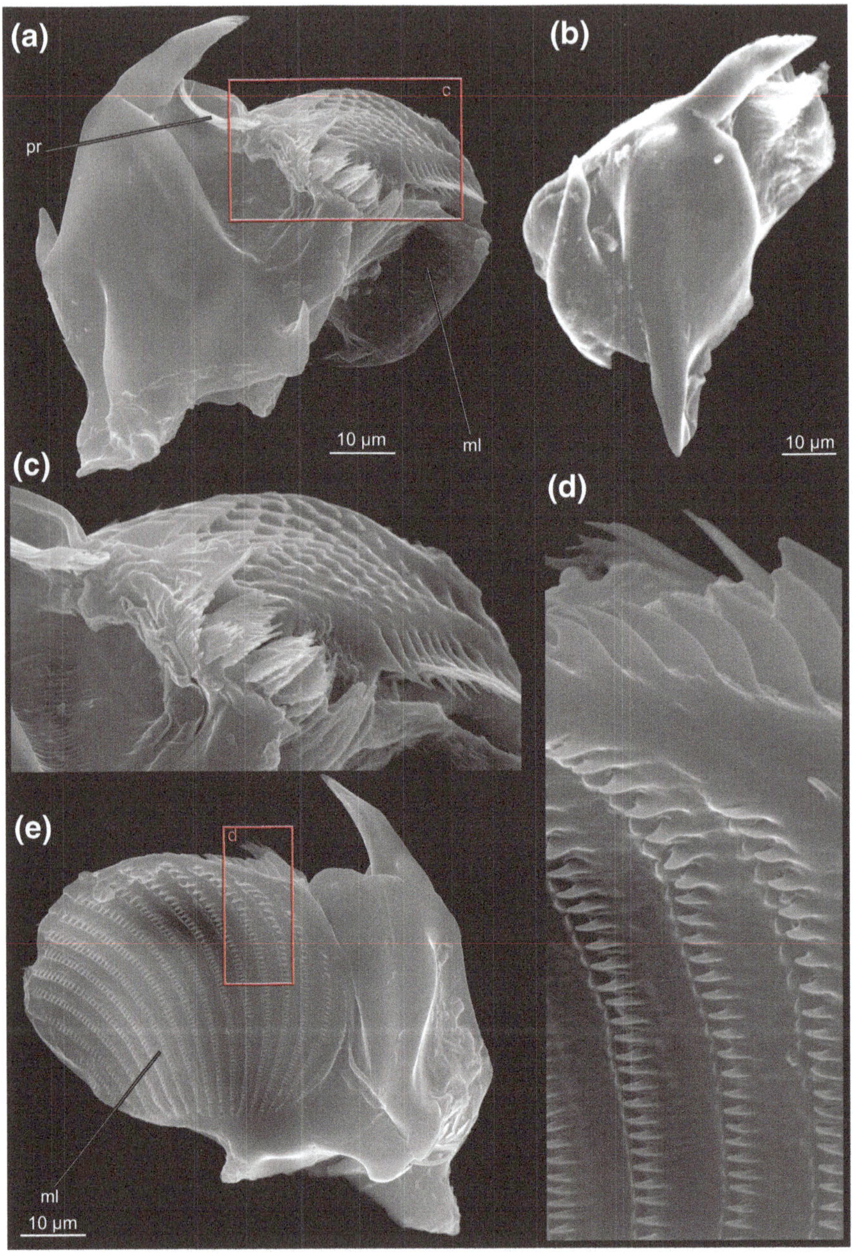

Fig. 5.7 Mandibles of adult ptiliids, SEM: **a**, **c**–**w** *Acrotrichis sericans*; **b** *Acrotrichis montandoni*; **a** dorsal view; **b** lateral view; **e** ventral view; *ml* mobile lobe, *pr* prostheca

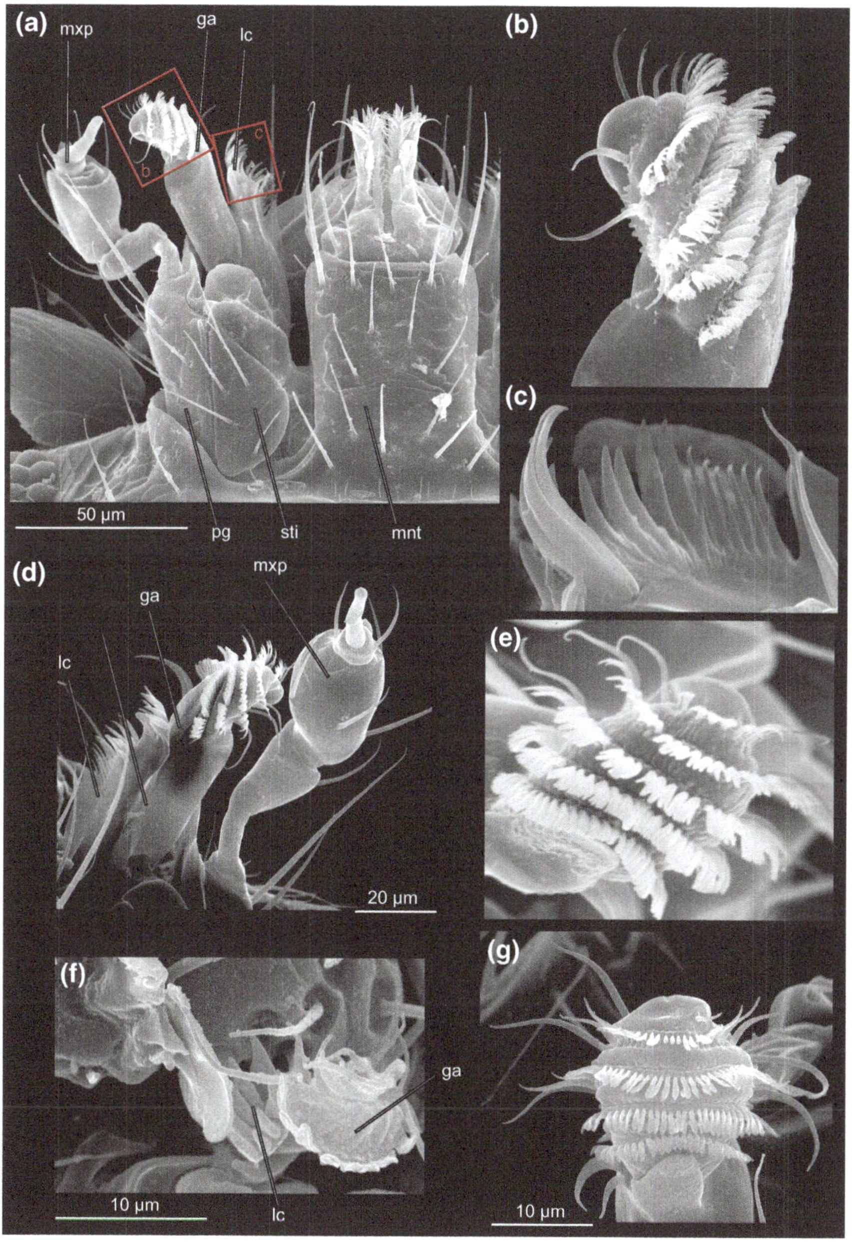

Fig. 5.8 Maxillae of adult ptiliids, SEM: **a–c** *Acrotrichis sericans*; **d** *Baeocrara variolosa*; **e** *Acrotrichis montandoni*, galea; **f** *Mikado* sp.; **g** *Ptiliolum sahlbergi*, galea; **a–d**, **f** ventral view; **e**, **g** medial view; *ga* galea, *lc* lacinia, *mnt* mentum, *mxp* maxillary palp, *pg* palpiger, *sti* stipes

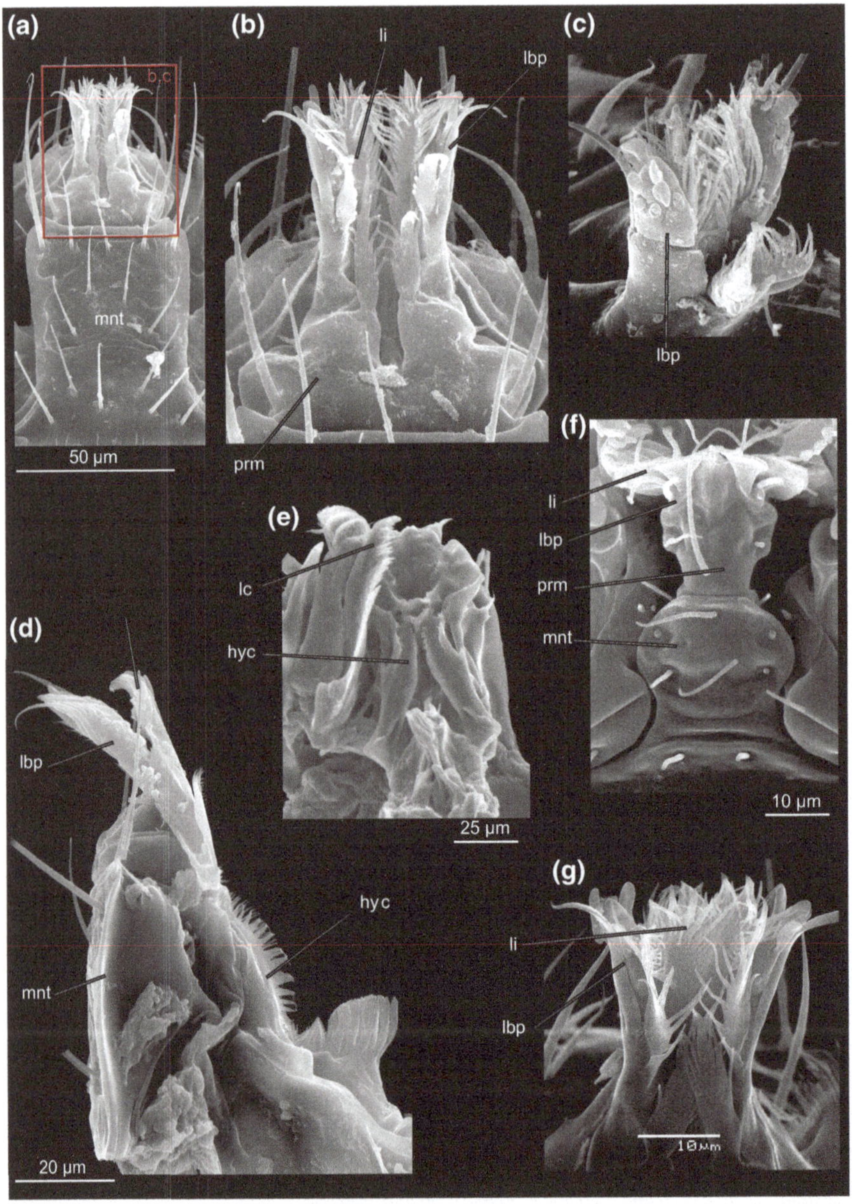

Fig. 5.9 Labium and hypopharynx of adult ptiliids, SEM: **a–d** *Acrotrichis sericans*; **e** *Acrotrichis montandoni*; **f** *Mikado* sp.; **g** *Baeocrara variolosa*; **a**, **b**, **f** ventral view; **c**, **e** dorsal view; **d** lateral view; *lbp* labial palp, *lc* lacinia, *li* ligula, *mnt* mentum, *hyc* hypopharyngeal comb

developed, symmetrical or almost symmetrical, short with wide base, generally similar in shape to mandibles of primitive Staphylinidae (*Proteinus* sp. and others) and Hydraenidae, but clearly distinguished by presence of spine on external surface and mobile lobe on dorsal surface of molar process. Dorsal surface of mobile lobe bearing regular rows of small teeth, together with ridges on surface of epipharynx participating in grinding spores and other small food particles. Surface of mobile lobe facing medial axis and surface of molar process with large teeth grinding large food particles. **Maxillae** rather uniform in representatives of family and generally similar in structure to those of other staphylinoids. Body of maxilla consisting of distinguished cardo, stipes, and palpiger. Maxillary palp 4–5-segmented, preapical palpomere strongly dilated, apical palpomere short, apically bearing basiconic sensillum (probably chemoreceptor). Lacinia and galea elongate, mobile, lacinia somewhat shorter, apically bearing about ten teeth, galea apically bearing 2–5 regular 'fan-like' rows of flattened setae. **Labium** consisting of well developed mentum, membranous prementum, reduced palps, and ligula (consisting of fused reduced glossae and paraglossae); submentum fused with cranium. Different groups of ptiliids differ in degree of reduction of labial palps and ligula, which distinguishes Ptiliidae from Staphylinidae and Hydraenidae, in which these structures are usually well developed. Hypopharynx with pair of longitudinal ridges bearing rows of setae. Similar structure found also in Staphylinidae.

Labium, maxillae, and partly labrum connected by membrane, forming cylindrical preoral cavity containing sunk mandibles. At rest, labrum and mentum joined, isolating preoral cavity containing sunk mandibles, galea, and lacinia, from external environment. In active state, galea and lacinia protruding beyond preoral cavity, and used by beetle for raking food (in most ptiliids, probably fungal, yeast, and bacterial films formed on surfaces of various decomposing plant substrates) into preoral cavity. Subsequently, this semiliquid substrate processed with help of mandibles and epipharynx, grinding fungal spores, yeast cells, and possibly bacteria.

Structure of thorax

The division of thorax is an important phylogenetic character considered for recognizing large taxa within the order Coleoptera. In most beetles, the prothorax is especially strongly modified (with particular parts reduced or fused); the metathorax is modified to a smaller degree. The metathorax is the most completely divided part of the thorax.

Prothorax in Ptiliidae, as in other beetles, covered dorsally with monolytic 'shield'—pronotum (Fig. 5.9). Shape of pronotum strongly varying among ptiliids. Division of prothorax considerably different from that of Staphylinidae and Hydraenidae: in most Ptiliidae all sclerites fused and sutures indiscernible, in some, e.g., *Acrotrichis*, with weakly discernible incomplete suture, tentatively interpreted as notosternal. Dorsal part of prothorax formed by shield of pronotum; ventral part formed by hypomeron. The shape of the procoxae and coxal cavities is an important taxonomic character of Coleoptera. In ptiliids, as in many staphylinoids, procoxae rounded, or slightly wider than long. In most Ptiliidae, as in Staphylinidae and

Hydraenidae, procoxal cavities open posteriorly, but in Nanosellini completely closed.

Mesothorax in ptiliids, as in other beetles, smaller than other segments of thorax and rather strongly modified (Fig. 5.10). In most Ptiliidae all sclerites of ventral surface of mesothorax fused, and sutures indiscernible, except in genus *Nossidium*, in which mesothorax with clearly distinguished pleurites. In some staphylinids sclerites of prothorax also fused or reduced, but in all staphylinids epimera separated by clearly discernible suture. Sclerites of tergal part of mesothorax in Ptiliidae reduced as in other beetles, membranous and covered under elytra, except for scutellum. Mesocoxae and coxal cavities rounded. In some representatives of Nanosellini (*Porophilla, Isolumpia*) lateral part prothorax with orifices, probably of glands of unknown function. Mesothoracic spiracles localized between pro- and mesothorax on membranous area, spiracle sclerites absent in ptiliids, in contrast to other staphylinoids.

Metathorax in ptiliids fully divided, as in other beetles. Metaventrite in all ptiliids separated from pleurite by clearly discernible suture (Fig. 5.9). Pleurites divided by weakly by discernible suture on narrow episterna and epimera. Tergal part of metathorax with usual for beetles division into scutum, scutellum, and postnotum. Scutellum in ptiliids peculiarly structured, bearing process (alacrista), strongly protruding posteriad beyond thorax, in some species reaching abdominal tergite 4 or 5. Dorsal surface of this process with medial furrow for receiving longitudinal protrusions of medial margin of elytra. Thus this process joining elytra, preventing wings from damage, as necessary for living in 'soil-like' substrates. Metacoxae of various shape, from transverse, with well developed femoral plates (*Nossidium*), to almost conical without such plates (*Ptenidium, Ptinella, Pteryx*): contiguous (*Nossidium*, Nanosellinae) or widely set apart (*Ptenidium, Micridium, Ptilium, Pteryx, Ptinella*).

Structure of **metendosternite** (metafurca), considered conserved character for purposes of classification of beetles (Crowson 1938, 1944), in ptiliids showing several variants (Fig. 5.11a–c). In *Nossidium* metendosternite consisting of two branches with shared base, in most ptiliids consisting of two branches widely set apart, and in many Nanosellini metendosternite strongly reduced. All variants share absence of common stem, in contrast to hydraenids and most staphylinids; absence of this stem probably explained by size of ptiliids and their widely set apart of coxae. Reduction of branches of metendosternite, found in Nanosellinae, probably explained by extremely small size, elongate body shape, and closely set coxae.

Elytra in ptiliids of various shape and microsculpture, without pronounced longitudinal striae. In most ptiliids elytra with narrow epipleura. Usually, elytra slightly shortened and not covering apical 2–3 segments of abdomen, less often not shortened and covering all abdomen or strongly shortened and covering only first abdominal segments. Along suture with lock joining elytra. In some Nanosellini (*Nanosella, Primorskiella*) internal surface of elytra and corresponding surface of metepimeron with complex structures very similar to stridulatory apparatus (Fig. 5.12). This apparatus represented by dentate zone of numerous cuticular folds on elytra and by plectrum of several cuticular ridges on epimeron of metathorax.

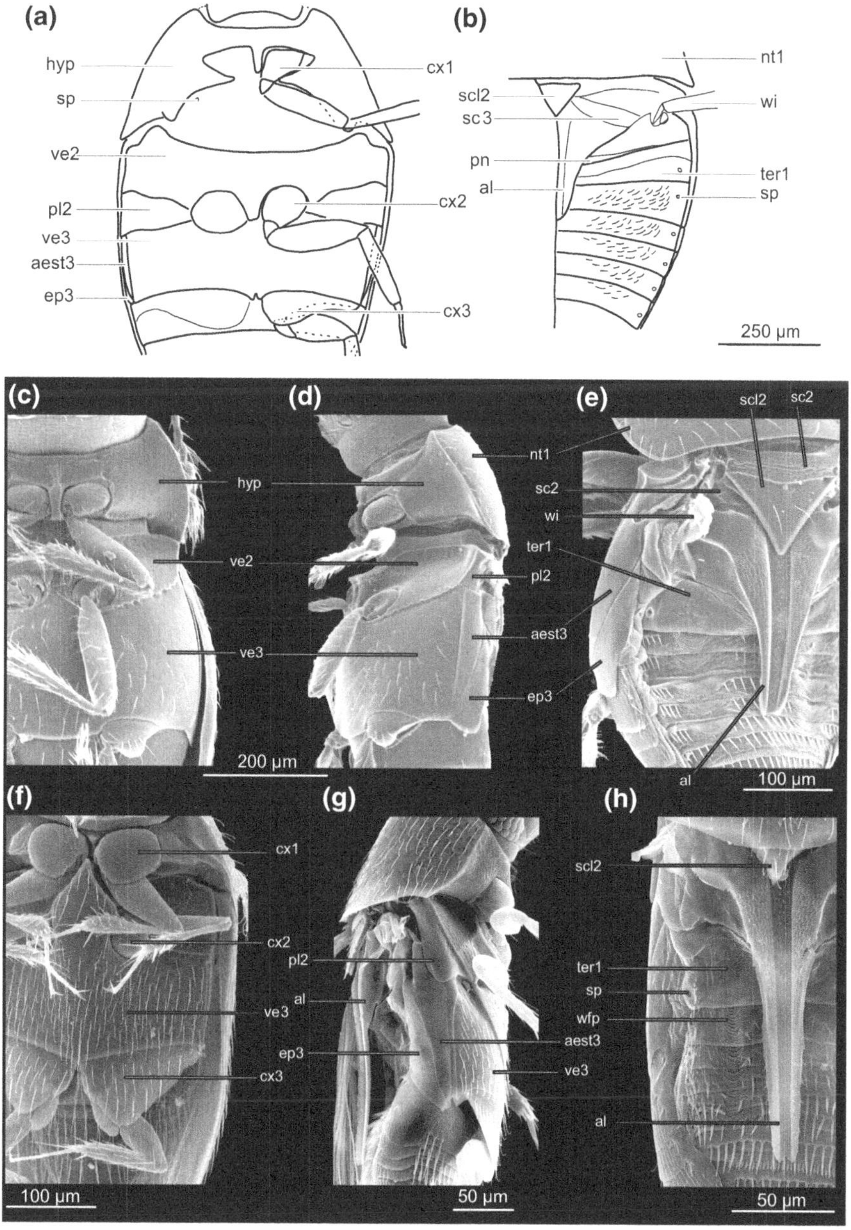

Fig. 5.10 Thorax of adult ptiliids: **a**, **b** *Nossidium pilosellum*, schematic; **b–h** SEM; **c–d** *Ptenidium formicetorum*; **f** *Mikado* sp.; **g**, **h** *Primorskiella ànodonta*; **a**, **c**, **f** ventral view; **b**, **e**, **h** dorsal view; **d**, **g** lateral view; *aest3* metepisternum, *al* alacrista, *cx1.2.3* pro-, meso-, and metacoxae, *el* elytron, *ep3* epimeron of metathorax, *hyp* hypomeron, *nt1* pronotum, *pl2* pleurite of mesothorax, *pn* postnotum, *sc2* mesoscutum, *sc3* metascutum, *scl2* mesoscutellum, *scl3* metascutellum, *sp* spiracle, *ter* tergite, *ve2.3* ventrite of meso- and metathorax, *wfp* wing folding patches, *wi* wing

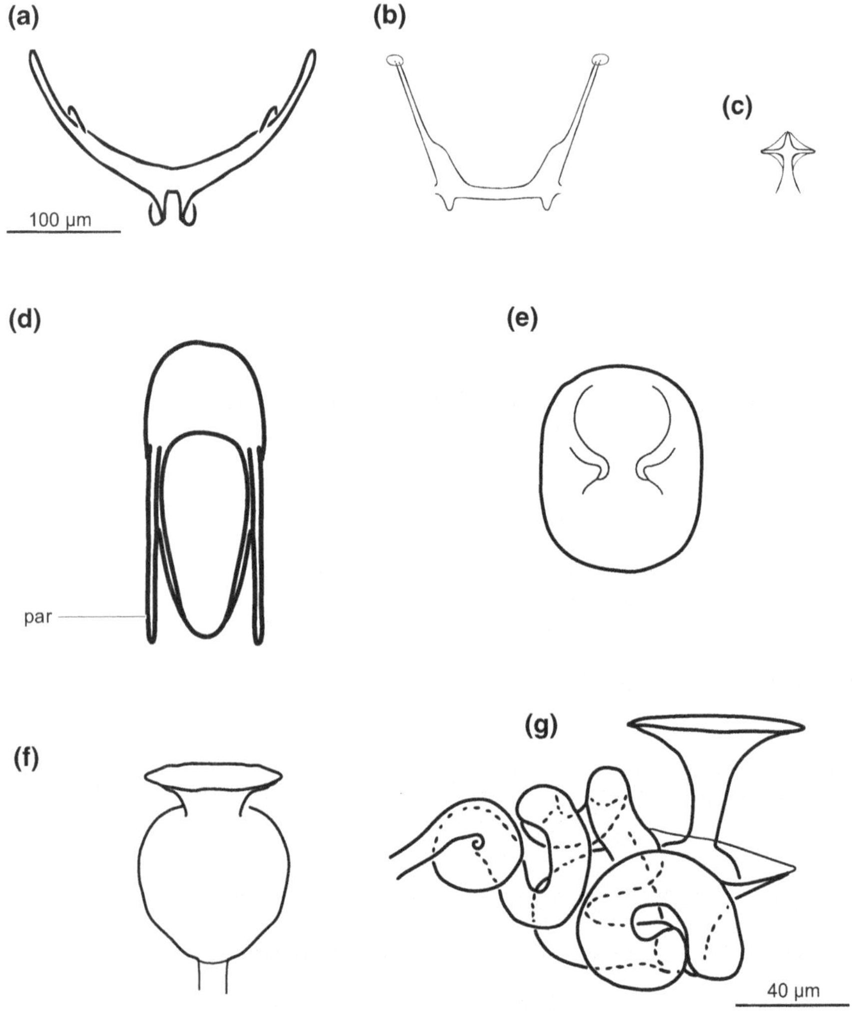

Fig. 5.11 Shapes of metendosternite and external copulatory apparatus in ptiliids: **a**, **d**, **f** *Nossidium pilosellum*; **b** *Acrotrichis montandoni*; **c** *Nanosella* sp.; **e**, **g** *Acrotrichis intermedia*; **a**–**c** metendosternite; **d**, **e** aedeagus; **f**, **g** spermatheca; *par* parameres

Hindwings in ptiliids typical 'feather-like' or absent (in wingless forms of some genera) (Fig. 5.13 and 5.14). Wing consisting of narrow blade with 2–3 discernible reduced veins and fringe of long setae on perimeter (along anterior and posterior margins). This type of wing is not found larger beetles, because in all others the wing blade is wide, and the fringe, if present, is situated only on the posterior margin of the wing and consists of short setae. Wing venation of staphylinoid type (crossveins absent), but strongly simplified because of narrowed wing blade. Homology of venation difficult to trace, but considering venation patterns of smaller

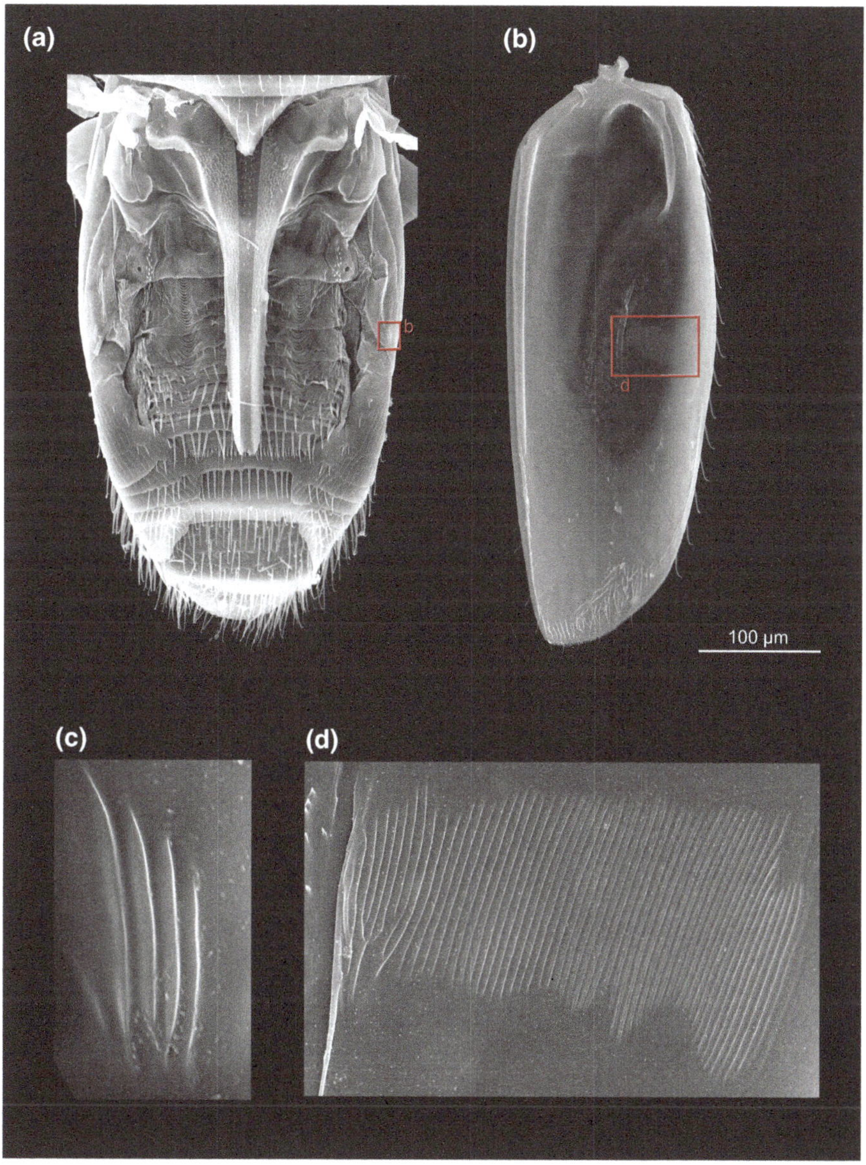

Fig. 5.12 Stridulatory apparatius of adult *Nanosella* sp., SEM: **a** thorax and abdomen, dorsal view; **b** elytron, ventral view; **c** plectrum; **d** dentate zone

staphylinids and hydraenids, wing blade of ptiliids probably containing parts of subcostal, radial, and median veins probably preserved and all others lost (Table 5.1).

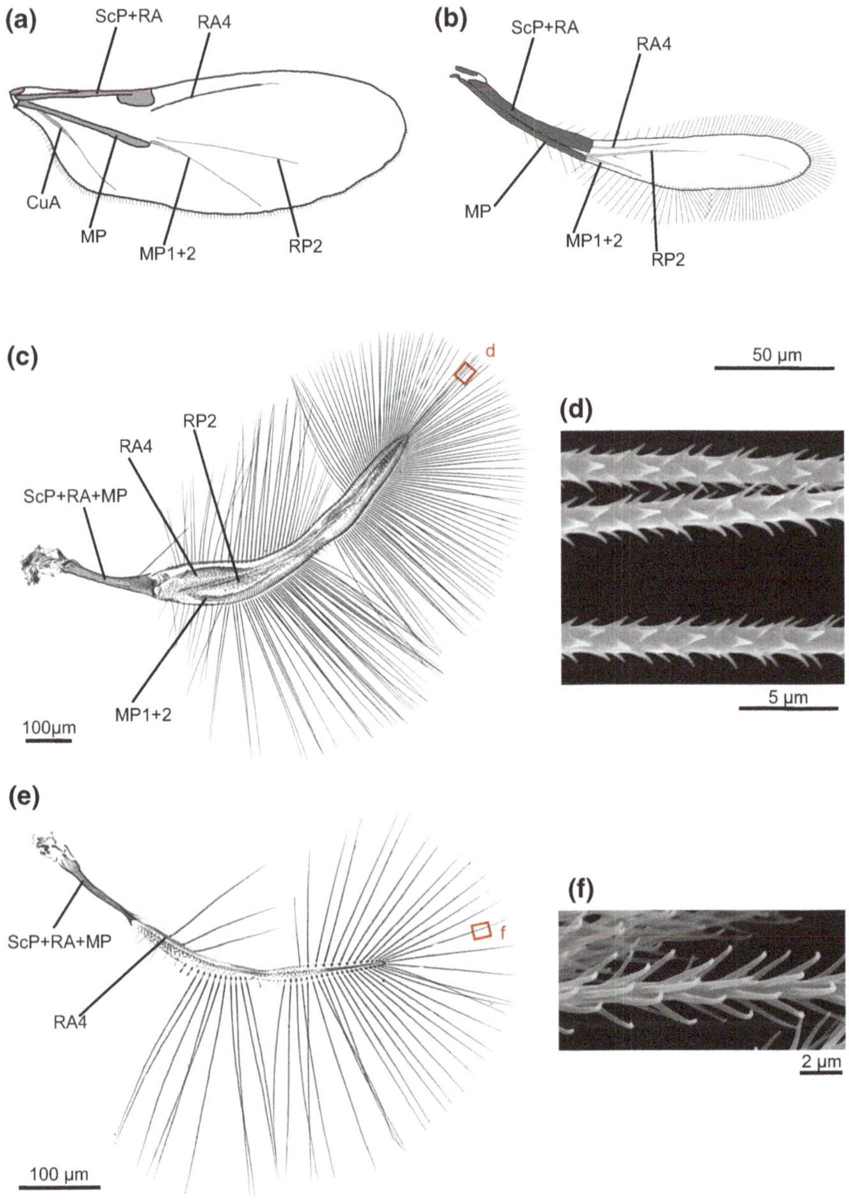

Fig. 5.13 Wings of Ptiliidae and Hydraenidae: **a** *Hydraena* sp.; **b** *Nossidium pilosellum*; **c**, **d** *Acrotrichis montandoni*; **e**, **f** *Primorskiella anodonta*. Venation see text

Base of wing represented by petiole consisting of one vein without wing membrane (most Ptiliidae, ScP + RA + MP) or of two almost contiguous veins (*Nossidium*, ScP + RA, and MP). Membranous part of wing with three veins (RA4,

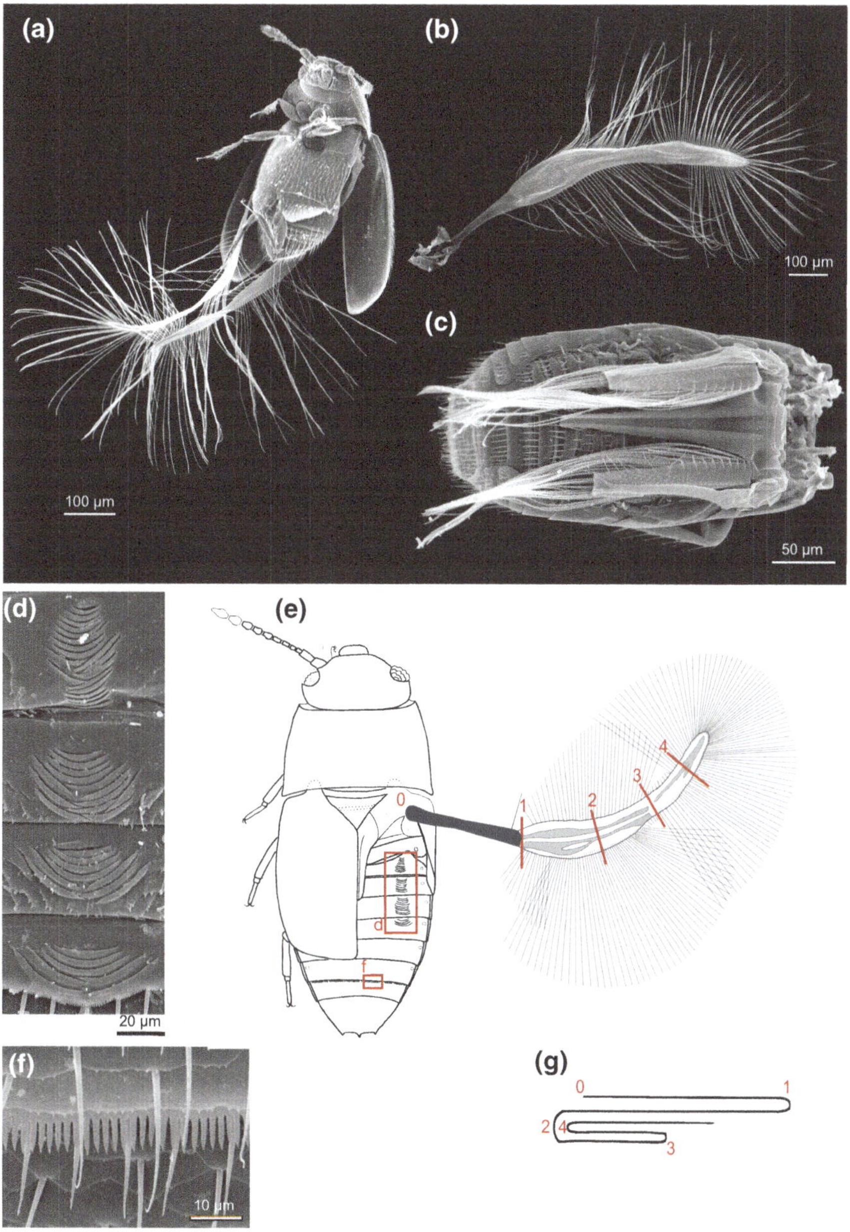

Fig. 5.14 Mechanism of wing folding in ptiliids: **a** *Ussurilumpia trichaptumi* with unfolded wings; **b** unfolded wing of *Acrotrichis montandoni*; **c** folded wings of *Primorskiella anodonta*; **d** wing folding patches; **e** schematic drawing of wing and folding apparatus; **f** cuticular 'comb'; **g** schematic drawing of folded wing; 0–4 planes of wing folding

Table 5.1 Homology of venation

Abbreviations used in this study	Naomi (1989a)	Kukalová-Peck and Lawrence (1993, 2004); Hansen (1997)	Fedorenko (2009)
ScP + RA	Sc + R	ScP, RA	ScP
RA4	R	RA4. RP1	SV
MP	Cu + PCu	MP1 + 2	CuA
RP2	M1	RP2	RMP
MP1 + 2	Cu1	MP1 + 2. RP3 + 4	MP3 + 4
CuA	A1 + A2	CuA	AAP

RP2, and MP1 + 2), and in smallest ptiliids number only with one vein (*Nanosella*, RA4). Wing in ptiliids, as in other staphylinoids, several times folded at rest, but in contrast to other staphylinoids folding along four lines strictly perpendicular to wing blade; veins at these lines narrowed and less strongly sclerotized. Folding of wing also accompanied by tucking down of setal fringe; in folded state all setae directed posteriad, parallel to wing blade (Fig. 5.14).

Legs ambulatorial. Tarsi 2–3-segmented, consisting of rather long apical tarsomere and 1–2 shortened basal tarsomeres (often weakly discernible). Apical tarsomere bearing two simple claws and simple empodium. In some representatives one claw reduced, markedly smaller than other claw (Sorensson 1997).

Structure of abdomen

Abdomen consisting of 10 segments, as in other beetles (Fig. 5.15). Number of visible sternites 6–7, because, as in other Staphylinoidea, sternite 1 absent, and sternites 9 and 10 modified and sunk into abdomen. In male ptiliids sternite 9 with long medial process directed anteriad; such structure of sternite 9 not found in other staphylinoids. This sternite probably participates in pulling out of the copulatory apparatus. Number of visible tergites 9–10, in contrast to Staphylinidae and Hydraenidae, which seldom have more than eight visible segments. Spiracles usually present on tergites 1–8, sometimes only six pairs of spiracles present (Nanosellini). Stigmata simple, ring-shaped, without filter apparatus, with spherical atrium. First 3–5 tergites in all ptiliids with cuticular structures shaped as regularly arranged patches (ridges) in most representatives or shaped as irregular folds (*Nossidium*). These structures probably participate in folding of the wing and fixation of the wing under elytra, because they are absent in wingless forms. In females secondary ovipositor shaped as short tube formed by tergite 10 and membranous sternites 9 and 10.

External male genitalia in most species represented by simple penis of various shape, only in species of genus *Nossidium* with parameres (Fig. 5.11d, e). One important diagnostic character is the structure of the **spermatheca**, very strongly varying within the family: from simple, spherical (*Nossidium*) or linear (*Ptenidium*), to complex, helically arranged (*Acrotrichis*) (Fig. 5.11f, g).

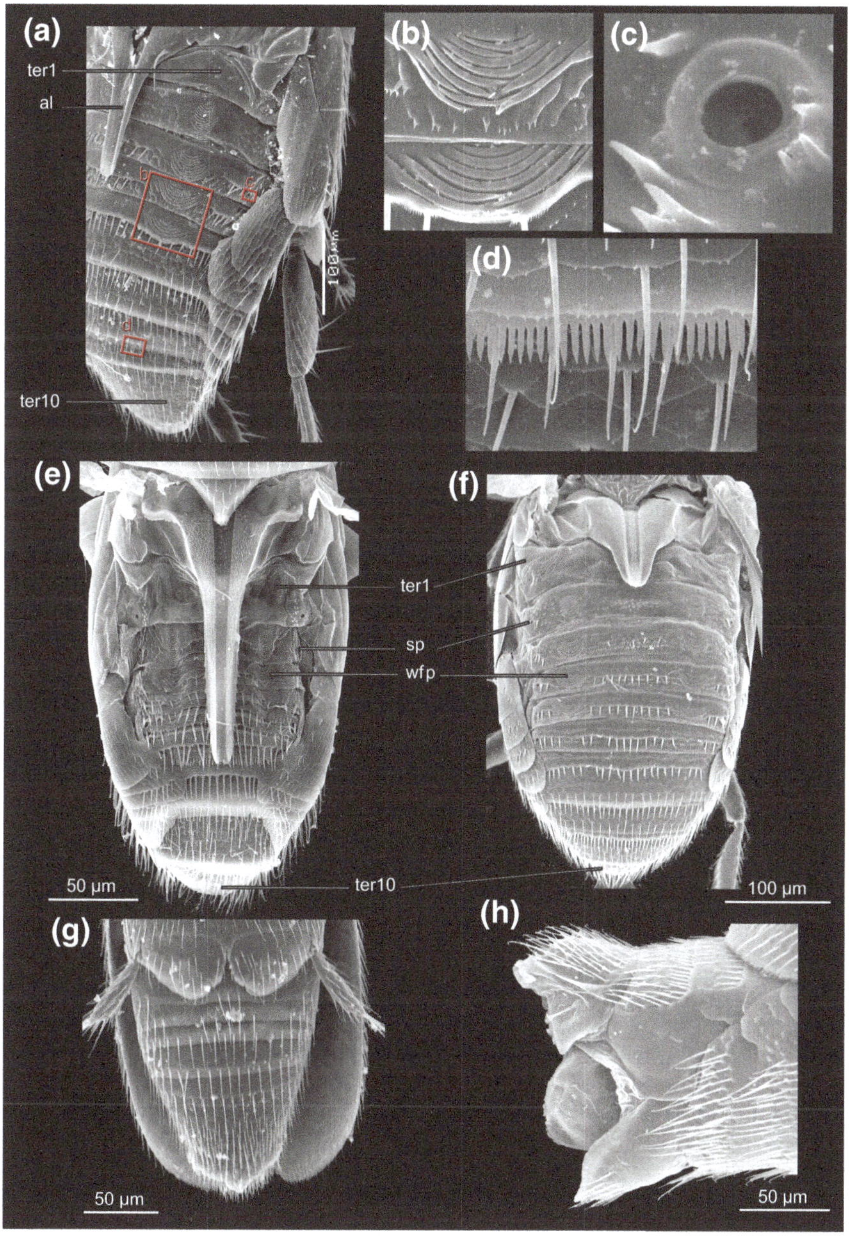

Fig. 5.15 Abdomen of adult ptiliids, SEM: **a–d** *Acrotrichis montandoni*: **e**, **g** *Primorskiella ànodonta*; **f**, **h** *Ptilium myrmecophilum*; **b** wing patch of tergite; **c** spiracle; **h** laying of egg; **a–f** dorsal view; **g** ventral view; **h** lateral view; *al* alacrista, *sp* spiracle, *ter1–10* tergite, *wfp* wing folding patch

5.3.2 Internal Morphology

General plan of internal structure: medial part of body cavity occupied by intestine, most of thorax occupied by musculature, space between all organs filled with fat body, one large free body cavity being egg cavity of females; nervous system strongly oligomerized and concentrated in thorax; reproductive system occupying considerable part of apical segments of abdomen; heart and diaphragms absent (Fig. 5.16).

Integument

Integument in ptiliids consisting of cuticle, hypoderm, and basal membrane. Cuticle thickness in *Acrotrichis grandicollis* 2.3–11.1 μm (M = 6.2; n = 80), in *Mikado* sp. 0.3–5.1 (M = 1.9; n = 80), in *Nanosella* sp. 0.8–6.1 (M = 2.9; n = 80). Cuticle consisting of epicuticle about 70 nm thick and weakly delimited exo- and endocuticle (Fig. 5.17a, b). Exocuticle and endocuticle lamellar, except for intersegmental areas, in which exocuticle and endocuticle homogeneous. Hypoderm consisting of flattened cells and up to 3 μm thick.

Digestive and excretory systems

Digestive canal divided into fore-, mid-, and hindgut (Fig. 5.18a, b). Fore- and hindgut with thin cuticular lining. Midgut without cuticular lining. Intestinal canal somewhat longer than body, forming loop in metathorax. Of salivary glands, only labial glands have been revealed in ptiliids.

Foregut divided into pharynx and oesophagus; crop or muscular stomach undeveloped. Pharynx of rather great diameter, with strong musculature, this structure of pharynx possibly indicating feeding on semiliquid food (Pavlovsky 1956). Oesophagus straight, with one-layer circular muscles, in contrast to other beetles, in which with both circular and longitudinal muscles present (Crowson 1981).

Midgut appearing as tube with numerous short diverticula, in cardial section these diverticula somewhat longer and directed anteriad. Wall consisting of unilayered epithelium, in cross-sections with 8–15 cells visible, much less than in other insects (Fig. 5.19). Slender unilayered peritrophic membrane present over entire length of midgut. Muscles of midgut not found; food possibly moved along intestine by somatic musculature. This character distinguishes Ptiliidae from the other beetles, in which the midgut has two layers of muscles (Crowson 1981).

Hindgut divided into small intestine and rectum. Small intestine rounded in section, rectum oval in section.

Boundary between mid- and hindgut bearing four **Malpighian tubules,** one pair directed anteriad, other pair directed posteriad.

Circulatory system and fat body

Circulatory system strongly reduced due to extremely small size, heart absent, or strongly reduced (in larger representatives), and hemolymph occupying extremely small part of body cavity, probably because of diffusion sufficing for transport of molecules in organisms of such small sizes. **Fat body** occupying almost all free body cavities, replacing hemolymph. Similar organization is found in many

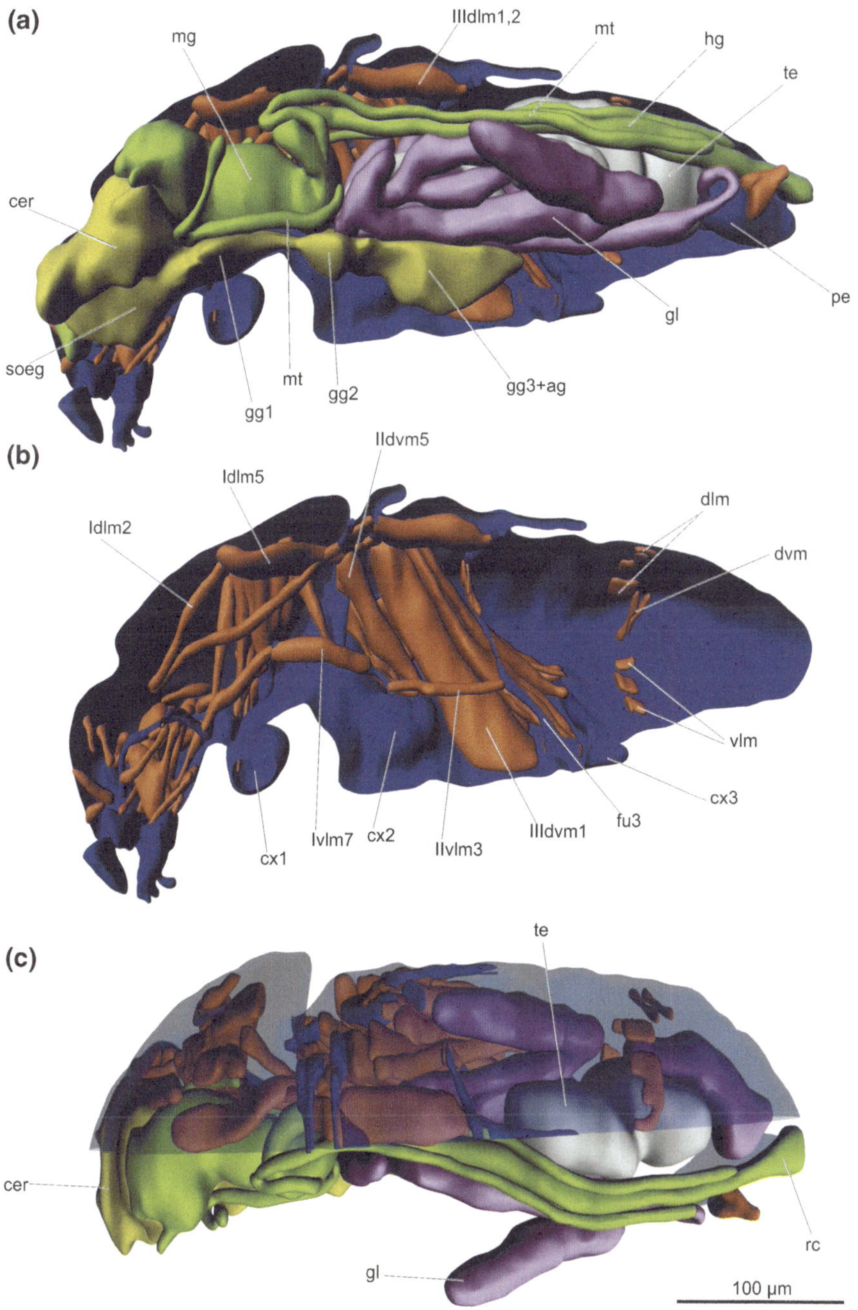

Fig. 5.16 Internal morphology of adult *Mikado* sp., male, 3D: **a**, **b** lateral internal view; **c** dorsal view; *ag* abdominal ganglion, *cer* cerebrum, *cx1.2* meso- and metacoxae, *fu3* metafurca, *gg3* metathoracic ganglion, *hg* small intestine, *mg* midgut, *mt* Malpighian tubules, *pe* penis, *rc* rectum, *soeg* suboesophageal ganglion, *te* testis. Musculature see text

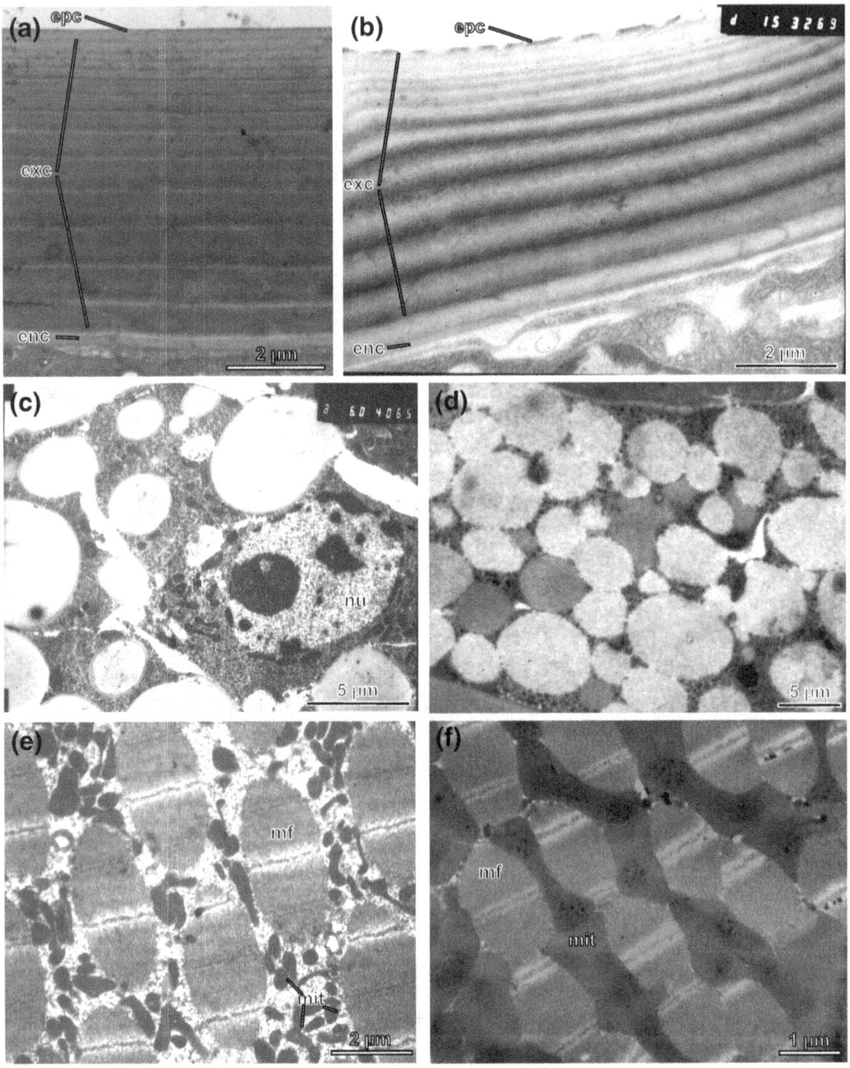

Fig. 5.17 Ultrastructure of adult ptiliids, TEM: **a–e** *Acrotrichis montandoni*; **f** *Primorskiella anodonta*; **a**, **b** cuticle, **c**, **d** fat body; **e**, **f** muscle; *enc* endocuticle, *epc* epicuticle, *exc* exocuticle, *mf* muscle fiber, *mit* mitochondrion, *nu* nucleus

representatives of Acari, especially smaller ones (Silvere and Shtein-Margolina 1976), in which hemolymph is completely replaced by parenchyma, which assumes the functions of both fat body and hemolymph. Fat body consisting of cells of various shape 8–20 µm long, containing electron transparent inclusions (Fig. 5.17c, d). Body cavities containing enocytes, present in many insects, but having still somewhat unclear functions.

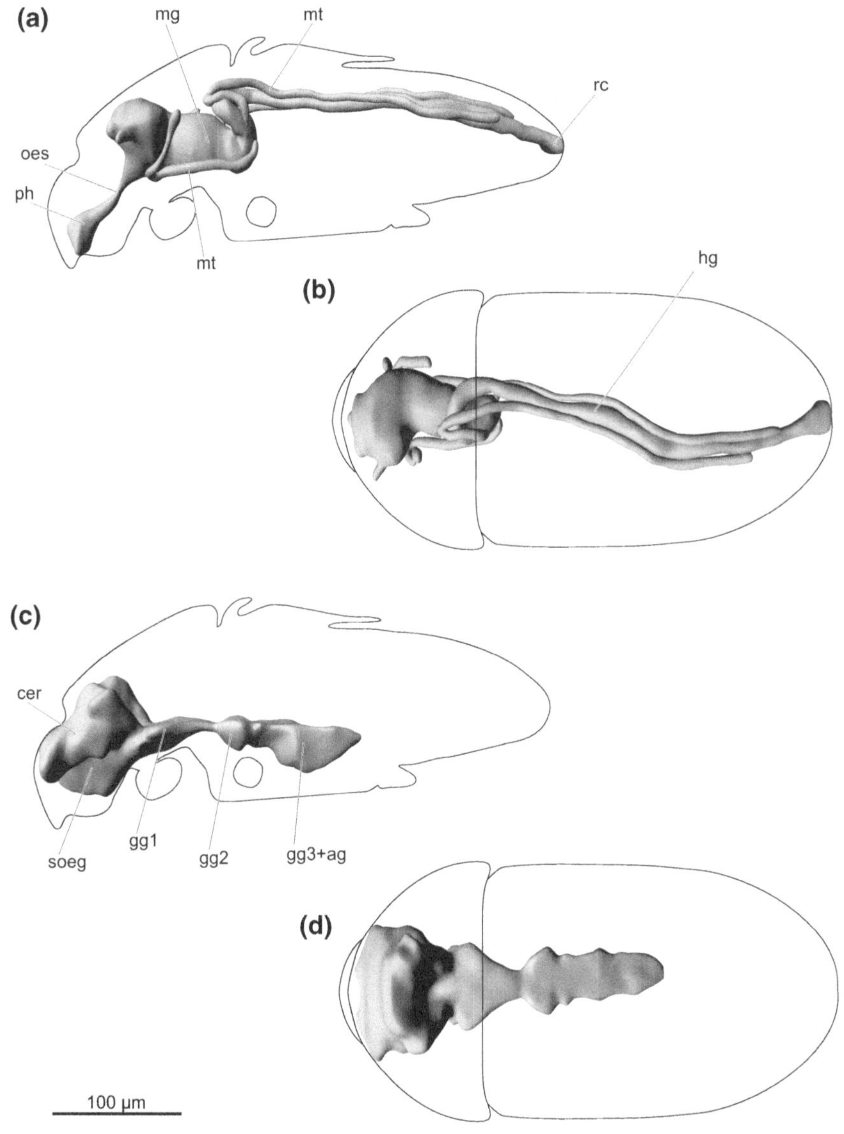

Fig. 5.18 Digestive and nervous systems of adult *Mikado* sp., 3D: **a**, **b** digestive system; **c**, **d** central nervous system; **a**, **c** lateral view; **b**, **d** dorsal view; *ag* abdominal ganglion, *cer* cerebrum, *gg1.2.3* pro-, meso-, and metathoracic ganglia, *hg* hindgut, *mg* midgut, *mt* Malpighian tubules, *oes* oesophagus, *ph* pharynx, *rc* rectum, *soeg* suboesophageal ganglion

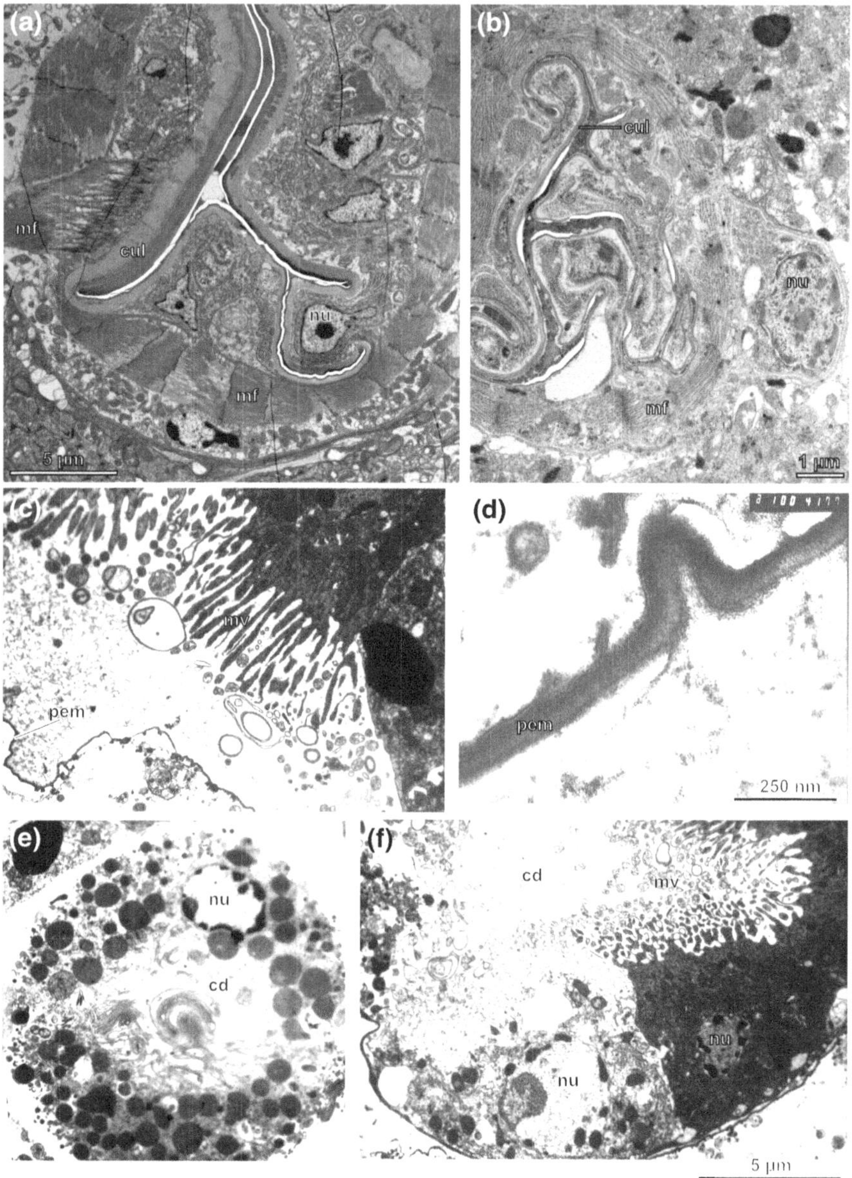

Fig. 5.19 Ultrastructure of digestive system in adult ptiliids, TEM: **a**, **c–f** *Acrotrichis montandoni*; **b** *Primorskiella anodonta*; **a**, **b** pharynx, **c** midgut; **d** peritrophic membrane; **e**, **f** diverticula of midgut; *cd* cavity of diverticulum, *cul* cuticular lining, *mf* muscle fiber, *mf* microfibers, *nu* nucleus, *pem* peritrophic membrane

Tracheal system

Tracheal system displaying features of strong simplification. Only small number of weakly branching tracheae present, connected with spiracles. Longitudinal and transverse stems and air sacs absent, probably because of diffusion sufficing for oxygen transport in organisms of such small sizes. Tracheae with structure typical of insects, consisting of hypoderm and intima, intima with helical thickenings—taenidia. Tracheoles (without taenidia, filled with hemolymph) not found, possibly because of capillary forces making tracheal control of respiration impossible at smaller diameters of tracheae.

Nervous system

Central nervous system displaying strong oligomerization and condensation of ganglia (Fig. 5.18c, d). In *Acrotrichis* supraoesophageal ganglion occupying almost entire volume of posterior half of head, in Nanosellini part of protocerebrum shifted into prothorax. Left half of protocerebrum in *Nanosella* with asymmetrical distal process. Suboesophageal ganglion partly (*Acrotrichis*) or completely (Nanosellini) shifted into prothorax and fused with prothoracic ganglion. Mesothoracic ganglion separate, abdominal ganglia fused with metathoracic ganglion.

Plan of organization of ganglia typical of insects. Cell bodies of neurons localized on perimeter; neuropil occupying central part. Extraneural sheath, formed by cells of fat body, represented not by single formation, but by scattered cells covering only some areas of surface of central nervous system. Ganglionic sheath consisting of two layers: external, noncellular neural lamella and internal, cellular perineurium. Thickness of neural lamella in *A. grandicollis* inconstant, about 0.3–0.6 μm, in *Nanosella* sp. 0.06–0.14 μm. Thickness of perineurium in *A. grandicollis* 0.4–1 μm, in *Nanosella* sp. 0.03–0.3 μm.

Size of cell bodies in cerebrum considerably smaller than in other insects, 2.52–5.68 μm (M = 3.36; n = 775) in *A. grandicollis* and 1.23–2.25 μm (M = 1.71; n = 336) in *Nanosella* sp. Nucleus occupying up to 90 % of volume of cell bodies; in smallest representatives of family chromatin strongly compacted. In spite of strongly reduced volume of cytoplasm, all organelles typical of neurons present in ptiliids: mitochondria, granular endoplasmic reticulum, Golgi complex, lysosomes, and vacuoles with inclusions (Figs. 5.20 and 5.21). Neuron nuclei often not regularly rounded, but uneven, sometimes even slightly lobe-shaped. Neuropil of cerebrum clearly discernible projections of neurons, their axolemmas, mitochondria, microtubules, and microfilaments. Minimum recorded diameter of neuron projections in *Nanosella* 0.078 μm. Neuropil with polarized and nonpolarized synaptic contacts. Most common contacts of divergent type, axo-dendritic. Cortical layer of cerebrum also with somato-somatic contacts, phenomenon rare among insects. Cerebral and cortical layers of cerebrum with strongly pronounced neuroglia, tightly adjacent to cell bodies of neurons (cortical glia) and their projections. Presence recorded of neuroglia embedding glomerular structures of mushroom bodies and comprising perineural layer of ganglionic sheath of cerebrum.

Number of cells in cerebrum in *A. grandicollis* about 26,000, in *Mikado* sp. 12,000, in *Nanosella* sp. 8000.

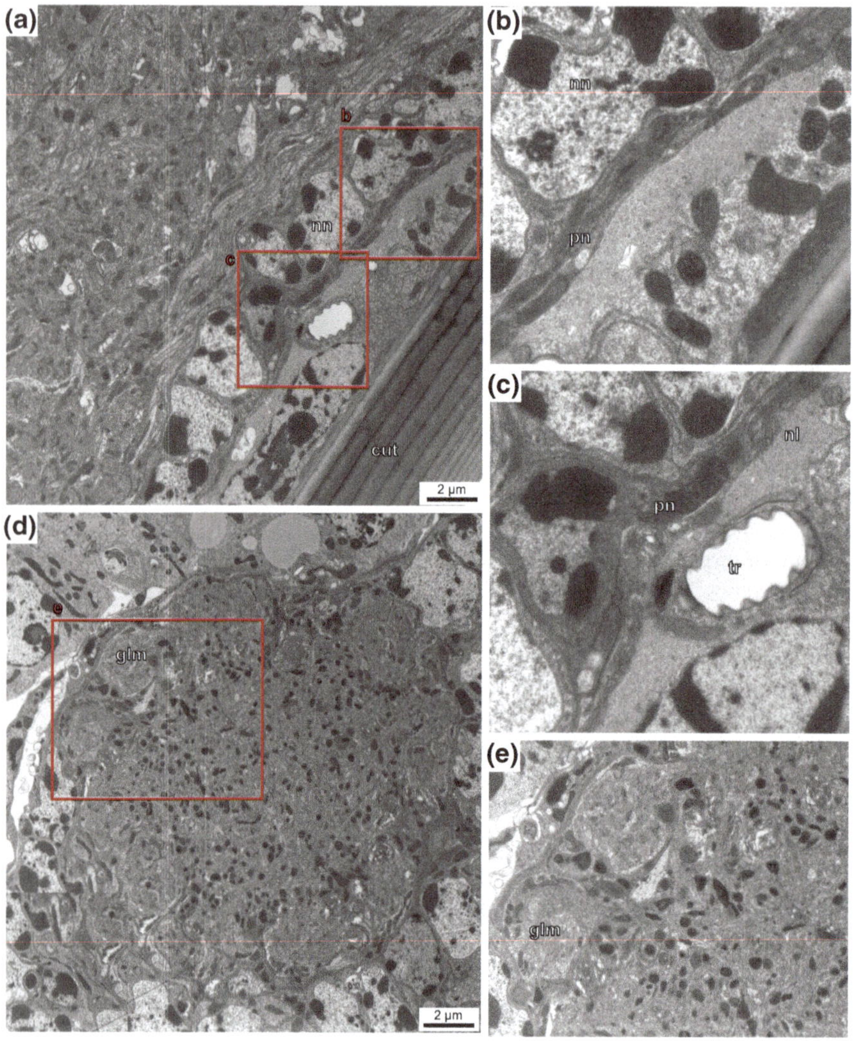

Fig. 5.20 Ultrastructure of cerebrum in adult *Acrotrichis grandicollis*, TEM: **a**–**c** structure of cerebral membranes; **d**, **e** ultrastructure of calyx of mushroom body. *glm* glomerule; *nn* nucleus of neuron; *nl* neural lamella; *pn* perineurium; *cut* cuticle; *tr* trachea

The protocerebrum, the largest region of the cerebrum, includes the most important association centers: the central body complex (central body and protocerebral bridge) and mushroom bodies, as well as optic lobes. The central body occupies intermediate position (above the pharynx and between the peduncles of mushroom bodies), has complex modular structure, and is differentiated into the dorsal region, which consists of the dorsal arch and fan-shaped body, ventral region (ellipsoid body), and noduli (Fig. 5.22 and 5.23). The protocerebral bridge is strongly pronounced, appearing as cord of commissural fibers. The mushroom

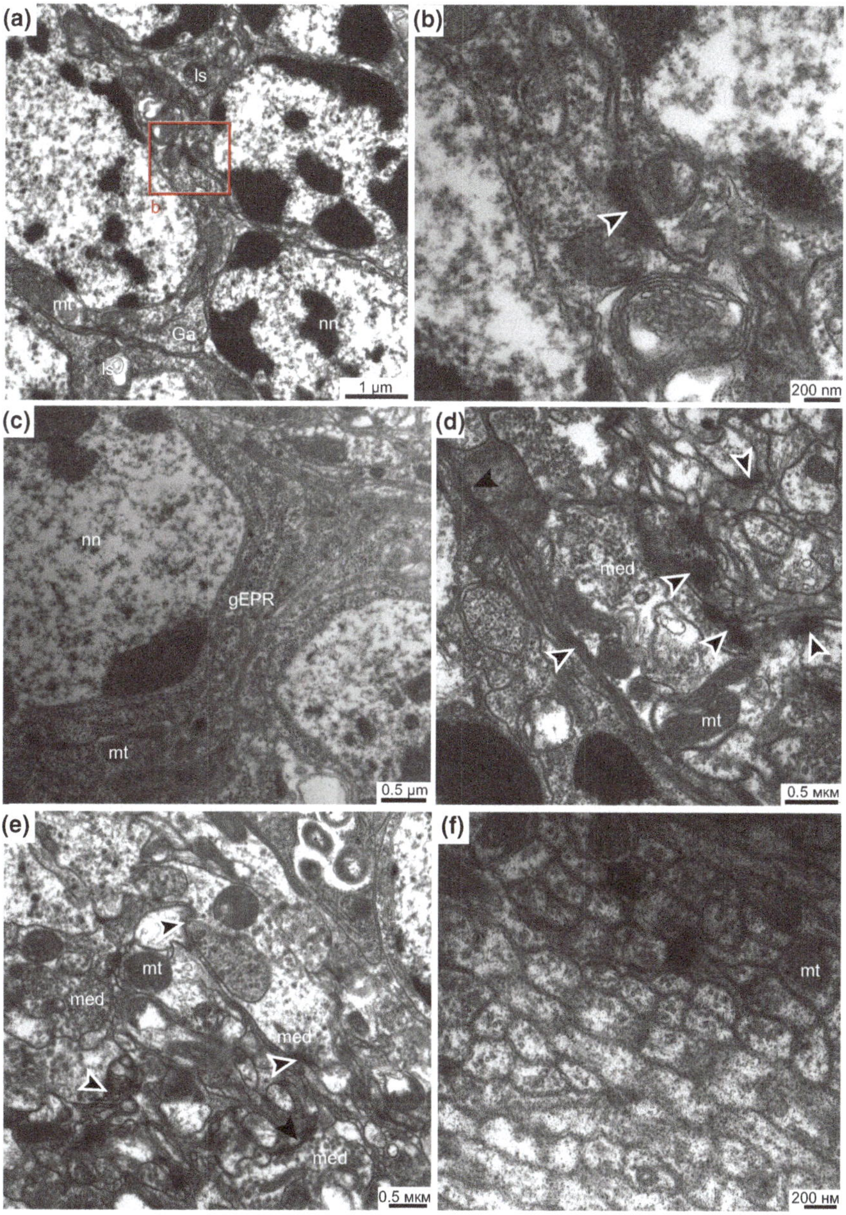

Fig. 5.21 Ultrastructure of cerebrum in adult *Acrotrichis grandicollis*, TEM: *Ga* Golgi apparatus; *gEPR* granular endoplasmic reticulum; *ls* lisosomes; *med* granules of mediator; *mt* mitochondrion; *nn* nucleus. *Arrows* indicate synaptic contacts

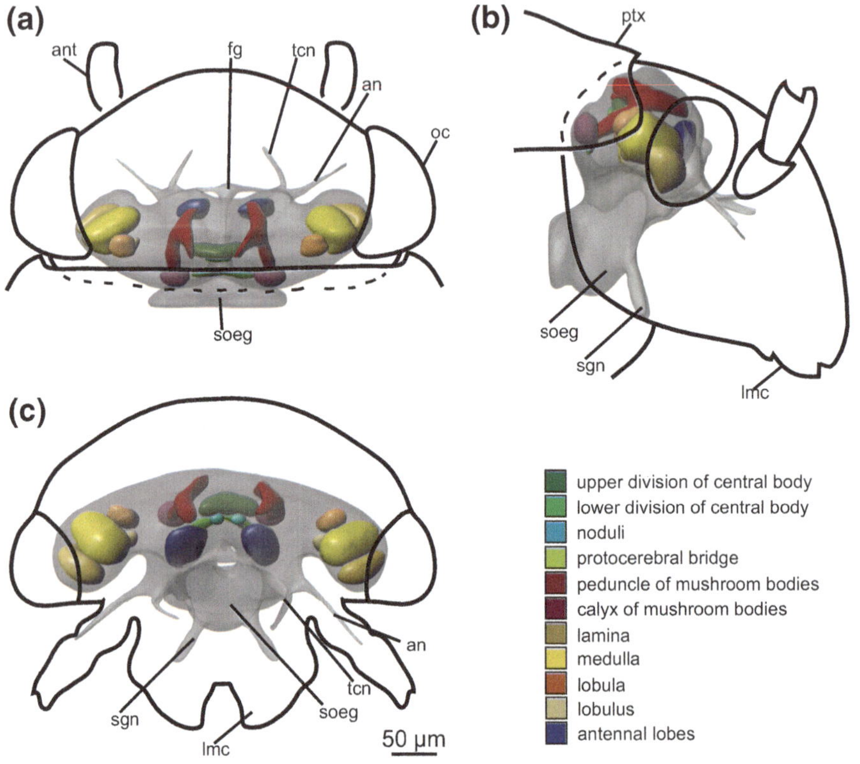

Fig. 5.22 3D reconstructions of cerebrum of adult *Acrotrichis grandicollis* (modified from Makarova and Polilov 2013): **a** dorsal view; **b** lateral view; **c** frontal view; *an* antennal nerves; *ant* antennae; *fg* frontal ganglion; *oc* compound eyes; *ptx* prothorax; *lmc* mouthparts; *sgn* nerves of suboesophageal ganglion, innervating mouthparts; *soeg* suboesophageal ganglion; *tcn* paired nerves of tritocerebrum

bodies are situated on both sides of central body in distal part of protocerebrum, differentiated into the peduncle and calyx. The peduncles of the mushroom bodies are well developed and display division into the α- and β-lobes; the α-lobe is elongated toward the calyx, and the β-lobe is directed toward the antennal lobes. The calyces of the mushroom bodies are single, in Nanosellini weakly developed.

The optic lobes in *Acrotrichis* are oriented perpendicular to the longitudinal body axis, as in other insects; in Nanosellini they are oriented almost longitudinally. The optic lobes are represented by three neuropils (optic ganglia): the lamina, medulla, and lobular complex, which is divided into the lobula and lobular plate. All three optic neuropils are close to each other.

The lateral protocerebrum is situated between the mushroom bodies and optic lobes. In contrast to the above-mentioned centers, it has no distinct structure or boundaries, and therefore it will not be included in comparative analysis as a separate zone of the cerebrum.

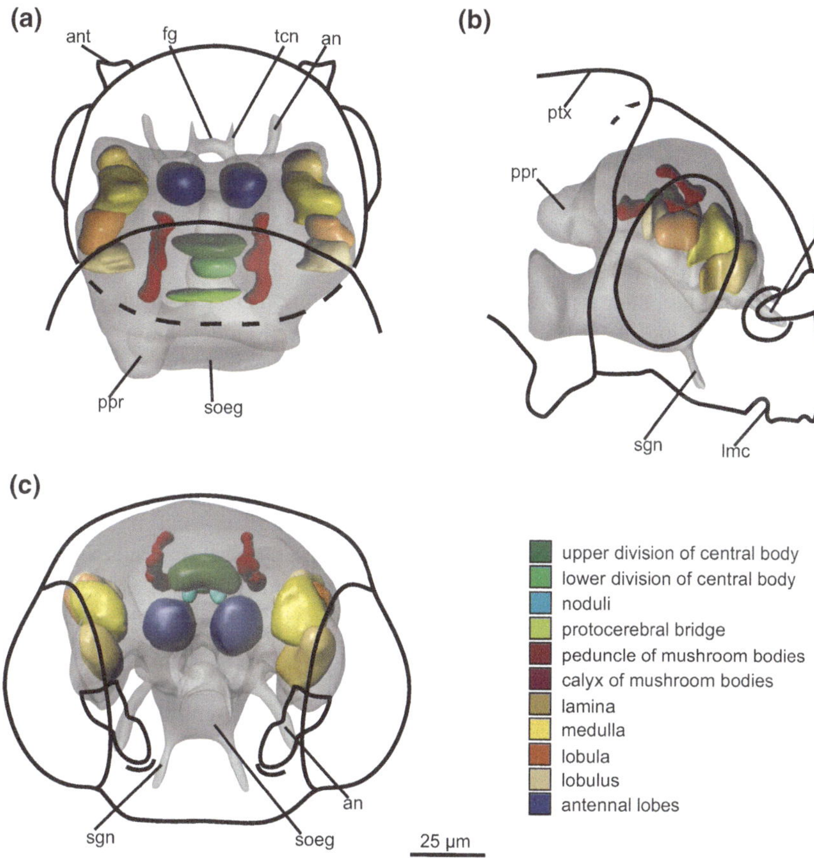

Fig. 5.23 3D reconstructions of cerebrum of adult *Nanosella* sp. (modified from Makarova and Polilov 2013): **a** dorsal view; **b** lateral view; **c** frontal view; *an* antennal nerves; *ant* antennae; *fg* frontal ganglion; *oc* compound eyes; *lmc* mouthparts; *ppr* posterior protocerebral process; *ptx* prothorax; *sgn* nerves of supraoesophageal ganglion innervating mouthparts; *soeg* suboesophageal ganglion; *tcn* paired nerves of tritocerebrum

Because of the strong condensation and proximity of all structures of the cerebrum, it is difficult to distinguish any other, smaller structures, such as commissures, tracts, chiasmata, etc.

In the deutocerebrum, antennal lobes can be distinguished. They are spherical; the neuropil is organized into glomerules. The glomerules are dense neuropilar structures; their approximate diameter and approximate number, respectively, are about 7–10 μm and 100–130 in *A. grandicollis*, 5–7 μm and 90 in *Mikado*, and 3–4 μm and 70 in *Nanosella*. The glomerules cannot be counted with precision because of unclear boundaries between many of them.

The tritocerebrum is the smallest region of the cerebrum, represented by a paired formation on the sides of the intestine. The tritocerebral commissure is strongly pronounced and runs under the intestine. The circum-oesophageal connectives are integrated into one ganglionic mass at the boundary between the cerebrum and suboesophageal ganglion and not clearly pronounced in histological sections. Paired connectives diverge from the tritocerebrum toward frontal ganglion.

The stomatogastric nervous system is represented by the strongly pronounced frontal ganglion, from which the recurrent nerve diverges. The occipital ganglion is not pronounced and is probably fused with the postcerebral endocrine complex.

Muscular system

The structure of the muscular system of ptiliids, in spite of their small size, complies with the general plan typical of insects. Sarcomere length in *Acrotrichis montandoni* is about 4 μm, about average for insects (Mandelshtam, 1983). The composition of musculature has been studied in *Mikado* sp. (Table 5.2).

Musculature of head (Fig. 5.24). 0an1 (M. tentorioscapalis anterior): O, anterior tentorial arms; I, anterior margin of scape. 0an2 (M. tentorioscapalis posterior): O, anterior tentorial arms; I, posterior margin of scape. 0an4 (M. tentorioscapalis medialis): O, anterior tentorial arms; I, medial margin of scape. 0lb5 (M. labroepipharyngealis): O, dorsal wall of labrum; I, wall of epipharynx. 0md1 (M. craniomandibularis internus), largest muscle in head: O, dorsal and lateral parts of cranium; I, adductor of mandibles. 0md3 (M. craniomandibularis externus): O, lateral part of cranium; I, abductor of mandible. 0mx1 (M. craniocardinalis): O, ventrolateral part of cranium; I, base of cardo. 0mx3 (M. tentoriocardinalis): O, anterior tentorial arms; I, ventral surface of cardo. 0mx2 (M. craniolacinialis): O, posterolateral part of cranium; I, base of lacinia. 0mx4 (M. tentoriostipitalis anterior): O, anterior tentorial arms; I, ventral surface of stipes. 0la5 (M. tentoriopraementalis): O, ventral part of cranium; I, posterior margin of prementum. 0la6 (M. tentorioparaglossalis): O, ventral part of cranium behind 0la5; I, posterior margin of prementum. 0hy1 (M. frontooralis): O, frons; I, lateral part of epipharynx. 0ci1 (M. clypeopalatalis): O, frontoclypeal region; I, dorsal part of epipharynx. 0bu2 (M. frontobuccalis anterior): O, anterior part of frons; I, dorsal surface of pharynx. 0bu3 (M. frontobuccalis posterior): O, frons behind 0hy1; I, dorsolateral part of pharynx. 0bu5 (M. tentoriobuccalis anterior): O, tentorial bridge; I, posterior part of hypopharynx. 0ph2 (M. tentoriopharyngealis): O, tentorial bridge; I, ventral wall of pharynx. 0st1 (M. annularis stomodaei) numerous. 0st2 (M. longitudinalis stomodaei) absent. Internal musculature of antennae and mouthparts not studied because of extremely small size.

Musculature of thorax. Prothorax (Fig. 5.25). Idlm1 (M. prophragma-occipitalis): O, medial part of prophragma; I, postoccipital area together with Idlm2. Idlm2 (M. pronoto-occipitalis): O, posterior part of pronotum; I, medial part postoccipital area together with Idlm1. Idvm2 (M. cervico-occipitalis): O, lateral part of sternite; I, lateral part postoccipital area. Idlm5 (M. pronoto-phragmalis anterior): O, prophragma; I, medial part of pronotum. Idvm6 (M. pronoto-cervicalis medialis): O, pronotum lateral to Idlm2; I, cervical region. Idvm10 (M. profurca-phragmalis): O, profurca; I, intersegmental membrane in front of mesepisternum. Idvm13 (M. pronoto-trochantinalis anterior): O, lateral part of pronotum; I, posterior margin of trochantin. Idvm17 (M. pronoto-coxalis posterior): O, lateral part of pronotum; I,

Table 5.2 Musculature of adults in Staphyliniformia and Cucujiformia, partly from literature

	Head muscles											
	0an1	0an2	0an3	0an4	0lb2	0lb5	0md1	0md3	0mx1	0mx2	0mx3	0mx5
	1	2	3	4	9	7	11	12	15	19	17	18
Mikado	+	+	–	+	–	+	+	+	+	+	+	+
Ochthebius	+	+	–	+	+	–	+	+	+	+	+	+
Hydraena	+	+	–	+	+	+	+	+	+	+	+	+
Necrophilus	+	+	–	+	–	+	+	+	+	+	+	+
Agathidium	+	+	–	+	–	+	+	+	+	+	+	+
Lesteva	+	+	–	+	+	+	+	+	+	+	+	+
Omalium	+	+	–	+	+	+	+	+	+	+	+	+
Tachyporus	+	+	–	+	+	+	+	+	+	+	+	+
Gyrophaena	+	+	–	+	+	+	+	+	+	+	+	+
Atheta	+	+	–	+	+	+	+	+	+	+	+	+
Oxytelus	+	+	–	+	+	+	+	+	+	+	+	+
Helophorus	+	+	–	+	+	+	+	+	+	+	+	+
Sericoderus	+	+	–	+	+	+	+	+	+	+	a	a
Coccinella	+	+	+	+	+	+	+	+	+	+	+	+
Lytta	+	+	+	+	+	+	+	+	+	+	+	+

	Head muscles														
	0la5	0la6	0la8	0la14	0hy1	0hy3	0hy9	0ci1	0bu2	0bu3	0bu5	0ph1	0ph2	0st1	0st2
	29	30	28	34	41	42	67	43	45	46	48	51	52	68	69
Mikado	+	+	–	–	+	+	–	+	+	+	–	–	+	+	–
Ochthebius	+	+	+	–	+	+	+	+	+	+	–	–	+	?	?
Hydraena	+	+	+	+	+	+	+	+	+	+	–	–	+	?	?
Necrophilus	+	+	+	+	+	+	?	+	+	+	–	?	?	?	?
Agathidium	+	+	+	+	+	+	?	+	+	+	–	?	?	?	?
Lesteva	+	+	+	+	+	+	?	+	+	+	–	?	?	?	?
Omalium	+	+	+	+	+	+	?	+	+	+	–	?	?	?	?

(continued)

Table 5.2 (continued)

	Head muscles														
	0la5	0la6	0la8	0la14	0hy1	0hy3	0hy9	0ci1	0bu2	0bu3	0bu5	0ph1	0ph2	0st1	0st2
Tachyporus	+	+	+	+	+	+	?	+	+	+	–	?	?	?	?
Gyrophaena	+	+	+	+	+	–	?	+	+	+	–	?	?	?	?
Atheta	+	+	+	+	+	–	?	+	+	+	–	?	?	?	?
Oxytelus	+	+	+	+	+	+	?	+	+	+	–	?	?	?	?
Helophorus	+	+	+	+	+	+	+	+	+	+	–	+	+	+	+
Sericoderus	b	b	–	–	+	–	–	+	+	+	–	–	+	+	–
Coccinella	+	+	–?	–	+	?	+?	+?	+?	+?	+?	+?	?	+	+
Lytta	?	+	–	–	+	?	+	+	+	+	+	–	?	+	+

	Thoracic muscles										
Taxon	Idlm1	Idlm2	Idlm3	Idlm5	Idvm2	Idvm5	Idvm6	Idvm8	Idvm10	Idvm13	Idvm17
Mikado	+	+	–	+	+	–	+	–	+	+	+
Ochthebius	+	–?	–	+	+	–	+	–	+	+	?
Creophilus	+	+	–	+	+	–	+	–	+	+	+
Nicrophorus	+	+	–	+	+	–	+	–	+	+	+
Hydrophilus	+	+	–	+	+	–	+	–	+	+	+
Margarinotus	+	+	–	+	+	–	+	–	+	+?	+
Aphodius	+	+	–	+	+	–	+	–	+	–	+
Cetonia	+	+	–	+	+	–	+	–	+	–	+
Sericoderus	**+**	**+**	–	**+**	**+**	–	**+**	–	**+**	**+**	**+**
Coccinella	+	+	–	+	+	+	+	–	+	–	+
Lytta	+	+	–	+	+	+	+	–	+	+	+
Meloe	+	+	–	+	+	+	+	–	+	+	+
Tenebrio	+	+	–	+	+	–	+	+	+	+	+
Cassida	+	+	–	+	+	–	+	–	+	+	+
Donacia	+	+	+	+	+	–	+	+	+	+	+

(continued)

Table 5.2 (continued)

	Thoracic muscles									
Taxon	Idvm18	Itpm3	Itpm6	Ipcm4	Ipcm8	Ivlm1	Ivlm3	Ivlm7	Ivlm9	Iscm2
	Thoracic muscles									
Taxon	Idvm18	Itpm3	Itpm6	Ipcm4	Ipcm8	Ivlm1	Ivlm3	Ivlm7	Ivlm9	Iscm2
Mikado	+	+	+	+	+	+	+	+	−	+
Ochthebius	?	+	+	?	+	+	+	+	−	+?
Creophilus	−	+	+	+	+	+	+	+	−	−
Nicrophorus	+	+	+	+	+	+	+	+	−	+
Hydrophilus	+	+	+	+	+	+	+	+	−	+
Margarinotus	−	+	+	+	+	+	+	+	−	+
Aphodius	+	+	+	+	+	+	+	+	−	−
Cetonia	+	+	+	+	+	+	+	+	−	−
Sericoderus	−	**+**	**+**	**+**	**+**	**+**	**+**	**+**	−	**+**
Coccinella	−	+	+	+	+	+	+	+	−	+
Lytta	+	+	+	+	+	+	+	+	+	−
Meloe	+	+	+	+	+	+	+	+	+	−
Tenebrio	−	+	+	+	+	+	+	+	−	−
Cassida	−	+	+	+	+	+	+	+	−	+
Donacia	−	+	+	+	+	+	+	+	−	−

Taxon	IIdlm1	IIdlm2	IIdvm4	IIdvm5	IIdvm6	IIdvm8	IItpm2	IItpm7	IItpm10	IIspm2	IIpcm2	IIpcm3	IIpcm5	IIpcm6	IIvlm3	IIscm1	IIscm2	IIscm6
Mikado	+	+	+	+	−	−	+	+	−	−	−	+	+	+	+	+	+	+
Ochthebius	+	+	+	?	−	−	+	?	?	−	+	−	−	+	+?	?	+	−
Creophilus	+	+	+	+	−	+	+	+	−	−	−	+	−	+	+	+	+	−
Nicrophorus	+	+	+	+	−	−	+	+	−	+	+	−	−	+	+	+	+	−
Hydrophilus	+	+	+	+	−	+	+	+	−	−	+	+	−	−	+	+	+	+
Margarinotus	+	+	+	+	**+**	+	+	+	−?	+	+	+	−	?	+	+	+	+
Aphodius	+	+	+	+	−	+	+	+	+	+	+	+	−	−	+	+	+	+
Cetonia	+	+	+	+	−	+	+	+	−	−	−	+	+	−	+	+	+	−
Sericoderus	+	+	+	+	−	−	+	+	−	+	+	+	+	−	−	+	−	+
Coccinella	+	+	+	+	−	−	+	+	+	+	+	+	−	+	+	−	+	+
Lytta	+	+	+	+	−	−	+	+	+	+	+	+	−	+	+	+	+	+
Meloe	+	+	+	+	−	−	+	+	−	−	+	−	−	−	+	+	+	+

(continued)

Table 5.2 (continued)

Taxon	IIdlm1	IIdlm2	IIdvm4	IIdvm5	IIdvm6	IIdvm8	IItpm2	IItpm7	IItpm10	IIspm2	IIpcm2	IIpcm3	IIpcm5	IIpcm6	IIvlm3	IIscm1	IIscm2	IIscm6
Tenebrio	+	+	+	+	−	−	+	+	+	+	+	+	−	+	+	+	−	+
Cassida	+	+	+	+	−	−	+	+	+	+	+	+	−	+	+	+	−	+
Donacia	+	+	+	+	−	−	+	+	+	+	+	+	−	+	+	+	−	+

Taxon	IIIdlm1	IIIdlm2	IIIdvm1	IIIdvm2	IIIdvm4	IIIdvm5	IIIdvm6	IIIdvm7	IIIdvm8	IIItpm2	IIItpm3	IIItpm7
Mikado	c	c	+	+	+	+	+	−	+	+	+	a
Ochthebius	+	+	+	−	+	+	+	+	+	+	+	a
Creophilus	+	+	+	+	+	+	+	−	+	+	+	a
Nicrophorus	+	+	+	+	+	+	+	−	+	+	+	a
Hydrophilus	+	+	+	−	+	+	+	−	+	+	+	a
Margarinotus	+	+	+	+	+	+	+	−	+	+	+	a
Aphodius	+	+	+	+	+	+	+	−	+	+	+	a
Cetonia	+	+	+	−	+	+	+	−	+	+	+	a
Sericoderus	+	+	+	+	+	+	+	−	+	+	+	a
Coccinella	+	+	+	+	−	+	−	−	+	+	+	a
Lytta	+	+	+	+	+	+	+	−	+	+	+	a
Meloe	−	−	−	−	−	+	−	−	+	−	−	a
Tenebrio	+	+	+	+	+	+	−	−	+	+	+	a
Cassida	+	+	+	+	−	+	−	−	+	+	+	a
Donacia	+	+	+	+	−	+	+	−	+	+	+	a

Taxon	IIItpm9	IIItpm10	IIIppm1	IIIspm1	IIIpcm3	IIIpcm4	IIIscm1	IIIscm4	IIIscm2	IIIscm6
Mikado	a	?	+	+	−	+	+	−	?	+
Ochthebius	a	+?	+	+	?	+	+	+	+	+
Creophilus	a	+	+	+	−	+	+	+	+	+
Nicrophorus	a	+	+	+	+	+	+	+	+	+
Hydrophilus	a	+	+	+	−	+	+	+	+	+
Margarinotus	a	+	+	+	+	+	+	+	+	+
Aphodius	a	+	+	+	+	+	+	+	+	+
Cetonia	a	+	−	+	−	+	+	+	+	+

(continued)

Table 5.2 (continued)

Taxon	IIItpm9	IIItpm10	IIIppm1	IIIspm1	IIIpcm3	IIIpcm4	IIIscm1	IIIscm4	IIIscm2	IIIscm6
Sericoderus	a	+	+	+	−	+	+	+	+	+
Coccinella	a	+	+	+	−	+	+	+	+	+
Lytta	a	+	+	+	+	+	+	+	+	+
Meloe	a	−	+	−	−	+	+	+	+	+
Tenebrio	a	+	−	+	+	+	+	+	+	+
Cassida	a	+	−	+	+	+	+	+	+	+
Donacia	a	+	−	+	−	+	+	+	+	+

Praghan (1938), Schneider (1981), Beutel and Haas (2000), Jach et al. (2000), Beutel et al. (2003), Beutel and Komarek (2004), Weide and Betz (2009)
+ present; − absent; *a*, *b* fused muscles, ? unknown

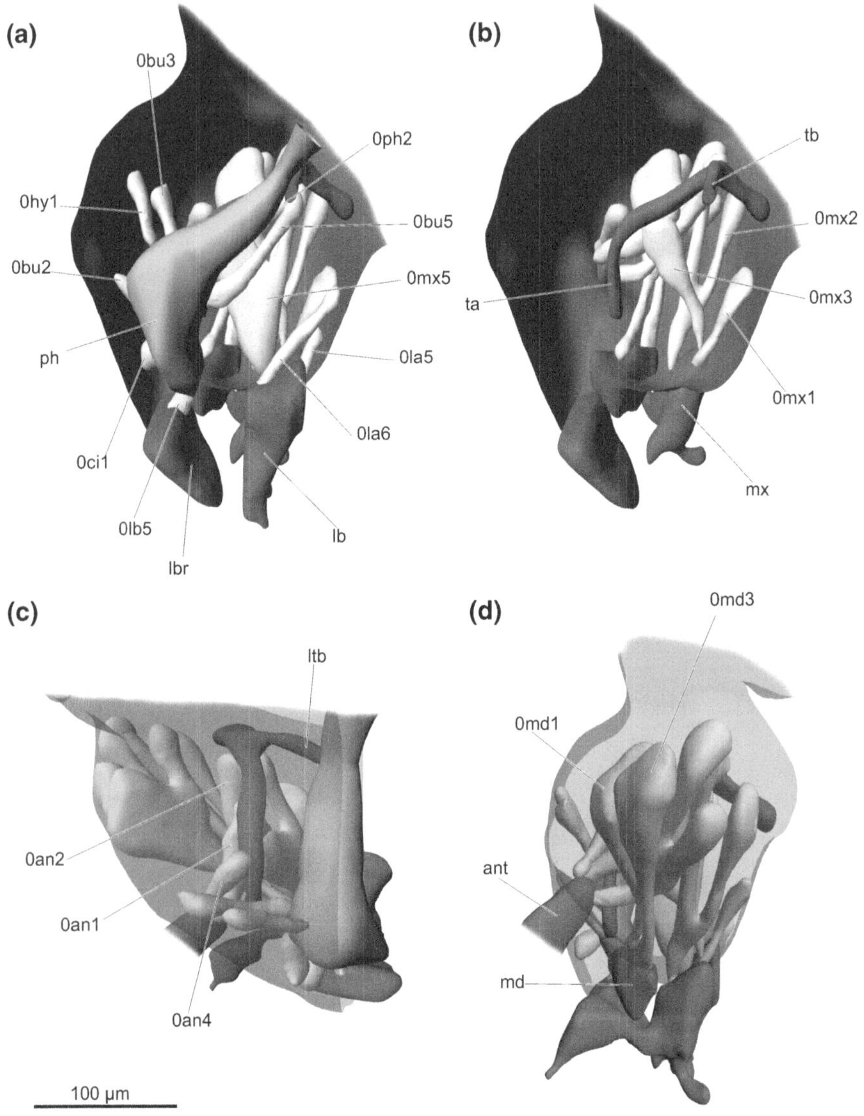

Fig. 5.24 Musculature of head in adult *Mikado* sp., 3D: **a**, **b** lateral view; **c** dorsal view; **d** lateral external view; *ant* antenna, *lb* labium, *lbr* labrum, *md* mandible, *mx* maxilla, *ph* pharynx, *ta* tentorial arms, *tb* tentorial bridge. Musculature see text

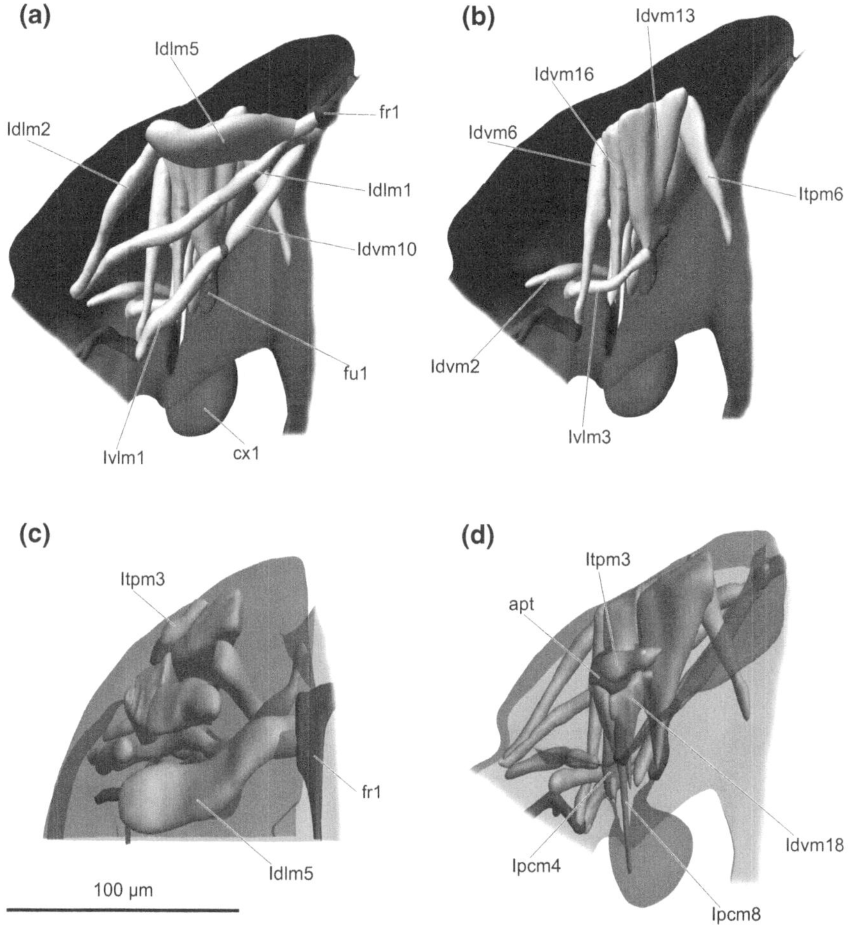

Fig. 5.25 Musculature of prothorax in adult *Mikado* sp., 3D: **a**, **b** lateral internal view; **c** dorsal view; **d** lateral external view; *apt* apodeme of trochantinopleura, *cx1* coxa of foreleg, *fr1* prophragma, *fu1* profurca. Musculature see text

anterolateral margin of base of coxa. Idvm18 (M. pronoto-coxalis lateralis): O, apodeme of trochantinopleura; I, posterior margin of base of coxa. Itpm3 (M. pronoto-pleuralis anterior): O, lateral part of pronotum; I, dorsal surface of apodemal plate of trochantinopleura. Itpm6 (M. pronoto-intersegmentalis): O, posterior part of pronotum lateral to Idlm2; I, intersegmental membrane anterior to mesepisternum. Ipcm4 (M. propleuro-coxalis superior): O, apodeme of trochantinopleura; I, anterior margin of base of coxa. Ipcm8 (M. propleuro-trochanteralis): O, apodeme of trochantinopleura; I, trochanter. Ivlm1 (M. profurca-cervicalis): O, profurca; I, cervical region. Ivlm3 (M. profurca-tentorialis): O, profurca; I, gular suture. Ivlm7 (M. profurca-mesofurcalis): O, profurca; I, mesofurca. Iscm2 (M. profurca-coxalis posterior): O, profurca; I, posterior margin of base of coxa. Internal musculature of legs not studied because of extremely small size.

Mesothorax (Fig. 5.26). IIdlm1 (M. prophragma-mesophragmalis): O, prophragma; I, mesophragma. IIdlm2 (M. mesonoto-phragmalis): O, medial suture of prophragma; I, lateral margin of mesophragma. IItpm2 (M. mesopleura-praealaris): O, pleural apodeme; I, anterolateral part of mesonotum. IItpm7 (M. mesanepisterno-axillaris) fused with IItpm9: O, episternum: I, third axillary plate. IItpm9

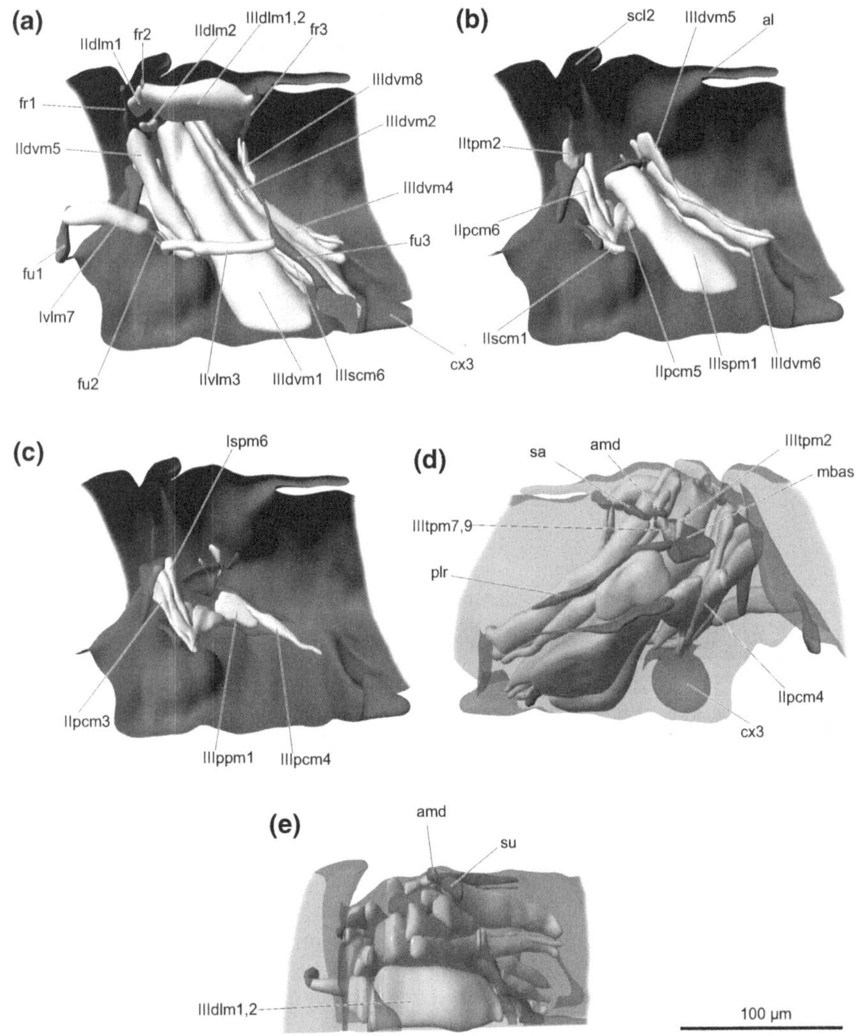

Fig. 5.26 Musculature of meso- and metathorax in adult *Mikado* sp., 3D: **a–c** lateral internal view; **d** lateral external view; **e** dorsal view; *al* alacrista, *amd* third axillary plate, *cx2.3* meso- and metacoxae, *fr1.2.3* pro-, meso-, and metaphragma, *fu1.2.3* pro-, meso-, and metafurca, *mbas* disk of basalare, *plr* pleural ridge, *sa* subalare. Musculature see text

(M. mesepimero-axillaris tertius): fused with IItpm7. IIdvm5 (M. mesonoto-coxalis posterior): O, prophragma; I, posterior margin of base of coxa. IIpcm3 (M. mesanepisterno-coxalis anterior): O, basalare; I, anterior margin of base of coxa, together with IIpcm4. IIpcm4 (M. mesanepisterno-coxalis posterior): O, episternum; I, anterior margin of base of coxa. IIpcm5 (M. mesanepisterno-trochanteralis): O, basalare; I, trochanter, together with IIpcm6. IIpcm6 (M. mesopleuratrochanteralis): O, episternum; I, trochanter. IIvlm3 (M. mesofurca-metafurcalis): O, mesofurca; I, metendosternite. IIscm1 (M. mesofurca-coxalis anterior): O, mesofurca; I, anterior part of coxa. IIscm2 (M. mesofurca-coxalis posterior): O, mesofurca; I, posterior part of coxa. Internal musculature of legs not studied because of extremely small size.

Metathorax (Fig. 5.26). IIIdlm1 (M. mesophragma-metaphragmalis) and IIIdlm2 (M. metanoto-phragmalis) fused: O, mesophragma and scutum; I, metaphragma. IIIdvm1 (M. metanoto-sternalis): O, anterior part of metanotum and mesophragma; I, ventrite of metathorax. IIIdvm2 (M. metanoto-trochantinalis anterior): O, metascutellum; I, process of trochantin. IIIdvm4 (M. metanoto-coxalis anterior): O, metanotum posterior to IIIdvm1; I, posterolateral margin of base of coxa. IIIdvm5 (M. metanoto-coxalis posterior): O, posterolateral part of metanotum; I, lateral part of coxa. IIIdvm6 (M. metacoxa-subalaris): O, coxa; I, subalare. IIIdvm8 (M. metafurca-phragmalis): O, apex of metendosternite; I, metaphragma. IItpm2 (M. metapleura-praealaris): O, prealar sclerite; I, pleural ridge. IIItpm3 (M. metanoto-basalaris): O, lateral margin of metanotum; I, disk of basalare. IIItpm7 (M. metanepisterno-axillaris): fused with IIItpm9. IIItpm9 (M. metepimero-axillaris tertius) fused with IItpm7: O, anterior part of pleural ridge; I, disk of third axillary plate. IIItpm10 (M. metepimero-subalaris) not found, possibly absent. IIIppm1 (M. metatransanapleuralis): O, pleural suture; I, lateral margin of ventrite. IIIspm1 (M. metapleura-sternalis): O, ventrite lateral to IIIdvm1; I, disc of basalare. IIIpcm4 (M. metanepisterno-coxalis posterior): O, episternum; I, laterally anterior margin of base of coxa. IIIscm1 (M. metafurca-coxalis anterior): O, anterior part metendosternite; I, anterior margin of base of coxa. IIIscm6 (M. metafurca-trochanteralis): O, metendosternite; I, trochanter. Internal musculature of legs not studied because of extremely small size.

Musculature of abdomen Dorsal longitudinal musculature (Mm. dorsales), several parallel fibers, O, anterior phragma; I, posterior phragma. Ventral longitudinal musculature (Mm. ventrales), several parallel fibers: O, anterior margin of segment; I, posterior margin of segment. Dorsoventral musculature, pair of urotergosternal muscles.

Reproductive system

Male reproductive system consisting of unpaired, judging by position, right testis, spermiduct, accessory glands, and copulatory apparatus (Fig. 5.27a, b). Spermatozoon of some species of Ptiliidae sometimes longer than body (Dybas and Dybas 1981, 1987; De Marzo 1992).

Female reproductive system consisting of unpaired, judging by position, right ovary, oviduct, spermatheca, and accessory glands (Fig. 5.27c, d). Ovary consisting

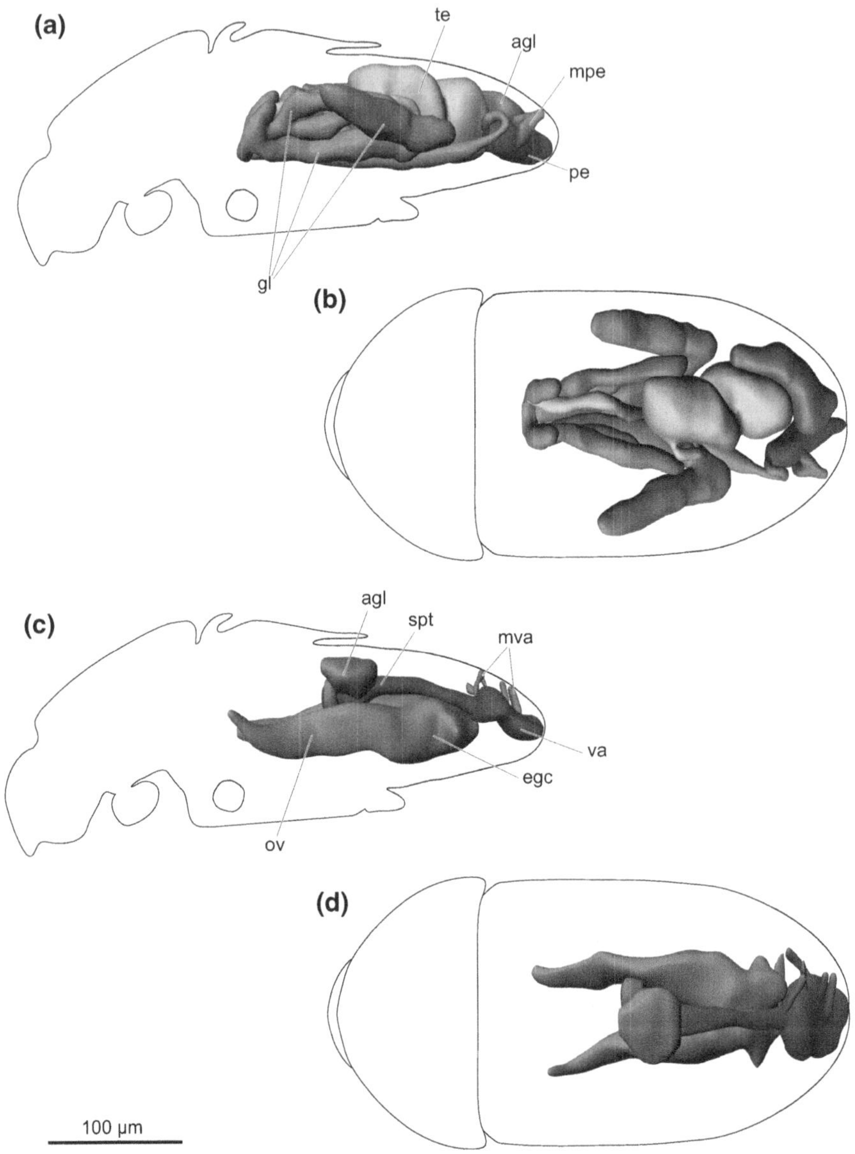

Fig. 5.27 Reproductive system of *Mikado* sp., 3D: **a**, b male; **c**, **d** female; **a**, **c** lateral view, **b**, **d** dorsal view; *agl* accessory glands, *egc* egg chamber, *mpe* musculature of penis, *mva* musculature of vagina, *ov* ovariole, *pe* penis, *spt* spermatheca, *te* testis, *va* vagina

of 2–4 teleotrophic ovarioles. Oviduct forming dilation, egg chamber, in which egg develops. Only one egg develops at a time, and egg ready for laying occupies up to half body length of the female.

5.4 Morphology of Larvae

5.4.1 External Morphology

Body length of first instar larva in *Mikado* sp. 390–450 μm (M = 430, n = 6), body length of last instar larva 790–920 (M = 840, n = 12).

Morphology of head

Head prognathous, approximately as long as wide, slightly flattened in dorsoventral direction (Fig. 5.28 and 5.29). **Cranium** solid, without sutures, distinguishing Ptiliidae from Staphylinidae and Hydraenidae, in which cranium with clearly discernible epicranial suture. Dorsal tentorial arms absent, anterior and posterior arms together forming long slender parallel structures, rounded in cross-section. Tentorial bridge strongly curved and reaching into thorax. **Ocelli** absent, except for one species of *Nossidium* (Dybas 1976). **Antennae**, as in most Staphylinoidea, 3-segmented. Antennomere 2 bearing sensorium (sensory cone) and 4–6 trichoid sensilla; apical antennomere bearing 3–5 trichoid sensilla and 2–4 cuticular apodemes of various length. **Labrum** movably joined with cranium and appearing as simple plate bearing 10–18 setae of various lengths. **Mandibles** well developed, symmetrical, or almost symmetrical (*Ptinella*), short, with wide base, well-developed molar process and prostheca, generally similar in shape to mandibles of Hydraenidae and primitive Staphylinidae (Betz et al. 2003). Structure of **maxillae** in Ptiliidae very similar to those of Hydraenidae, Leiodidae, and Agyrtidae; based on this character and some other characters these families used to be united in the so-called leptinid association (Boving and Craighead 1931). Body of maxilla consisting of cardo and stipes. Stipes bearing 3-segmented palp. Palomere 1 without setae; palpomere 2 with two trichoid sensilla and apically with sensillum ampullaceum; palpomere 3 apically with several basiconical sensilla, probably functioning as contact chemoreceptors. Lacinia and galea short, weakly separated. Lacinia with several teeth, galea bearing two 'fan-shaped' rows of long setae. In Nanosellini fan-shaped structures on galea absent. **Labium** membranous, weakly sclerotized. Mentum and submentum fused with cranium; prementum separated from mentum by weakly discernible suture. Labial palps 1–2-segmented, with ligula between them.

Morphology of thorax

Thorax consisting of three separate segments, subequal in size, slightly wider than abdomen (Fig. 5.29). Thoracic segments with only weakly sclerotized non-pigmented dorsal sclerites, whereas in most staphylinoids with strongly sclerotized, pigmented dorsal, lateral, and ventral sclerites. Pair of spiracles present, as in other beetles, between pro- and mesothorax. **Legs** consisting of coxa, trochanter, femur, fused tibia and tarsus, and claw, as in larvae of other staphylinoids.

Morphology of abdomen

Abdomen consisting of 10 segments. All segments without sclerites, in contrast to Staphylinidae (in most staphylinids segments with dorsal, pleural, and ventral sclerites. Larvae of most Ptiliidae (except for *Nossidium* and *Nephanes*) without

Fig. 5.28 Larva of *Acrotrichis montandoni*, SEM: **a** habitus in lateral view; **b** head, ventral view; **c** head, frontal view; **d** apex of abdomen, lateral view; **e** galea; **f** antenna; **g** maxillary palp; *ah* anal hook, *ant* antenna, *as* segment of abdomen, *ga* galea, *lbp* labial palp, *lbr* labrum, *md* mandible, *mnt* mentum, *mx* maxilla, *mxp* maxillary palp, *sen* sensorium, *ug* urogomphi

Fig. 5.29 Larva of *Mikado* sp. SEM: **a–c** first instar larva, lateral view; **d–g** last instar larva; **d** thorax, ventral view; **e** head, dorsal view; **f** apex of abdomen, lateral view; **g** head, ventral view; *ant* antenna, *as* segment of abdomen, *cx2* mesocoxa, *lbr* labrum, *md* mandible, *mx* maxilla, *pyg* pygopod, *ca* cardo, *lb* labium, *mxp* maxillary palp, *prm* prementum, *sti* stipes

abdominal spiracles, present in staphylinids and hydraenids on segments 1–8. Segment 9 with 1-segmented **urogomphi**, in some representatives (Nanosellini, *Ptenidium*) absent. Segment 10 with anal hooks (in some Nanosellinae absent) and membranous anal vesicle (homology of which somewhat unclear, Dybas 1976: anal vesicle, Grebennikov and Beutel 2002: pygopod).

5.4.2 Internal Morphology

General plan of internal structure: most of body cavity occupied by intestine, considerable part of thorax occupied by musculature, space between all organs filled with fat body; diaphragms and heart absent.

Particular organ systems or functional groups of systems are described in more detail in special sections below.

Integument

Cuticle thickness in *Mikado* 0.08–1.19 μm (M = 0.67; n = 80) in first instar larvae and 0.4–4.9 μm (M = 1.0; n = 80) in last instar larvae. Cuticle of abdominal segments 8–10 several times as thick as on other parts of body, consisting of cuticle, hypoderm, and basal membrane.

Digestive and excretory systems

Structure of digestive system in ptiliid larvae, as in adults, generally complying with structure typical of insects (Fig. 5.30a–d).

Foregut divided into pharynx and oesophagus, crop and muscular stomach undeveloped. Oesophagus straight, rounded in cross-section. Oesophagus with one thin layer of circular muscles, distinguishing ptiliids from other beetles (Crowson 1981).

Midgut appearing as tube, in cardial section with two lateral diverticula directed anteriad. Wall consisting of unilayered epithelium, with 8–14 cells visible in cross-section, much fewer than in other insects. Entire midgut with slender unilayered peritrophic membrane. Muscles of midgut not found.

Hindgut divided into small intestine and rectum.

Boundary between mid- and hindgut bearing four **Malpighian tubules,** one pair directed anteriad, other pair directed posteriad.

Circulatory system and fat body

In ptiliid larvae, as in adult, **circulatory system** reduced, heart absent, and hemolymph occupying extremely small part of body cavity volume, probably because of diffusion sufficing for transport of molecules in organisms of such small sizes. **Fat body** in ptiliid larvae occupying almost all free body cavities, replacing hemolymph.

Tracheal system

In larvae of most species of Ptiliidae only one pair of spiracles present, between pro- and mesothorax, and small number of tracheae in prothorax and head. In other segments of thorax and in abdomen tracheae absent. Judging by this fact, as well as

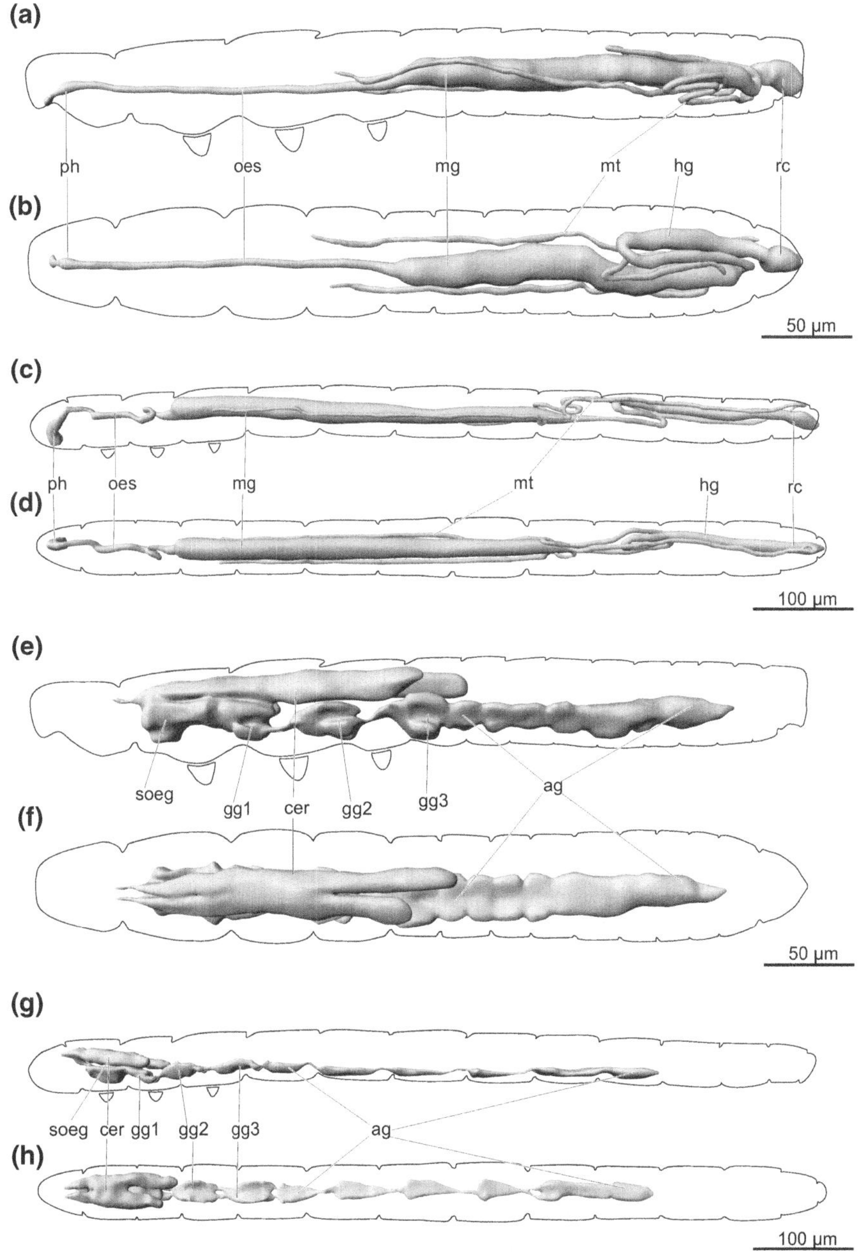

Fig. 5.30 Digestive and nervous systems in larva of *Mikado* sp. SEM: **a**, **b**, **e**, **f** first instar larva; **c**, **d**, **g**, **h** last instar larva; **a–d** digestive system; **e–h** nervous system; **a**, **c**, **e**, **g** lateral view; **b**, **d**, **f**, **h** dorsal view; *ag* abdominal ganglion, *cer* cerebrum, *gg1.2.3* pro-, meso-, and metathoracic ganglia, *hg* hindgut, *mg* midgut, *mt* Malpighian tubules, *oes* oesophagus, *ph* pharynx, *rc* rectum, *soeg* suboesophageal ganglion

by the fact that ptiliid larvae have thin, weakly sclerotized integument, it can be suggested that they employ mixed, tracheal-cutaneous respiration.

Nervous system

Central nervous system of ptiliid larvae different from that of adults in weakly pronounced condensation and compaction of ganglia, especially in last instar larvae (Fig. 5.30e–h). Cerebrum elongate, completely shifted into thorax, and in first instar larvae of *Mikado* sp. posterior protocerebral lobes of cerebrum asymmetrical and reaching abdominal segment 2. Deutocerebrum with elongate lobes (directed into cranium). Single neuropilar mass of cerebrum appearing homogeneous and not differentiated into cerebral centers. Suboesophageal ganglion fused with prothoracic ganglion. Mesothoracic and metathoracic ganglia separated. Ganglia of abdominal segments 1–5 in late instar larvae separate, ganglia of segments 6–8 fused and situated in segment 6. Abdominal ganglia in first instar larvae of *Mikado* sp. very closely set, resembling elongate synganglion. Ultrastructure of central nervous system as in adult.

Cell size in nervous system 1.19–1.98 μm (M = 1.35; n = 23) om first instar larvae. Cerebrum of first instar larvae of *Mikado* sp. containing about 6700 cells.

Elements of stomatogastric nervous system not revealed, possibly fused with central nervous system.

Muscular system

Ptiliid larvae, in spite of their small sizes, have all main groups of musculature and almost all particular muscles typical of beetle larvae. Musculature has been studied in detail in first and last instar larvae of *Mikado* sp. (Figure 5.31 and 5.32); published data on *Ptinella tenella* (Grebennikov and Beutel 2002) have also been used (Table 5.3).

Musculature of head. M1 (M. tentorioscapalis anterior): O, anterior tentorial arms; I, anterior margin of base of scape. M2 (M. tentorioscapalis posterior): O, anterior tentorial arms; I, posterior margin of base of scape. M4 (M. tentorioscapalis medialis), in first instar larvae absent, in last instar larvae: O, anterior tentorial arms; I, base of scape. M11 (M. craniomandibularis internus), largest muscle in head: O, posterior and lateral part of cranium; I, adductor of mandible. M12 (M. craniomandibularis externus): O, posterior and lateral part of cranium; I, abductor of mandible. M15 (M. craniocardinalis): O, ventrolateral part of cranium; I, base of cardo. M17 (M. tentoriocardinalis): O, tentorial arms; I, ventral surface of cardo. M18 (M. tentoriostipitalis): O, tentorial arms near M17; I, ventral surface of stipes. In last instar larvae M17 and M18 fused. M19 (M. craniolacinialis): O, lateral part of cranium; I, base of galea. M29 (M. tentoriopraementalis inferior): O, tentorial arms; I, lateral part of prementum. M30 (M. tentoriopraementalis superior): O, tentorial arms; I, lateral part of prementum. M41 (M. frontohypopharyngalis): O, frons; I, lateral and posterior margin of epipharynx. M42 (M. tentoriohypopharyngalis): O, tentorial bridge; I, hypopharynx. M43 (M. clypeopalatalis): O, frontoclypeal part of cranium; I, dorsal surface of epipharynx. M45 (M. frontobuccalis anterior): O, frontal part of cranium; I, dorsal and lateral surface of pharynx. M 46 (M. frontobuccalis posterior): O, frons behind base of M41; I, dorsal and lateral surface of pharynx. M 48 (M. tentoriobuccalis anterior): O, tentorial

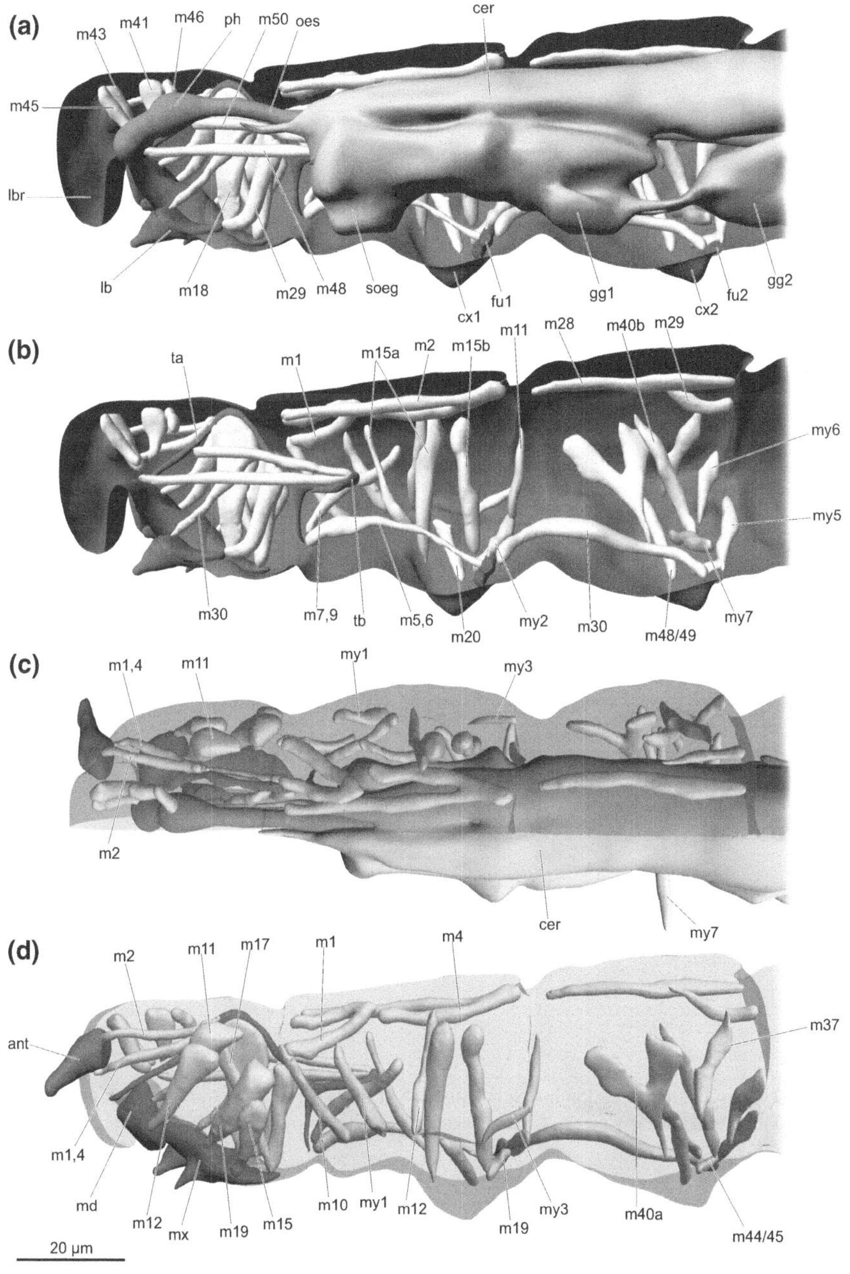

Fig. 5.31 First instar larva of *Mikado* sp., 3D: **a**, **b** lateral internal view; **c** dorsal view; **d** lateral external view; *cer* cerebrum, *cx1.2* meso- and metacoxae, *fu1.2* pro- and mesofurca, *gg1.2* pro- and mesothoracic ganglia, *lb* labium, *lbr* labrum, *mg* midgut, *ph* pharynx, *oes* oesophagus, *ta* tentorial arms, *tb* tentorial bridge, *soeg* suboesophageal ganglion. Musculature see text

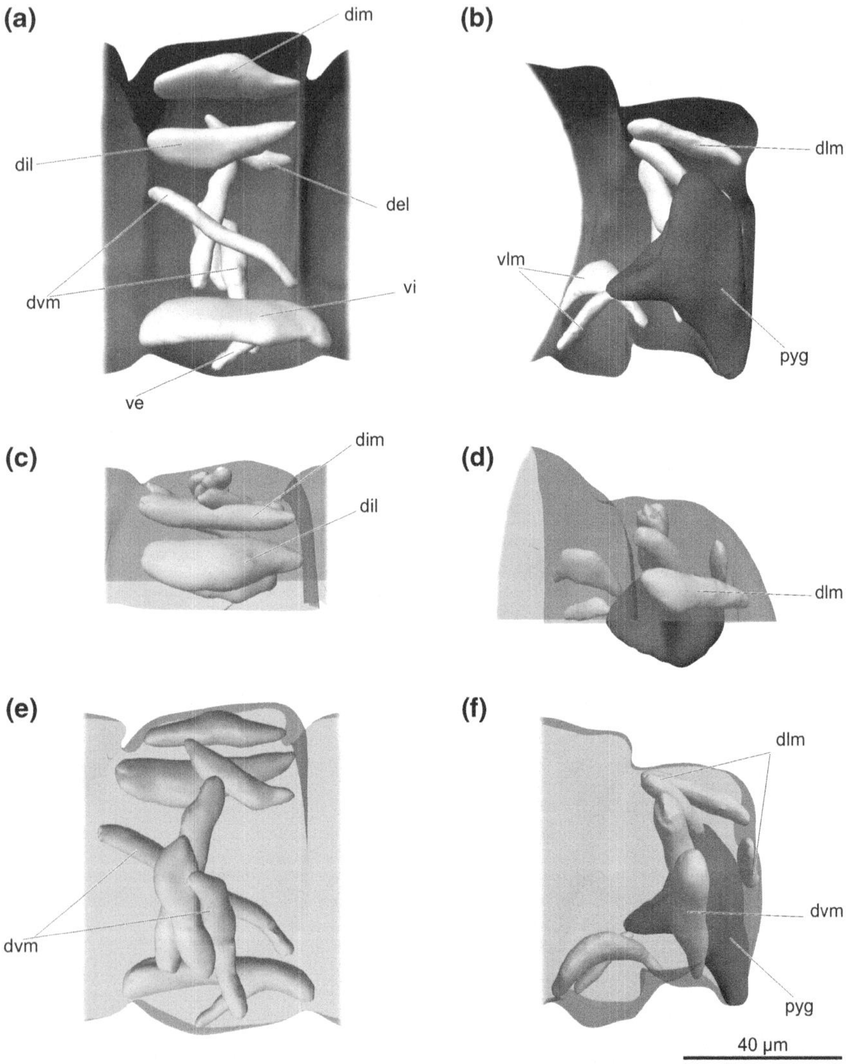

Fig. 5.32 Musculature of abdominal segments in first instar larva of *Mikado* sp., 3D: **a**, **b** lateral internal view; **c**, **d** dorsal view; **e**, **f** lateral external view; *pyg* pygopod. Musculature see text

bridge; I, posterior boundary of hypopharynx. M 50 (M. tentoriobuccalis posterior): O, tentorial bridge; I, anterior part of pharynx. Internal musculature of antennae and mouthparts not studied because of extremely small size.

Musculature of thorax. Prothorax. M1 (M. pronoti primus): O, pronotum; I, cervical membrane. M2 (M. pronoti secundus): O, central part of prophragma; I, dorsal part of postoccipital suture. M3 (M. pronoti tertius) in first instar larvae absent or fused with M2. y last instar larvae: O, medial part of prophragma; I, dorsal

Table 5.3 Musculature of larvae in Staphyliniformia and Cucujiformia, partly from literature

	Head muscles											
	1	2	4	7	9	11	12	15	17	18	19	28
Mikado, first ins.	+	+	−	−	−	+	+	+	+	+	+	−
Mikado, last ins.	+	+	+	−	−	+	+	+	+	+	+	−
Ptinella, last ins.	+	+	+	+	−	+	+	+	+	+	+	+
Hydraena	+	+	+	−	+	+	+	+	+	+	+	+
Catops	+	+	+	+	+	+	+	+	+	+	+	+
Tachinus	+	+	+	−	+	+	+	+	+	+	+	+
Staphylinus	+	+	+	−	−	+	+	+	+	+	+	+
Sepedophilus	+	+	+	−	+	+	+	+	+	+	+	+
Silpha	+	+	+	−	+	+	+	+	+	+	+	+
Hydrochara	+	+	+	−	−	+	+	+	+	+	+	+
Oryctes	+	+	+	+	+	+	+	+	+	+	+	−
Sericoderus, first ins.	−	+	+	+	+	+	+	+	a	a	+	+
Sericoderus, last ins.	+	+	+	+	+	+	+	+	a	a	+	+
Sphindus	+	+	+	+	+	+	+	+	+	+	+	+
Ericmodes	+	+	+	−	+	+	+	+	+	+	+	+
Prostomis	+	+	+	−	−	+	+	−	+	+	+	+

	Head muscles									
	29	30	41	43	44	45	46	48	50	52
Mikado, first ins.	+	+	+	+	−	+	+	+	+	−
Mikado, last ins.	+	+	+	+	−	+	+	+	+	−
Ptinella, last ins.	+	+	+	+	−	+	+	+	+	+
Hydraena	+	+	+	+	−	+	+	+	?	+
Catops	+	+	+	+	−	+	+	+	?	+
Tachinus	+	+	+	+	−	+	+	+	?	−?
Staphylinus	+	+	+	+	−	+	+	+	?	+
Sepedophilus	+	+	+	+	−	+	+	+	+	+
Silpha	+	+	+	+	−	+	+	+	?	+
Hydrochara	+	+	+	+	−	+	+	+	?	−
Oryctes	+	+	+	+	+	+	+	+	?	+
Sericoderus, first ins.	+	−	+	+	−	+	+	−	−	−
Sericoderus, last ins.	+	−	+	+	−	+	+	−	−	−
Sphindus	−	−	+	+	−	+	+	+	−	+
Ericmodes	+	−	+	+	−	+	+	+	−	+
Prostomis	−	−	+	+	−	+	+	+	−	+

	Thoracic muscles									
	1	2	3	4	5	6	7	8	9	10
Mikado, first ins	+	+	−	+	a	a	b	−	b	+
Mikado, last ins	+	+	+	+	a	a	+	−	+	+
Ptinella, last ins	+	+	+	+	+	+	+	+	+	+
Sericoderus, first ins	+	+	+	+	a	a	b	−	b	+
Sericoderus, last ins	+	+	+	+	+	+	b	−	b	+

	Thoracic muscles											
	11	12	15	19	20	Y1	Y2	Y3	Y8	Y9	Y10	Y11
Mikado, first ins	+	+	+	+	+	+	+	+	−	−	−	−
Mikado, last ins	+	+	+	+	+	+	+	+	−	−	−	−

(continued)

Table 5.3 (continued)

	Thoracic muscles											
	11	12	15	19	20	Y1	Y2	Y3	Y8	Y9	Y10	Y11
Ptinella, last ins	+	+	+	+	+	+	–	–	–	–	–	–
Sericoderus, first ins	+	+	+	+	+	+	+	+	+	+	+	+
Sericoderus, last ins	+	+	+	+	+	+	+	+	+	+	+	+

	28	29	30	37	40	44/45	48/49	Y4	Y5	Y6	Y7	Y12	Y13	Y14	Y15	Y16
Mikado, first ins	+	+	+	+	+	+	+	–	+	+	+	–	–	–	–	–
Mikado, last ins	+	+	+	+	+	+	+	+	+	+	+	–	–	–	–	–
Ptinella, last ins	+	+	+	+	+	+	+	–	–	–	–	–	–	–	–	–
Sericoderus, first ins	+	+	+	+	+	+	+	+	+	+	–	+	+	+	+	+
Sericoderus, last ins	+	+	+	+	+	+	+	+	+	+	–	+	+	+	+	+

Beutel and Haas (1998), Leschen and Beutel (2001), Beutel and Splinski (2001), Grebennikov and Beutel (2002), Schunger et al. (2003)
\+ present; – absent; *a, b* fused muscles, ? unknown

part of postoccipital suture. M4 (M. pronoti quartus): O, prophragma, together with M2; I, pronotum. M5 (M. prosterni primus): O, profurca; I, lateral part of postoccipital suture. M6 (M. prosterni secundus): absent or fused with M5. M7 (M. dorsoventralis primus): O, pronotum; I, ventral part of postoccipital suture. M8 (M. dorsoventralis secundus; MI7b): absent. M9 (M. dorsoventralis tertius): in first instar larvae absent or fused with M7, y last instar larvae: O, pronotum; I, ventrolateral part of postoccipital suture. M10 (M. dorsoventralis quartus): O, sternite prothorax; I, lateral part of postoccipital suture. M11 (M. dorsoventralis quintus): O, prophragma; I, posterior part of pleurite. M12 (M. noto-pleuralis): O, lateral part of pronotum; I, pleurite. M14 (M. noto-trochantinalis): not found, possibly absent. M15a (M. noto-coxalis anterior): O, pronotum lateral to M4; I, anterior margin of base of coxa. M15b (M. noto-coxalis posterior): O, posterior part of pronotum; I, posterior margin of base of coxa. M 19 (M. furca coxalis): O, profurca; I, lateral part of base of coxa. M 20 (M. pleuratrochanteralis): O, pleurite; I, tendon of trochanter. Three muscles of unclear homology: MY1 (Speyer 1922: 10c?; Grebennikov and Beutel 2002: y): O, anterior part of pronotum; I, anterior part of sternite. MY2 (Speyer 1922: 4f?): O, profurca; I, laterally between pro- and mesothorax. MY3 (Speyer 1922: 10g?): O, posterior part of pronotum; I, posterior part of sternal and pleural region. Internal musculature of legs not studied because of extremely small size.

Mesothorax. Musculature identical to prothoracic, except for absence of cervical musculature and presence of one transverse muscle (MY7) of unclear homology running from ventral margin of pleurite on one side of body to ventral part of pleurite of other side of body.

Metathorax. Musculature identical to mesothoracic.

Musculature of abdomen. Segments I–VIII. Dorsal longitudinal musculature: dim (Mm. dorsales interni medialis): O, anterior phragma; I, posterior phragma. dil (Mm. dorsales interni lateralis): O, anterior phragma; I, posterior phragma. del (Mm. dorsales interni lateralis) O, tergite; I, posterior phragma. Ventral longitudinal musculature: vi (Mm. ventrales interni): O, anterior margin of segment; I, posterior margin of segment. Ve (Mm. ventrales externi): O, sternite; I, posterior margin of segment. Dorsoventral musculature (dvm): three pairs of tergosternal muscles, homology unknown. Segment IX. Dorsal longitudinal muscle (dlm): ventral longitudinal muscle (vlm): dorsoventral muscle (dvm). Segment X: one dorsolateral muscle (dlm). Pygopod, pair of dorsal muscles and one ventral longitudinal muscle, diverging from anterior margin of segment IX.

References

Araujo, J., & Pasteels, J. M. (1985). Ultrastructure de la Glande Defensive de *Drusilla canaliculata* (Fab.) (Coleoptera, Staphylinidae) *Archives de biologie (Bruxelles) 96*, 81–89.

Betz, O., Thayer, M. K., & Newton, A. F. (2003). Comparative morphology and evolutionary pathways of the mouthparts in spore-feeding Staphylinoidea (Coleoptera). *Acta Zoologica, 84* (3), 179–238.

Beutel, R. G., Anton, E., & Jäch, M. A. (2003). On the evolution of adult head structures and the phylogeny of Hydraenidae (Coleoptera, Staphyliniformia). *Journal of Zoological Systematics and Evolutionary Research, 41*(4), 256–275.

Beutel, R. G. (1994). Phylogenetic analysis of Hydrophiloidea based on characters of the head of adults and larvae (Coleoptera, Staphyliniformia). *Koleopterologische Rundschau, 64*, 103–131.

Beutel, R. G., & Haas, A. (1998). Larval head morphology of *Hydroscapha natans* LeConte 1874 (Coleoptera, Myxophaga, Hydroscaphidae) with special reference to miniaturization. *Zoomorphology, 118*(2), 103–116.

Beutel, R. G., & Haas, A. (2000). Phylogenetic releationships of the suborders of Coleoptera (Insecta). *Cladistics, 16*, 103–141.

Beutel, R. G., & Jäch, M. A. (1995). Untersuchungen unber die Propleura der Hydraenidae (Coleoptera: Staphyliniformia). *Zeitschrift der Arbeitsgemeinschaft österreichischer Entomologen, 47*, 59–63.

Beutel, R. G., & Komarek, A. (2004). Comparative study of thoracic structures of adults of Hydrophiloidea and Histeroidea with phylogenetic implications (Coleoptera, Polyphaga). *Organisms Diversity & Evolution, 4*, 1–34.

Beutel, R. G., & Molenda, R. (1997). Comparative Morphology of Selected Larvae of Staphilinoidea (Coleoptera, Polyphaga) with Phylogenetic Implication. *Zoologischer Anzeiger, 236*, 31–67.

Beutel, R. G., & Splinski, A. (2001). Comparative study of head structures of larvae of Sphindidae and Protocucujidae (Coleoptera: Cucujoidea). *European Journal of Entomology, 98*, 219–232.

Boving, A. G., & Craighead, F. C. (1931). *An illustrated synopsis of the principal larvae forms of the Coleoptera*. New York: The Brooklyn Entomological Society.

Carstens, S., & Storch, V. (1980). Beeinflussung der Ultrastruktur von Fettkorper und Mitteldarm des Staphyliniden Atheta fungi (Grav.) durch Umwelteinflusse. *Zoologische Jahrbuecher Abteilung für Anatomie und Ontogenie der Tiere, 103*, 73–84.

Crowson, R. A. (1938). The metendosternite in Coleoptera: a comparative study. *Transactions of the Royal Entomological Society of London, 87*, 397–415.

Crowson, R. A. (1944). Futher studies on the metendosternite in Coleoptera. *Transactions of the Royal Entomological Society of London, 94*, 273–310.

Crowson, R. A. (1981). *The biology of the Coleoptera*. London: Academic Press.

De Coninck, E., & Coessens, R. (1981). Life cycle and reproductive pattern of *Acrotrichis intermedia* (Coleoptera: Ptiliidae) in experimental conditions. *Journal of Natural History, 15* (6), 1047–1055.

De Marzo, L. (1992). Osservationi anatomiche sui genitali interni maschili in alcuni ptilidi (Coleoptera). *Entomologica Bari, 27*, 107–115.

De Marzo, L. (2000). Larve di coleotteri in detriti vegetali di orgine agricola: lineamenti morfologici e presenza stagionale (Polyphaga: 20 familglie). *Entomologica Bari, 2000*(34), 65–131.

De Marzo, L. (2001). Coleotterofauna dei depositi di Podisonia: morfologia larvale in alcune specie caratteristiche (Sphaerididae, Histeridae, Ptiliidae, Staphylinidae). *Estratto dagli Annali del Museo Civico di Storia Naturale "G.Doria" 93*(3), 461–471.

Delgado, J. A., & Soler, A. G. (1996). Morphology and chaetotaxy of the first-instar larva of *Hydraena (Phothydraena) hernandoi* Fresneda & Lagar (Coleoptera: Hydraenidae). *Koleopterologische Rundschau, 66*, 147–154.

Delgado, J. A., & Soler, A. G. (1997a). Morphology and Chaetotaxy of Larval Hydraenidae (Coleoptera) I: The Genus *Limnebius* Leach, 1815. Based on a Description of *Limnebius cordobanus* d'Orchymont. *Aquatic Insects, 19*(1), 37–49.

Delgado, J. A., & Soler, A. G. (1997b). Morphology and Chaetotaxy of Larval Hydraenidae (Coleoptera) II: The Subgenus *Ochthebius* s. str. Leac. *Bulletin de l'Institut Royal des Sciences Naturelles de Belgique Entomologie, 67*, 45–55.

Delgado, J. A., & Soler, A. G. (1997c). Morphology and Chaetotaxy of Larval Hydraenidae (Coleoptera) III: The Genus *Calobius* Wollaston, 1854. *Aquatic Insects, 19*(3), 165–175.

Dybas, H. S. (1976). The larval characters of featherwing and limulodid betles and family relationships in the Staphylinoidea. *Fieldiana Zoology 70*(3), 29–78.

Dybas, H. S. (1978). Polimorphism in Featherwing Beetles, with a Revision of the Genus Ptinelloides (Coleoptera: Ptiliidae). *Annals of the Entomological Society of America, 71*, 695–714.

Dybas, L. K., & Dybas, H. S. (1981). Coadaptation and taxanomic differentiation of sperm and spermathecae in featherwing beetles. *Evolution, 35*(1), 168–174.

Dybas, L. K., & Dybas, H. S. (1987). Ultrastructure of mature spermatozoa of a minute featherwing beetle from Sri Lanka (Coleoptera, Ptiliidae: Bambara). *Journal of Morphology, 191*, 63–76.

Fedorenko, D. N. (2009). *Evolution of the beetle hind wing, with special reference to folding (Insecta, Coleoptera)*. Sofia, Moscow: Pensoft Publishers.

Ganglbauer L (1895) Familienreihe Staphyloidea. In Die Käfer von Mitteleuropa 2. Gerold, Wien.

Grebennikov, V. V., & Beutel, R. G. (2002). Morphology of the minute larva of *Ptinella tenella*, with special reference to effects of miniaturisation and the systematic position of Ptiliidae (Coleoptera: Staphylinoidea). *Arthropod Structure & Development, 31*(2), 157–172.

Gillmeister, C. J. F. (1845). Trichopterygia, Beschreibung and Abbilidung der haarflugeligen Kafer. *Deutschlands Insekten Kafer, 17*, 1–98.

Hall, W. E. (1999). Generic revision of the tribe Nanosellini (Coleoptera: Ptiliidae: Ptiliinae). *Transactions of the American Entomological Society 125*(1/2), 39–126.

Hammond, P. M. (1976). A review of the genus *Anotylus* C. G. Thomson (Coleoptera: Staphylinidae). *Bulletin of the Natural History Entomology 33*(2), 141–187.

Hansen, M. (1997). Phylogeny and classification of the staphyliniform beetle families (Coleoptera). *Biologiske Skrifter, 48*, 1–339.

Herman, L. H. (1970). Phylogeny and reclassification of the Genera of the rove-beetle Subfamily Oxyteline of the World (Col., Staph.). *Bulletin of the American Museum of Natural History, 142*, 343–454.

Hunt, T., Bergsten, J., Levkanicova, Z., et al. (2007). A comprehensive phylogeny of beetles reveals the evolutionary origins of a super-radiation. *Science, 318*, 1913–1916.

Jach, M. A., Beutel, R. G., Dıaz, J. A., & Kodada, J. (2000). Subgeneric classification, description of head structures, and world check list of *Hydraena* Kugelann (Insecta: Coleoptera: Hydraenidae). *Annalen des Naturhistorischen Museums Wien, 102*, 177–258.
Kukalová-Peck, J., & Lawrence, J. F. (1993). Evolution of the hind wing in Coleoptera. *Canadian Entomologist, 125*(2), 181–258.
Kukalová-Peck, J., & Lawrence, J. F. (2004). Relationships among coleopteran suborders and major endoneopteran lineages: evidence from hind wing characters. *Journal of Entomology, 101*, 95–144.
Lawrence, J. F., Ślipiński, A., Seago, A. E., Thayer, M. K., Newton, A. F., & Marvaldi, A. E. (2011). Phylogeny of the Coleoptera based on morphological characters of adults and larvae. *Annals of Zoology, 61*(1), 1–217.
Lawrence, J. F., & Newton, A. F. (1982). Evolution and classification of beetles. *Annual Reviews of Ecology and Systematics, 13*, 261–290.
Lawrence, J. F., & Newton, A. F. (1995). Families and subfamilies of Coleoptera (with selected genera, notes, references and data family-group names) In: Lawrence JF, Newton AF Biology, Phylogeny and Classification of Coleoptera: Papers Celebrating the 80th Birthday of R. A. Crowson. Muzeum i Instytut Zoologii PAN (pp. 779–1006), Warzawa.
Leschen, R. A. B., & Beutel, R. G. (2001). Pseudotracheal tubes, larval head, and mycophagy in *Sepedophilius* (Coleoptera: Staphylinidae: Tachyporinae). *Journal of Zoological Systematics and Evolutionary Research, 39*(1/2), 25–35.
Makarova, A. A., & Polilov, A. A. (2013). Peculiarities of the brain organization and fine structure in small insects related to miniaturization. 1. The smallest Coleoptera (Ptiliidae). *Entomological Review, 93*(6), 703–713.
Mandelshtam, U. E. (1983). *Нейрон и мышца насекомого (Insect Neuron and Muscle)*. Leningrad: Nauka.
Matthews, A. (1872). *Trichopterygia illustrata et descripta*. Janson: London, E.W.
Matthews, A. (1900). *Trichopterygia illustrata et descripta*. Janson: London, E.W.
Naomi, S.-I. (1985). The phylogeny and higher classification of the Staphylinidae and their allied groups (Coleoptera, Staphilinoidea). *Eskaria, 23*, 1–27.
Naomi, S.-I. (1987a). Comparative Morphology of the Staphylinidae and the Allied Groups (Coleoptera, Staphylinoidea). I. Introduction, Head Sutures, Eyes and Ocelli. *Kontyu, 55*(3), 450–458.
Naomi, S.-I. (1987b). Comparative Morphology of the Staphylinidae and the Allied Groups (Coleoptera, Staphylinoidea). II. Cranial Structure and Tentorium. *Kontyu, 55*(4), 666–675.
Naomi, S.-I. (1988a). Comparative Morphology of the Staphylinidae and the Allied Groups (Coleoptera, Staphylinoidea). III. Antennae, Labrum and Mandibles. *Kontyu, 56*(1), 67–77.
Naomi, S.-I. (1988b). Comparative Morphology of the Staphylinidae and the Allied Groups (Coleoptera, Staphylinoidea). IV. Maxillae and Labium. *Kontyu, 56*(2), 241–250.
Naomi, S.-I. (1988c) Comparative Morphology of the Staphylinidae and the Allied Groups (Coleoptera, Staphylinoidea). V. Cervix and Prothorax. *Kontyu 56*(3), 506–513.
Naomi, S.-I. (1988d). Comparative Morphology of the Staphylinidae and the Allied Groups (Coleoptera, Staphylinoidea). VI. Masothorax and Metathorax. *Kontyu, 56*(4), 727–738.
Naomi, S.-I. (1989a). Comparative Morphology of the Staphylinidae and the Allied Groups (Coleoptera, Staphylinoidea). VII. Metendosternite and Wings. Jpn. *Journal of Entomology, 57* (1), 82–90.
Naomi, S.-I. (1989b). Comparative Morphology of the Staphylinidae and the Allied Groups (Coleoptera, Staphylinoidea). VIII. Thoracic Legs. *Japanese Journal of Entomology, 57*(2), 269–277.
Naomi, S.-I. (1989c). Comparative Morphology of the Staphylinidae and the Allied Groups (Coleoptera, Staphylinoidea). IX. General Structure, Lateral Plates, Stigmata and 1st to 7st Segments of Abdomen. *Japanese Journal of Entomology, 57*(3), 517–526.
Naomi, S.-I. (1989d). Comparative Morphology of the Staphylinidae and the Allied Groups (Coleoptera, Staphylinoidea). X. Eighth to 10th Segments of Abdomen. *Japanese Journal of Entomology, 57*(4), 720–733.

Naomi, S.-I. (1990). Comparative Morphology of the Staphylinidae and the Allied Groups (Coleoptera, Staphylinoidea). XII. Abdominal Glands, Male Genitalia Female Spermatheca. *Japanese Journal of Entomology, 58*(1), 16–23.

Newton, A. F., & Thayer, M. K. (1988). A critique on Naomi's phylogeny and higher classification of Staphylinidae and Allies (Coleoptera). *Entomologia Generalis, 14*(1), 63–72.

Newton, A. F., & Thayer, M. K. (1992). Current classification and family-group names in Staphyliniformia (Coleoptera). *Fieldiana Zoology, 67*, 1–92.

Panov, A. A. (1986). The cerebral neurosekretory cells and Retrocerebral endocrine complex in several representatives of Staphyliniformic Beetles (Coleoptera, Staphyliniformia). *Journal fur Hirnforschung, 27*(4), 409–421.

Paulian, R. (1941). Les premiers etats des Staphylinoidea (Coleoptera). Etude de morphologie comparee. *Mémoires du Muséum national d'histoire naturelle, 15*, 1–361.

Pavlovsky, E. N. (1956). К функциональной анатомии паразита выхухоли — жука *Silphopsyllus desmanae* Ols. (Coleoptera, Leptinidae) (On the functional anatomy of the beetle *Silphopsyllus desmanae* Ols. (Coleoptera, Leptinidae), parasite of desman). *Entomologiceskoe obozrenie 35*(3), 518–529.

Perkins, P. D. (1980). Aquatic beetles of the family Hydraenidae in the Western Hemisphere: classification, biogeography and inferred phylogeny (Insecta: Coleoptera). *Quaestiones Entomologicae, 16*, 3–554.

Perkins, P. D. (1997). Life on the effective bubble: exocrine secretion delivery systems (ESDS) and the evolution and classification of beetles in the family Hydraenidae (Insecta: Coleoptera). *Annals of the Carnegie Museum, 66*, 89–207.

Polilov, A. A. (2005). Anatomy of the feather-winged beetles *Acrotrichis montandoni* and *Ptilium myrmecophilum* (Coleoptera, Ptiliidae). *Entomological Review, 85*(5), 467–475.

Polilov, A. A. (2008). Anatomy of the smallest of the Coleoptera, feather-winged beetles from tribe Nanosellini (Coleoptera, Ptiliidae) and limits to insect miniaturization. *Entomological Review, 88*(1), 26–33.

Polilov, A. A., & Beutel, R. G. (2009). Miniaturization effects in larvae and adults of *Mikado* sp. (Coleoptera: Ptiliidae), one of the smallest free-living insects. *Arthropod Structure & Development, 38*(3), 247–270.

Pototskaya, V. A. (1967). *Определитель личинок коротконадкрылых жуков (Staphylinidae) европейской части СССР (Guide to Rove Beetle (Staphylinidae) Larvae of the European USSR)*. Moscow: Nauka.

Pototskaya, V. A. (1971). Морфо-экологические типы личинок (Staphylinidae, Coleoptera) (Morphoecological types of larvae (Staphylinidae, Coleoptera)). *Zoologicheskii Zhurnal, 50* (11), 1665–1675.

Praghan, S. (1938). Neuromuscular study of the mouthparts of *Coccinella septempunctata* L. with a comparison of the mouthparts in carnivorous and herbivorous Coccinellids. *Records of The Indian Museum, 40*, 341–358.

Schneider, W. (1981). Zur Kopfmorphologie der Imago des Olkaafers *Lytta vesicatora* (Coleoptera: Meloidea). *Entomologia Generalis, 7*, 69–87.

Schunger, I., Beutel, R. G., & Britz, R. (2003). Morphology of immature stages of Prostomis mandibularis (Coleoptera: Tenebrionidae: Prostomidae). *European Journal of Entomology, 100*, 357–370.

Seevers, C. H., & Dybas, H. S. (1943). A synopsis of Limulodidae: A new family proposed mirmecophilies of the subfamilies Lemulodinae. *Annals of the Entomological Society of America, 36*, 546–586.

Silvere, A. P., & Shtein-Margolina, V. V. (1976). *Tetrapodili — четырехногие клещи: Электронномикроскопическая анатомия, проблемы эволюции и взаимоотношения с возбудителями болезней растений (Tetrapodili: Fore-legs Mites*. Evolution Problems and Mutual Relations with Plant Pathogenic Organisms) Valgus, Tallinn: Electron Microscopic Anatomy.

Sorensson, M. (1997). Morphological and taxonomical novelties in the world's smallest beetles, and the first Old World records of Nanosellini. *Systematic Entomology, 22*, 257–283.

Speyer, W. (1922). Die Muskulatur der Larve von *Dytiscus marginalis* L. Ein Beitrag zur Kenntnis des Insektenko¨rpers. *Zeitschrift für wissenschaftliche Zoologie, 119*, 423–492.

Tikhomirova, A. L. (1973). *Морфоэкологические особенности и филогенез стафилинид: (С каталогом фауны СССР) (Peculiar Morphoecological Features and Phylogeny of Staphylinids (With a Catalogue of the Staphylinids of Russia))*. Moscow: Nauka.

Tikhomirova, (1974). Строение личинок стафилинид (Coleoptera, Staphylinidae) в свете явления эмбрионизации и дезэмбрионизации (Morphology of rove beetle larvae (Coleoptera, Staphylinidae) in the light of the phenomena of embryonization and desembryonization). *Zoologicheskii Zhurnal, 53*(8), 1187–1195.

Weide, D., & Betz, O. (2009). Head morphology of selected Staphylinoidea (Coleoptera: Staphyliniformia) with an evaluation of possible groundplan features in Staphylinidae. *Journal of Morphology, 270*(12), 1503–1523.

Weide, D., Thayer, M. K., Newton, A. F., & Betz, O. (2010). Comparative morphology of the head of selected sporophagous and non-sporophagous aleocharinae (Coleoptera: Staphylinidae): musculature and hypopharynx-prementum complex. *Journal of Morphology, 271*(8), 910–931.

Weide, D., Thayer, M. K., & Betz, O. (2014). Comparative morphology of the tentorium and hypopharyngeal–premental sclerites in sporophagous and non-sporophagous adult Aleocharinae (Coleoptera: Staphylinidae). *Acta Zoologica, 95*(1), 84–110.

Chapter 6
Structure of the Principal Groups of Microinsects. IV. Hooded Beetles (Coleoptera: Corylophidae)

6.1 Introduction

Corylophidae is a rather small family of small beetles. Representatives of this family are distributed worldwide, but data on their biology are very scant. Most of them are found in the litter, under bark, or on polypores and considered sporophagous. The morphology of adults was described in one outdated study (Matthews 1899) and several modern taxonomic publications (Paulian 1950; Bowestead 1999; Ślipiński et al. 2009; Robertson et al. 2013). The general morphology of larvae was described only for a few genera (Boving and Craighead 1931; Paulian 1950; De Marzo 2000; Ślipiński et al. 2009), structure of the head was described in detail in larvae of *Holopsis* (Yavorskaya et al. 2014). Detailed descriptions of the external and internal morphology of adults and larvae became available only very recently (Polilov and Beutel 2010; Polilov 2011; Yavorskaya and Polilov 2016).

According to modern studies on the taxonomy and phylogeny of Coleoptera, Corylophidae belong to the superfamily Cucujoidea (Lawrence and Newton 1995; Hunt et al. 2007; Lawrence et al. 2011). The best studied of all related families is the family Coccinellidae. The morphology of adults and larvae of Coccinellidae has been well studied and summarized in reviews (Savoiskaya 1983; Seago et al. 2011). Internal morphology has been described in detail for *Coccinella* (Ibrahim 1948) and *Radolia* (Priore 1966) and treated in some studies on particular organs and tissues (Verhoeff 1895; Landis 1936; Praghan 1938, 1939, 1942, 1953; Amourig 1960). The anatomy of larvae has been less thoroughly studied than that of adults (Amourig 1963; Priore 1966).

A.A. Polilov, *At the Size Limit—Effects of Miniaturization in Insects*,
DOI 10.1007/978-3-319-39499-2_6

6.2 Brief Description of the Structure of Corylophidae

Small beetles, adults 0.5–2.3 mm long (on average about 1.5 mm). Body shape varying, from almost spherical (*Orthoperus*) to elongate or strongly flattened (*Sericoderus*) (Figs. 6.1 and 6.2). Coloration usually monotonous, from yellow to black, in some representatives of Periptycinae with spotted pattern. Structure of integument strongly varying within this family, surface smooth or sparsely punctate (*Orthoperus, Holopsis*), or with various microsculpture (*Corylophus*). Most corylophidae sparsely pubescent or without pubescence. **Head** prognathous or hypognathous, usually sunk into pronotum. Cranium solid, without sutures. Compound eyes consisting of 60–100 ommatidia each. Ocelli absent. Antennae 9–11-segmented, with 3-segmented club. Mouthparts consisting of well-developed mandibles, maxillae, labrum, and labium. **Thorax** consisting of detached prothorax and fused meso- and metathorax. Pronotum in most corylophids wide, dilating toward posterior margin. Elytra usually narrowing toward posterior margin, sometimes shortened and not covering apical abdominal segments (*Sericoderus*), without longitudinal striae, with well-developed epipleura. Hindwings with strongly depleted venation and fringe of long setae on perimeter of wing, at rest folded and tucked under elytra. Legs with normal division into coxa, trochanter, femur, tibia, tarsus, and pretarsus. Tarsi 3-segmented. **Abdomen** usually with six visible sternites and eight visible tergites. External male genitalia represented by simple penis and tegmen. Spermatheca usually simple, from spherical to elongate. External sexual dimorphism weakly pronounced. Wingless forms present in many species.

Larvae usually elongate, flattened, body from parallel-sided (*Arthrolips*) to broad discoid (Rypobiini), body length 1–3 mm (last instar). Integument weakly sclerotized, weakly pigmented, usually with stellate microsculpture and covered with several types of setae. **Head** prognathous. Cranium solid, without sutures. Antennae 3-segmented, less often 2-segmented (*Sericoderus*) or strongly shortened (*Orthoperus*). Mouthparts consisting of well-developed mandibles, maxillae, labrum, and labium. **Thorax** consisting of three separate segments, slightly wider than abdomen and only with weakly sclerotized nonpigmented dorsal sclerites. Pair of spiracles present between pro- and mesothorax. Legs consisting of coxa, trochanter, femur, fused tibia and tarsus, and claw. **Abdomen** 10-segmented. All segments without pronounced sclerites. Spiracles present on abdominal segments 1–7. Urogomphi absent. Abdominal segment 10 with pygopod.

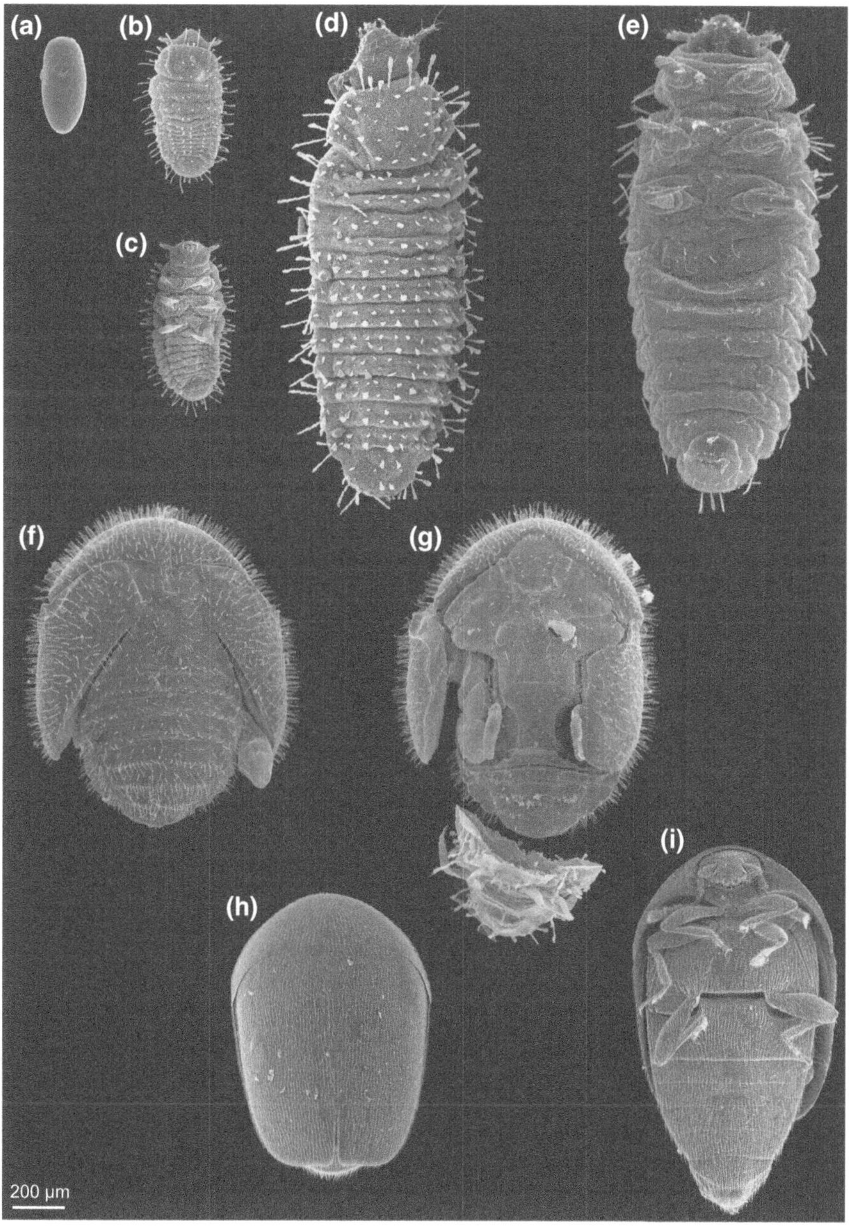

Fig. 6.1 External morphology of *Sericoderus lateralis*, SEM: **a** egg, **b**, **c** first instar larva; **d**, **e** last instar larva; **f**, **g** pupa; **h**, **i** adult; **b**, **d**, **f**, **h** dorsal view; **c**, **e**, **g**, **i** ventral view

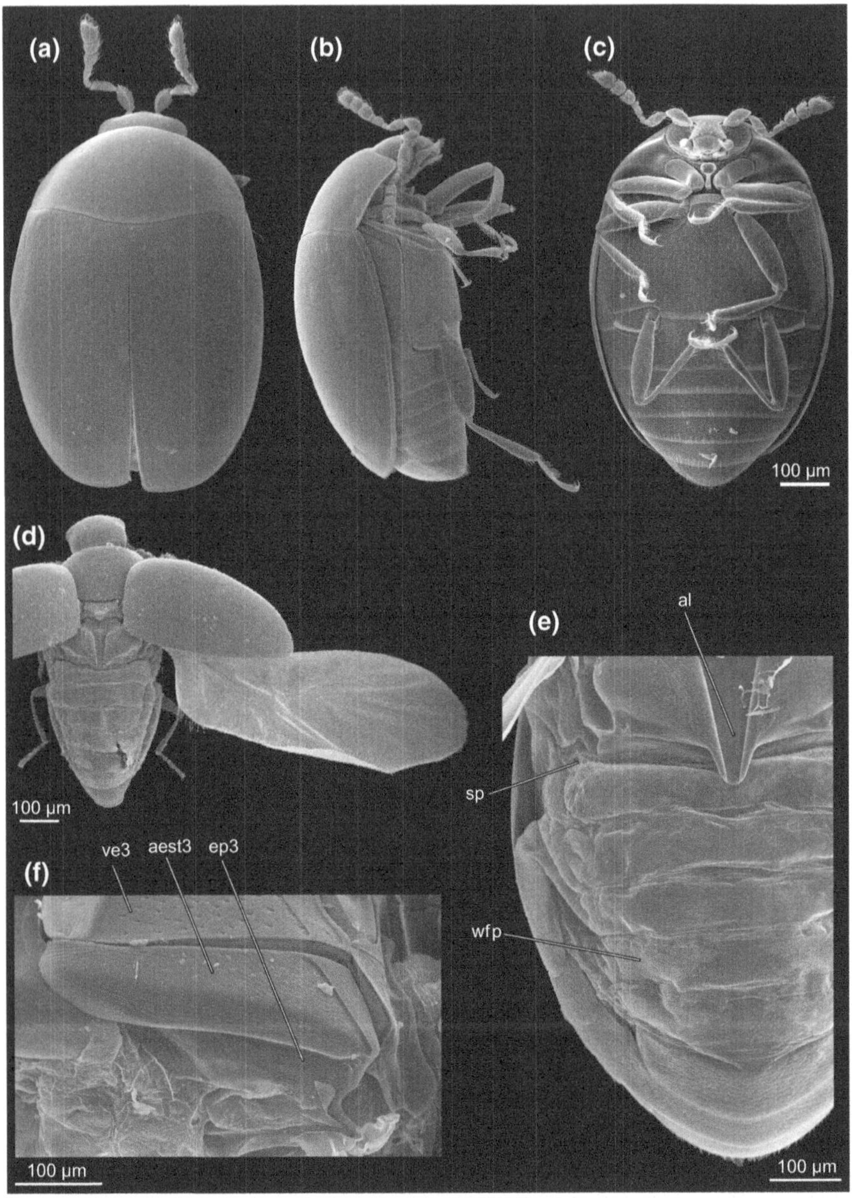

Fig. 6.2 External morphology of adult in *Orthoperus atomus*, SEM: **a**, **d**, **e** dorsal view; **b**, **f** lateral view; **c** ventral view; *aest3* metepisternum, *al* alacrista, *ep3* metepimeron, *sp* spiracle, *ve3* metaventrite, *wfp* cuticular folds for folding of wing

6.3 Morphology of Adult Corylophidae

6.3.1 External Morphology

Adult body length 950–1200 μm (M = 1060, n = 10) in *Sericoderus lateralis*, 767–815 μm (M = 795, n = 6) in *Orthoperus atomus*.

Structure of head

Cranium solid, without sutures. Head connected with prothorax only via articular membrane; cervical sclerites absent. In contrast to most other Cucujoidea, in Corylophidae, clypeus fused with frons (frontoclypeal suture absent), gular sutures and, accordingly, gular plate also absent. Tentorium strongly simplified, laminitentorium, dorsal and anterior arms absent. Posterior arms slender, rounded in cross section. Tentorial bridge well developed.

Compound eyes consisting of 60–100 ommatidia (in larger representatives of related groups of beetles eyes consisting of tens of times as many ommatidia). Ocelli absent.

Antennae in most representatives of Corylophidae 11-segmented, but in some species 10-segmented (*Sericoderus*), 9-segmented (*Orthoperus*), or even 8-segmented (*Cleidostethus*). Antenna with several types of sensilla. Trichoid sensilla (probably mechanoreceptors) present on all antennomeres (4–30 on each). Placoid sensilla (probably olfactory) present on two apical antennomeres (2–6 on each). Clavate sensilla (8–10) localized apically on apical antennomere.

Mouthparts different in representatives of different genera (Figs. 6.3 and 6.4). Labrum movably joined with cranium and appearing as simple plate. Internal surface of labrum and dorsal surface of oral cavity with complex epipharynx bearing tufts of setae. Mandibles well developed, symmetrical or almost symmetrical, very different in different genera, from compact (*Sericoderus*) to narrow and elongate (Rypobiini), with well developed mola (*Orthoperus*) or without it (*Corylophus*), with one or several large teeth. Prostheca also of various shape. Maxillae strongly varying, consisting of cardo, stipes, lacinia, and 3-segmented palp. Galea absent. In representatives of different genera maxillary palp and lacinia very strongly differ in shape and armature. Lacinia sometimes with apical teeth (*Sericoderus*) or with fan-shaped rows setae (*Corylophus*) Labium consisting of well developed mentum, membranous prementum, 2-segmented palps and ligulae; submentum fused with cranium.

Structure of thorax

Shape of **pronotum** strongly varying within this family, but in most genera transverse, dilating toward posterior angles, with rounded anterior margin. Hypomeron well developed. Trochantinopleura not visible from without, with well-developed apodeme. Notosternal sutures absent. Coxal cavities closed.

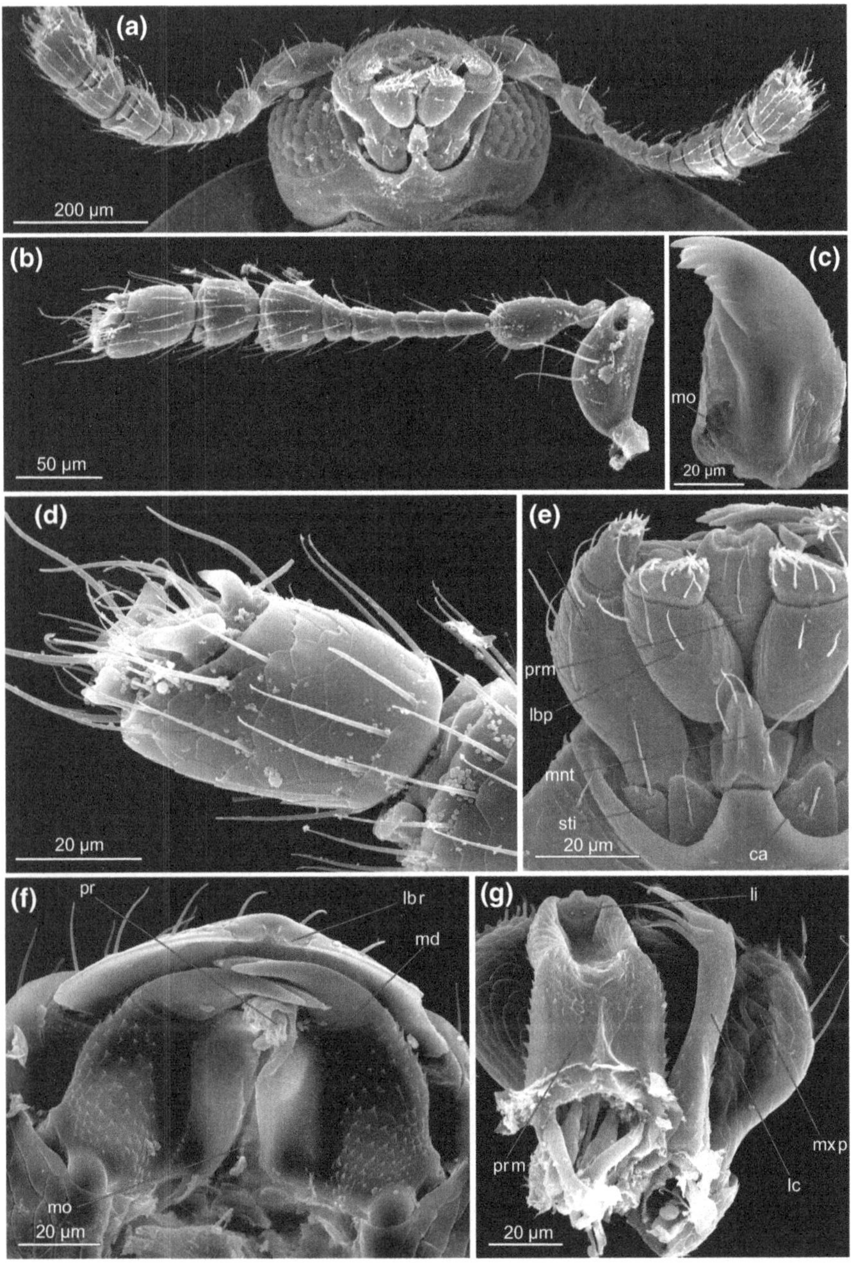

Fig. 6.3 Head and mouthparts of adult *Sericoderus lateralis*, SEM: **a** head, ventral view; **b**, **d** antenna; **e** mandible, dorsal view; **e** maxilla and labium, ventral view; **f** mandibles, ventral view; **g** maxilla and labium, dorsal view; *ca* cardo, *lb* labium, *lbp* labial palp, *lbr* labrum, *lc* lacinia, *li* ligula, *md* mandible, *mnt* mentum, *mo* mola, *mxp* maxillary palp, *pr* prostheca, *prm* prementum, *sti* stipes

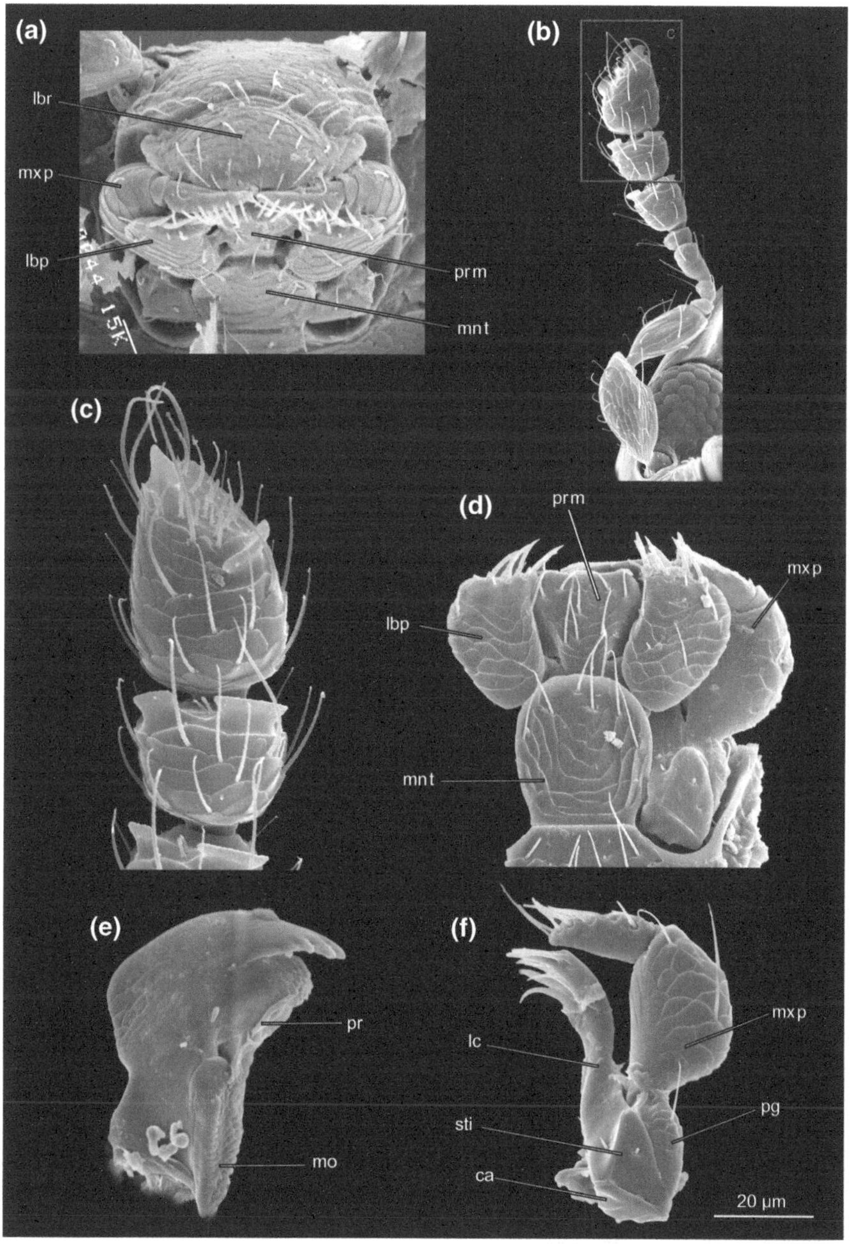

Fig. 6.4 External morphology of adult *Orthoperus atomus*, SEM: **a** mouthparts, frontal view; **b**, **c** antenna; **d** labium and maxilla; **e** mandible; **f** maxilla; *ca* cardo, *lbp* labial palp, *lbr* labrum, *lc* lacinia, *mnt* mentum, *mo* mola, *mxp* maxillary palp, *pg* palpiger, *pr* prostheca, *prm* prementum, *sti* stipes

Mesothorax in corylophids, as in other beetles, smaller than other segments of thorax (Fig. 6.5). Pleural part separated from of ventrite well marked suture. Episternum and epimeron fused. Tergal part of mesothorax membranous and hidden under elytra, except for scutellum. Mesocoxae and coxal cavities rounded. Mesofurca well developed, with long anterior and lateral lobes. Elytra usually narrowing toward posterior margin, sometimes shortened and not covering apical

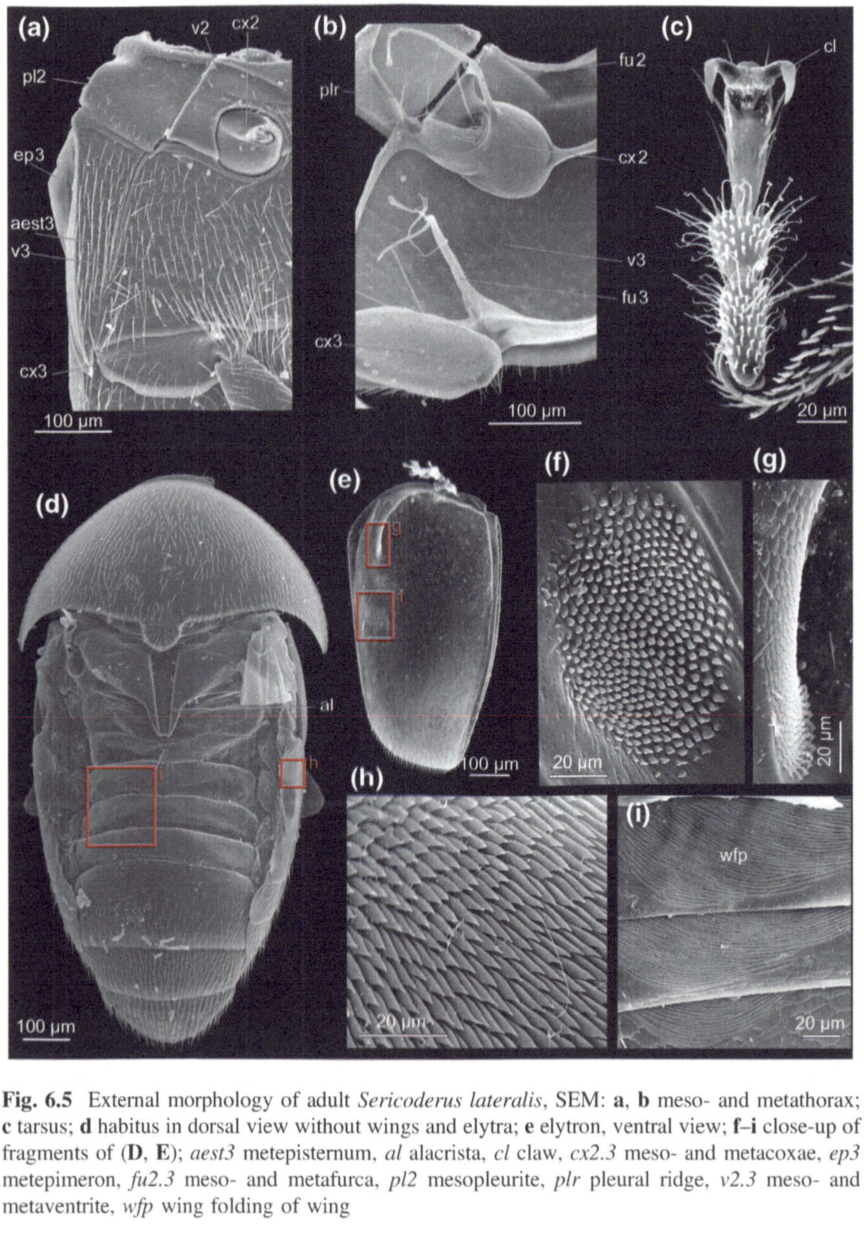

Fig. 6.5 External morphology of adult *Sericoderus lateralis*, SEM: **a**, **b** meso- and metathorax; **c** tarsus; **d** habitus in dorsal view without wings and elytra; **e** elytron, ventral view; **f–i** close-up of fragments of (**D**, **E**); *aest3* metepisternum, *al* alacrista, *cl* claw, *cx2.3* meso- and metacoxae, *ep3* metepimeron, *fu2.3* meso- and metafurca, *pl2* mesopleurite, *plr* pleural ridge, *v2.3* meso- and metaventrite, *wfp* wing folding of wing

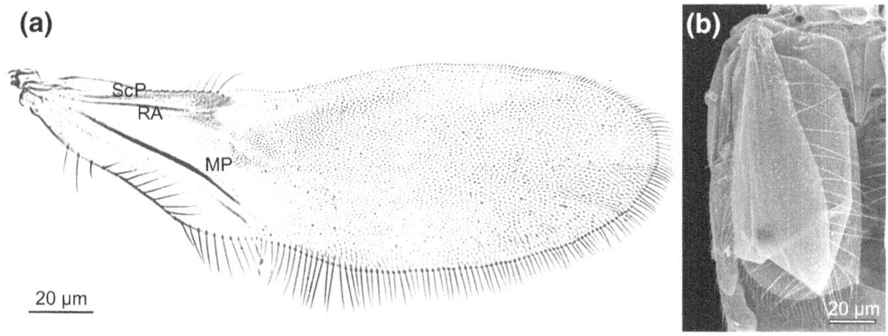

Fig. 6.6 Wing of *Sericoderus lateralis*: **a** unfolded; **b** folded; *MP* posterior medial vein, *RA* anterior radial vein, *ScP* posterior subcostal vein

abdominal segments (*Sericoderus*), without longitudinal striae, with well-developed epipleura. Internal surface of elytra with two areas of characteristic microrelief of large spinules, also with areas of spinules on corresponding areas of tergite of metathorax and abdominal segment 1; these structures probably participating in joining and fixing of elytra.

Metathorax completely divided, as in other beetles (Fig. 6.5). Ventrite separated from pleurite by clearly discernible suture. Pleurites divided by on episterna and epimera. Tergal part of metathorax with usual for beetles division into scutum, scutellum, and postnotum. All part of tergite membranous, except for scutellum, bearing well sclerotized allocrista. Metacoxae of various shape, wider than long, widely set apart. Metendosternite without common stem, represented by long lateral arms; anterior arms undeveloped. Hindwings well developed, appearing as rather narrow plate with fringe of setae on perimeter of wing (Fig. 6.6). Venation strongly depleted. Three shortened veins present: subcostal (ScP), anterior radial (RA), and posterior medial (RP) (Bowestead 1999: costal, subcostal, and cubital). Wing several times folded at rest. Legs ambulatorial. Tarsi 3-segmented. Apical tarsomere bearing two simple claws.

Structure of abdomen

Abdomen usually with six visible sternites and eight visible tergites. Proximal tergites membranous and weakly pigmented. Tergites 4–7 tergites bearing folds participating in folding of wing. Spiracles usually present on tergites 1–7, sometimes only six pairs of spiracles present (*Orthoperus*). Stigmata simple, ring-shaped, without filter apparatus, with spherical atrium. Females with secondary ovipositor shaped as short tube, formed by segments 9 and 10. External male genitalia represented by simple penis and tegmen. Spermatheca usually simple, from spherical to elongate.

6.3.2 Internal Morphology

General plan of structure in Corylophidae complying with general plan of larger representatives of related groups of insects. Medial part of body cavity occupied by intestine, most of thorax occupied by musculature, nervous system strongly oligomerized and concentrated in thoracic segments; reproductive system occupying considerable part of abdomen; heart present; diaphragms absent (Fig. 6.7).

Integument

Integument consisting of cuticle, hypoderm, and basal membrane. Cuticle thickness 1.7–17.6 μm thick (M = 6.1; n = 80) in *Sericoderus*, 2.6–13.7 μm (M = 6.7; n = 80) in *Orthoperus*. Cuticle consisting of epicuticle about 70 nm thick and weakly delimited exo- and endocuticle (Fig. 6.8A). Hypoderm consisting of flattened cells, 1–4 μm thick.

Digestive and excretory systems

Digestive canal divided into fore-, mid-, and hindgut (Fig. 6.9a, b). Fore- and hindgut with thin cuticular lining. Intestinal canal somewhat longer than body, forming loop in abdomen. Digestive glands not found.

Foregut divided into pharynx, oesophagus, and muscular stomach. Pharynx U-shaped, ventral wall strongly sclerotized. Oesophagus straight, rounded in cross section, with one layer of circular muscles (Fig. 6.8b), in contrast to other beetles, in which oesophagus with both circular and longitudinal muscles (Crowson 1981). Muscular stomach with thick cuticular lining and strong musculature.

Midgut appearing as short wide tube. Wall consisting of unilayered epithelium, in cross section 15–25 cells visible, cylindrical in anterior part and flattened in posterior part of gut. Entire midgut with slender unilayered peritrophic membrane. Muscles of midgut not found; food possibly moved along intestine by somatic musculature. This character distinguishes Corylophidae from other beetles, in which midgut with two layers of muscles (Crowson 1981).

Hindgut divided into small intestine and rectum. Small intestine rounded in section, rectum oval in section. Rectum with rectal ampulla.

Boundary between mid- and hindgut bearing four **Malpighian tubules** forming several loops.

Circulatory system

Circulatory system strongly reduced, heart appearing as slender tube, other vessels absent. **Fat body** consisting of cells of various shapes, containing electron transparent inclusions (Fig. 6.8c, d).

Tracheal system

Tracheal system strongly simplified. Only small number of weakly branching tracheae, connected to spiracles, and longitudinal stems in thoracic segments present. Transverse stems and air sacs reduced, probably because of diffusion sufficing for transport of molecules in organisms of such small sizes. Tracheae with structure typical of insects, consisting of hypoderm and intima, intima with helical thickenings—taenidia. Tracheoles (without taenidia, filled with hemolymph) not found.

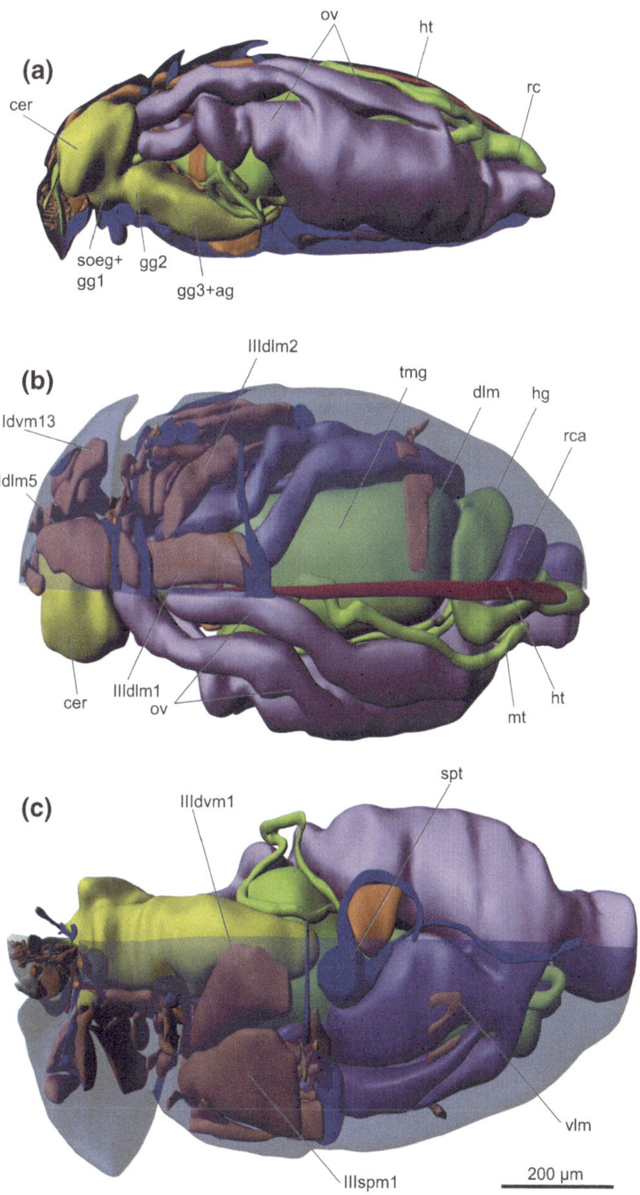

Fig. 6.7 Internal morphology of *Sericoderus lateralis*, 3D: **a** lateral internal view; **b** dorsal view; **c** ventral view; *ag* abdominal ganglion, *cer* cerebrum, *dlm* dorsal longitudinal muscle (shown only in one segment), *gg1.2.3* pro-, meso-, and metathoracic ganglia, *hg* hindgut, *ht* heart, *mg* midgut, *mt* Malpighian tubules, *ov* ovariole, *rc* rectum, *rca* rectal ampulla, *soeg* suboesophageal ganglion, *spt* spermatheca, *vlm* ventral longitudinal muscle (shown only in one segment). Colors: *blue* cuticle, *green* digestive system, *yellow* central nervous system, *brown* musculature, *purple* reproductive system. Musculature see text

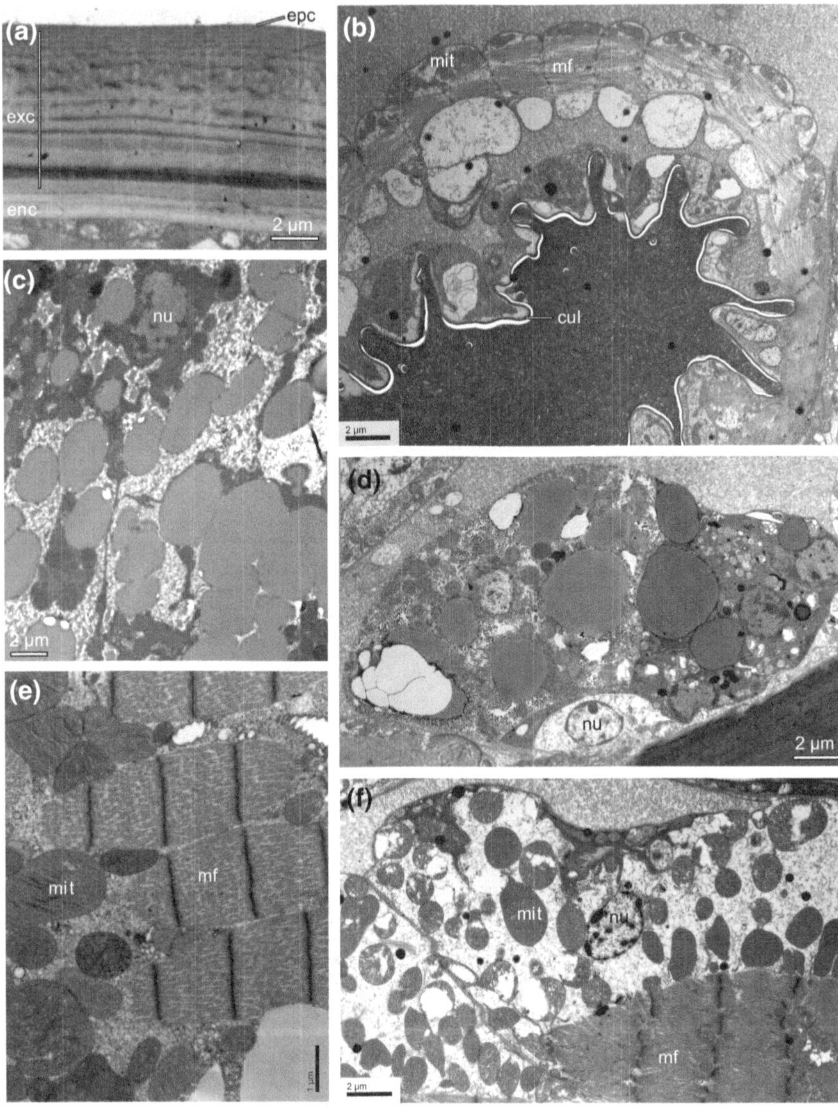

Fig. 6.8 Ultrastructure of adult *Sericoderus lateralis*, TEM: **a** cuticle; **b** oesophagus; **c**, **d** fat body; **e**, **f** muscle; *cul* cuticular lining, *enc* endocuticle, *epc* epicuticle, *exc* exocuticle, *mf* muscle fiber, *mit* mitochondrion, *nu* nucleus

Nervous system

Central nervous system in adult corylophids showing high degree of oligomerization and condensation of ganglia. All ganglia very closely set and appearing as one formation (Fig. 6.9c, d). Cerebrum and suboesophageal ganglion completely shifted into prothorax. Only frontal ganglion of stomatogastric nervous

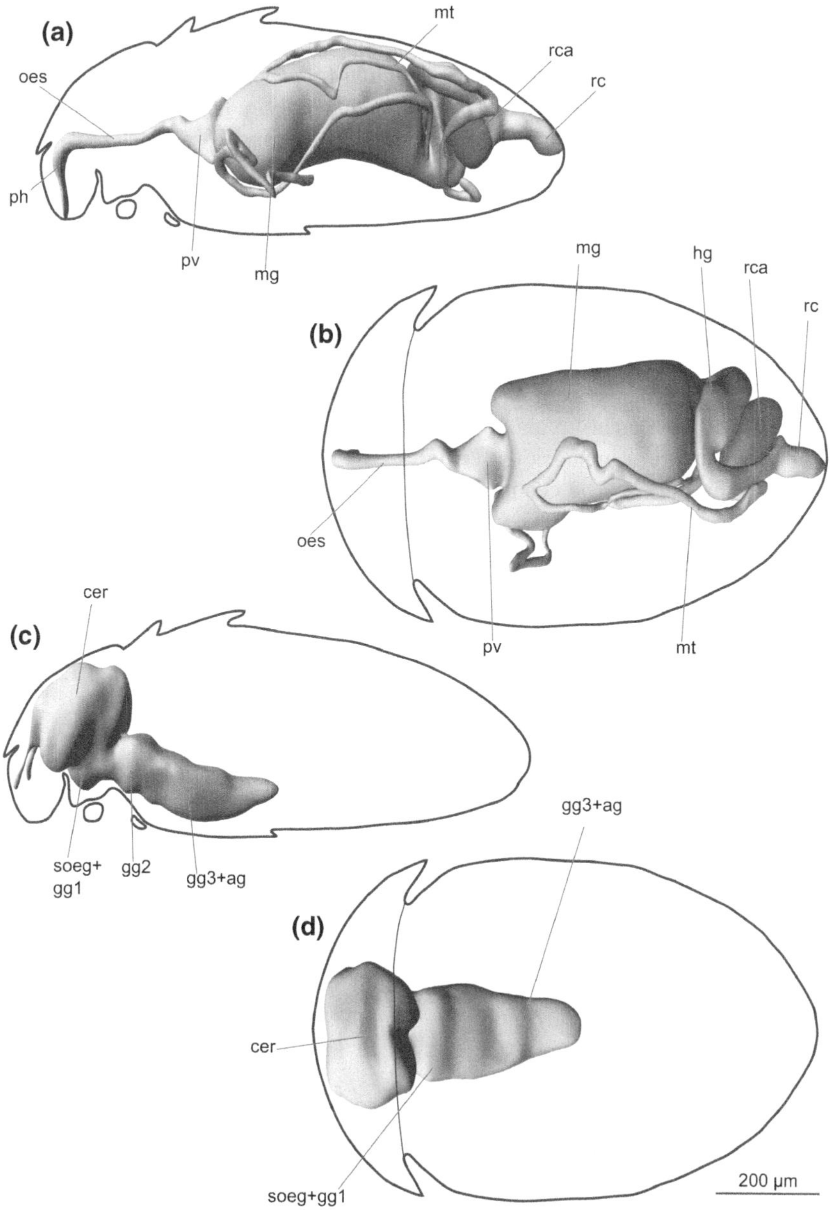

Fig. 6.9 Internal morphology of adult *Sericoderus lateralis*, 3D: **a**, **b** digestive system; **c**, **d** central nervous system; **a**, **c** lateral view; **b**, **d** dorsal view; *ag* abdominal ganglion, *cer* cerebrum, *gg1.2.3* pro-, meso-, and metathoracic ganglia, *hg* hindgut, *mg* midgut, *mt* Malpighian tubules, *oes* oesophagus, *ph* pharynx, *rca* rectal ampulla, *soeg* suboesophageal ganglion

system remaining in cranium. Antennal nerves, optic nerves, and nerves running from tritocerebrum and suboesophageal ganglion strongly elongated and stretched through entire cranium. Suboesophageal ganglion fused with prothoracic ganglion.

Cells of central nervous system, in spite of markedly reduced volume of cytoplasm, containing all necessary cellular organelles (Figs. 6.10 and 6.11). Neuron nuclei regularly oval. Most of chromatin in lax state. Thickness of neural lamella about 0.19–0.3 μm. Thickness of perineurium from 0.49 to 1 μm. Glia represented by same elements as in Ptiliidae. Mitochondria and microtubules visible in projections. Minimum diameter of examined projections 0.08 μm. Longitudinal and transverse synaptic contacts and connections of 'terminal button' type and dirvergent type recorded in common neuropilar mass. Cortical layer of cerebrum with somato-somatic contacts.

Cell size in nervous system of *S. lateralis* 2.21–5.78 μm (M = 3.11; n = 686). Number of cells in cerebrum about 18,000.

The protocerebrum includes the central body complex, mushroom bodies, and optic lobes (Fig. 6.12). The central body complex is differentiated into the central

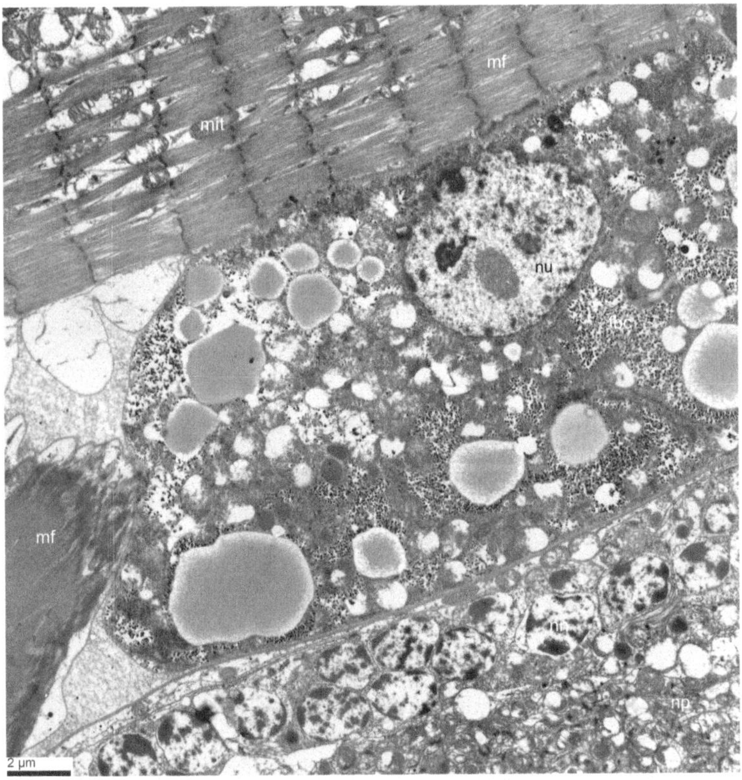

Fig. 6.10 Ultrastructure of adult *Sericoderus lateralis*, TEM: *fbc* cell of fat body, *mf* muscle fiber, *mit* mitochondrion, *nn* nucleus of neuron, *np* neuropil, *nu* nucleus

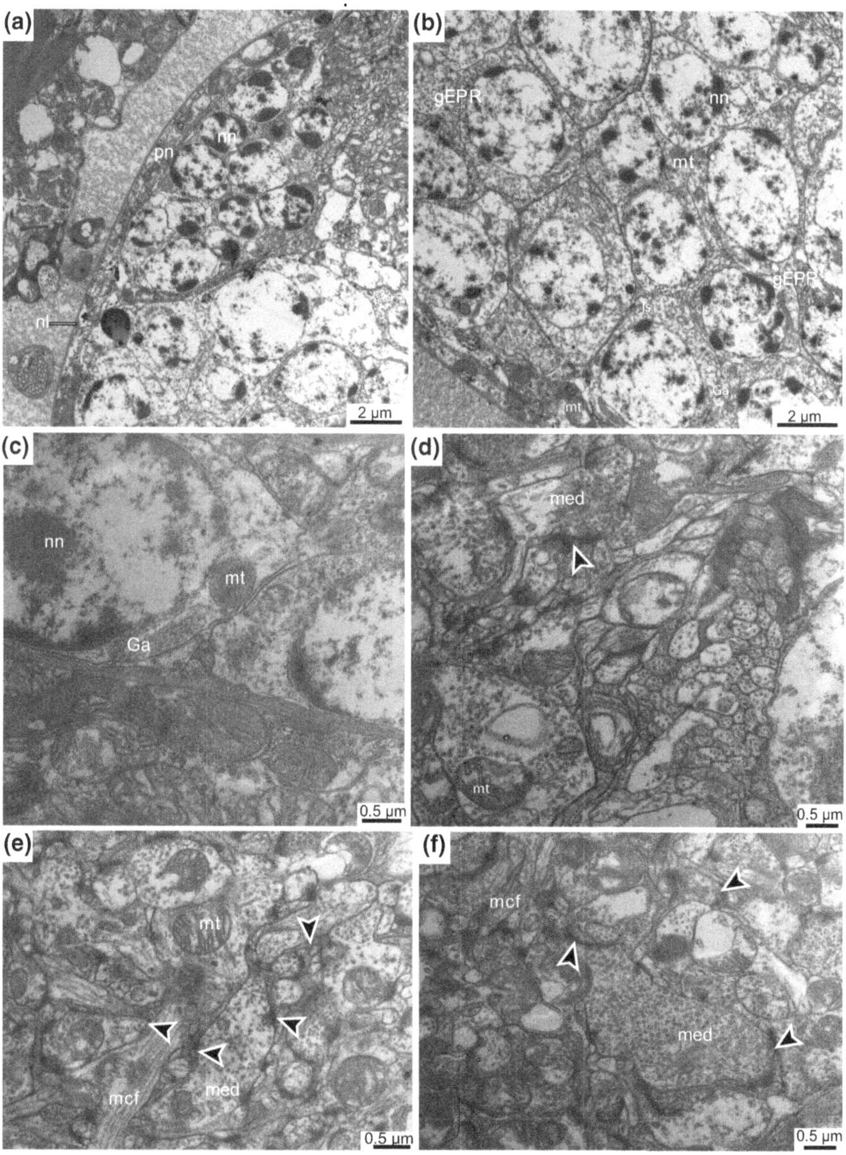

Fig. 6.11 Ultrastructure of cerebrum adult *Sericoderus lateralis*, TEM: *Ga* Golgi apparatus; *med* granules of mediator; *mt* mitochondrion; *nn* nucleus; *nl* neural lamella; *pn* perineurium; *syn* synaptic contact; **d**–**f** *arrows* indicate synaptic contacts

body (consisting of the dorsal region, which includes the dorsal arch and fan-shaped body, ventral region, or ellipsoid body, and noduli), and protocerebral bridge. The mushroom bodies are well developed. The peduncles of the mushroom bodies include strong, clearly pronounced α-lobes and weakly pronounced β-lobes. Large

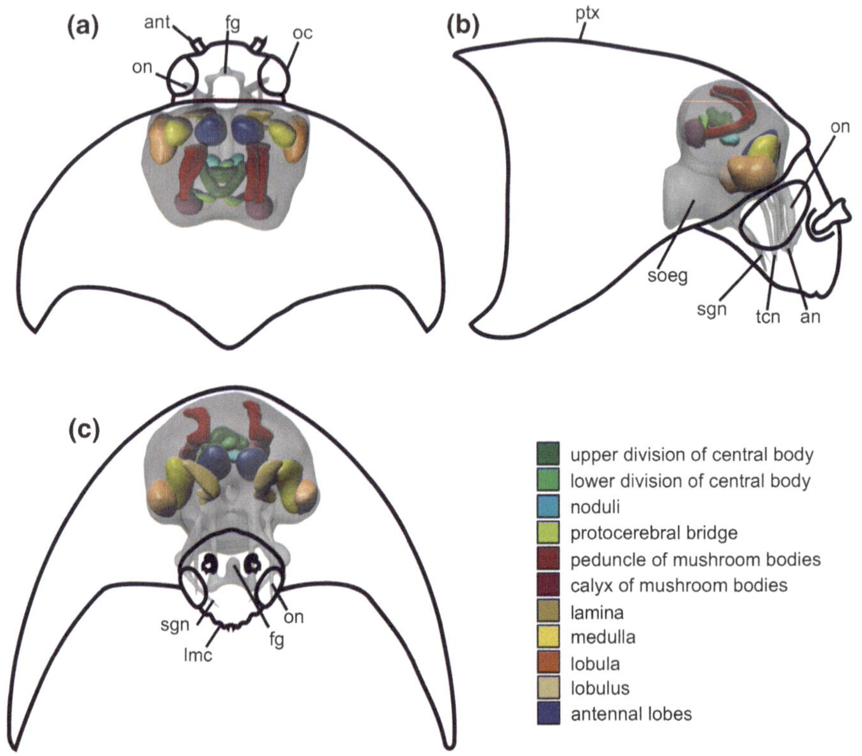

Fig. 6.12 3D reconstructions of cerebrum adult *Sericoderus lateralis*: **a** dorsal view; **b** lateral view; **c** frontal view; *an* antennal nerves; *ant* antennae; *fg* frontal ganglion; *oc* compound eyes; *on* optic nerves; *ptx* prothorax; *lmc* mouthparts; *sgn* nerves of suboesophageal ganglion, innervating mouthparts; *soeg* suboesophageal ganglion; *tcn* paired nerves of tritocerebrum

spherical calyces of the mushroom bodies on the perimeter are formed by glomerular structures; the central zone is represented by dense neuropilar mass.

The optic lobes are represented by three neuropils. All optic neuropils are arranged in a row in the anterior part of the protocerebrum, perpendicular to the longitudinal body axis, an orientation opposite to that found in other insects; the lamina is situated closer than the other optic neuropils to the central axis of the cerebrum, and lobula is situated on the periphery. The lamina is crescent-shaped. The other optic neuropils are oval.

The deutocerebrum is situated very close to the optic lobes; it includes large spherical antennal lobes with clearly discernible glomerules. The diameter of the glomerules is 8–11 μm. Lobes contain about 160 glomerules.

The tritocerebrum is very small, represented by paired formations connected with the tritocerebral commissure. Long, slender connectives link the tritocerebrum to the frontal ganglion. The frontal ganglion is rather small and situated in the cranium, between the compound eyes. The recurrent nerve is strongly elongated, because the cerebrum is shifted into the thorax. A minute formation, possibly the result of fusion of the occipital ganglion and postcerebral endocrine complex, is situated between the oesophagus and cerebrum.

Muscular system

Structure of muscular system in Corylophidae, studied in *Sericoderus lateralis* (Figs. 6.13, 6.14 and 6.15, Table 5.2).

Musculature of head. 0an1 (M. tentorioscapalis anterior): O, dorsal part of cranium; I, dorsal part of scape. 0an2 (M. tentorioscapalis posterior) and 0an4 (M. tentorioscapalis medialis): O, posterior tentorial arms, I, base of scape. 0lb2 (M. frontoepipharyngalis): O, frons; I, lateral part of epipharynx. 0lb5 (M. labroepipharyngealis): O, ventral wall of labrum; I, dorsal wall of epipharynx. 0md1 (M. craniomandibularis internus): largest muscle in head: O, posterior and lateral part of cranium; I, adductor of mandible. 0md3 (M. craniomandibularis

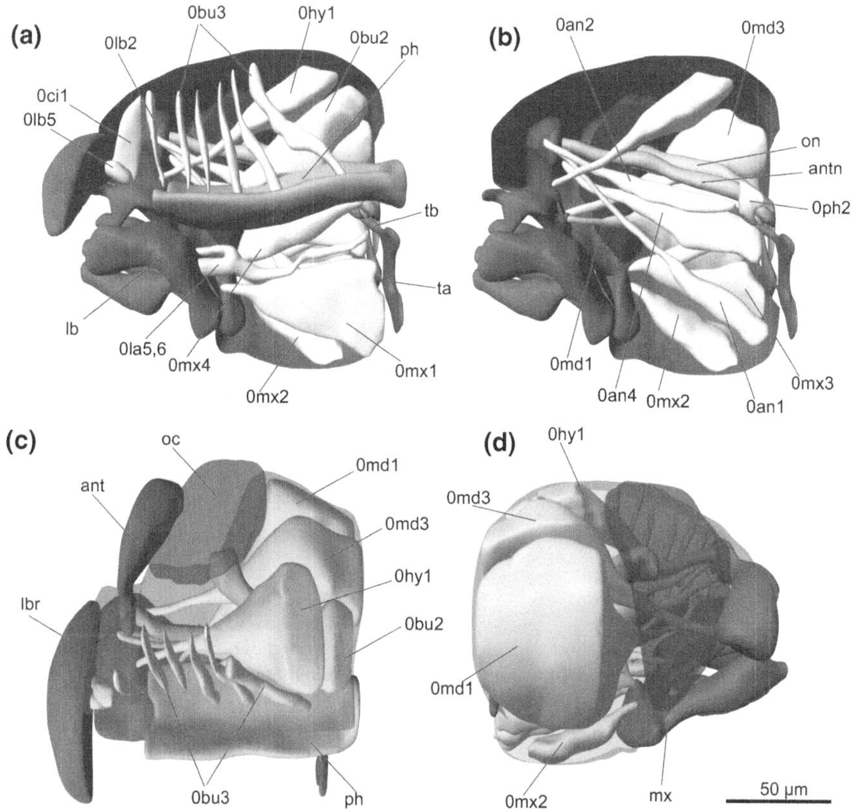

Fig. 6.13 Musculature of head in adult *Sericoderus lateralis*, 3D: **a**, **b** lateral internal view; **c** dorsal view; **d** lateral external view; *ant* antenna, *antn* antennal nerve, *lb* labium, *lbr* labrum, *mx* maxilla, *on* optic nerve, *oc* eye, *ph* pharynx, *ta* tentorial arms, *tb* tentorial bridge. Musculature see text

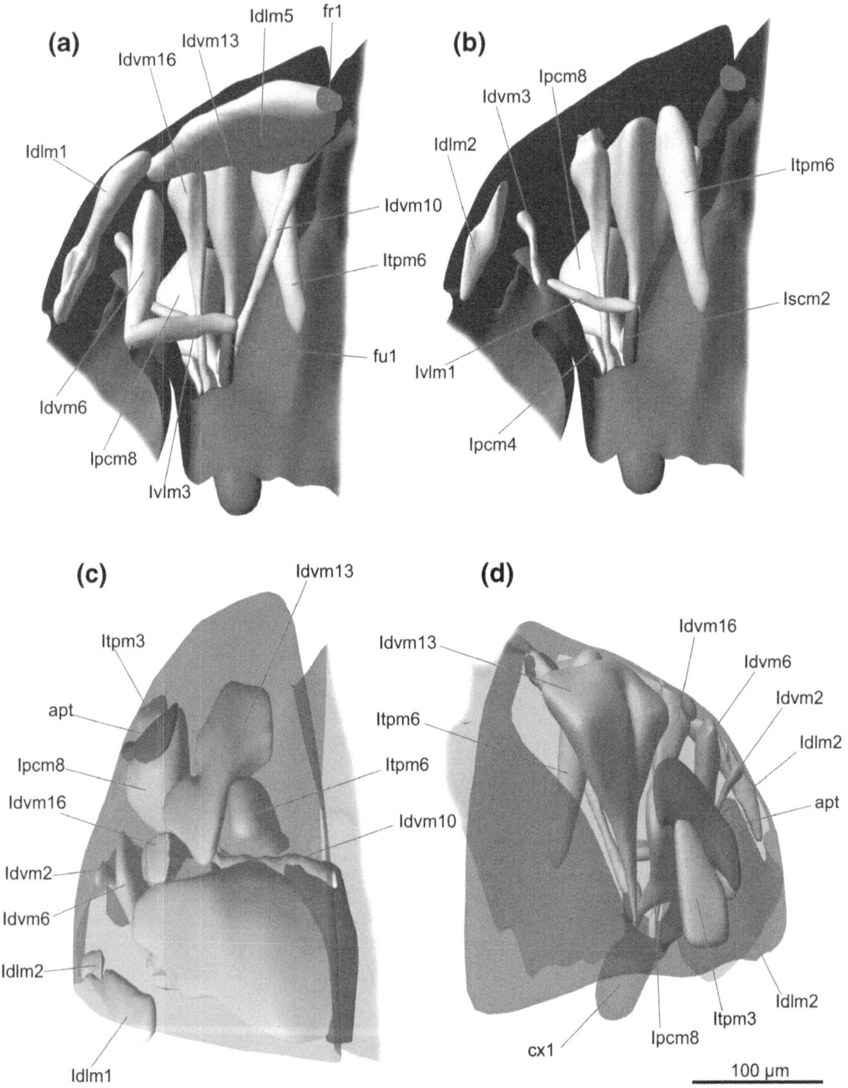

Fig. 6.14 Musculature of prothorax in adult *Sericoderus lateralis*, 3D: **a**, **b** lateral internal view; **c** dorsal view; **d** lateral external view; *apt* apodeme of trochantinopleura, *cx1* coxa of foreleg, *fr1* prophragma, *fu1* profurca. Musculature see text

externus): O, dorsolateral part of cranium; I, abductor of mandible. 0mx1 (M. craniocardinalis): O, posterolateral part of cranium; I, ventrolateral part of cardo. 0mx2 (M. craniolacinialis): O, posterior part of cranium; I, base of lacinia. 0mx3 (M. tentoriocardinalis) and 0mx4 (M. tentoriostipitalis anterior) fused: O, tentorial arms; I, ventral surface of base of maxilla. 0la5 (M. tentoriopraementalis) and 0la6

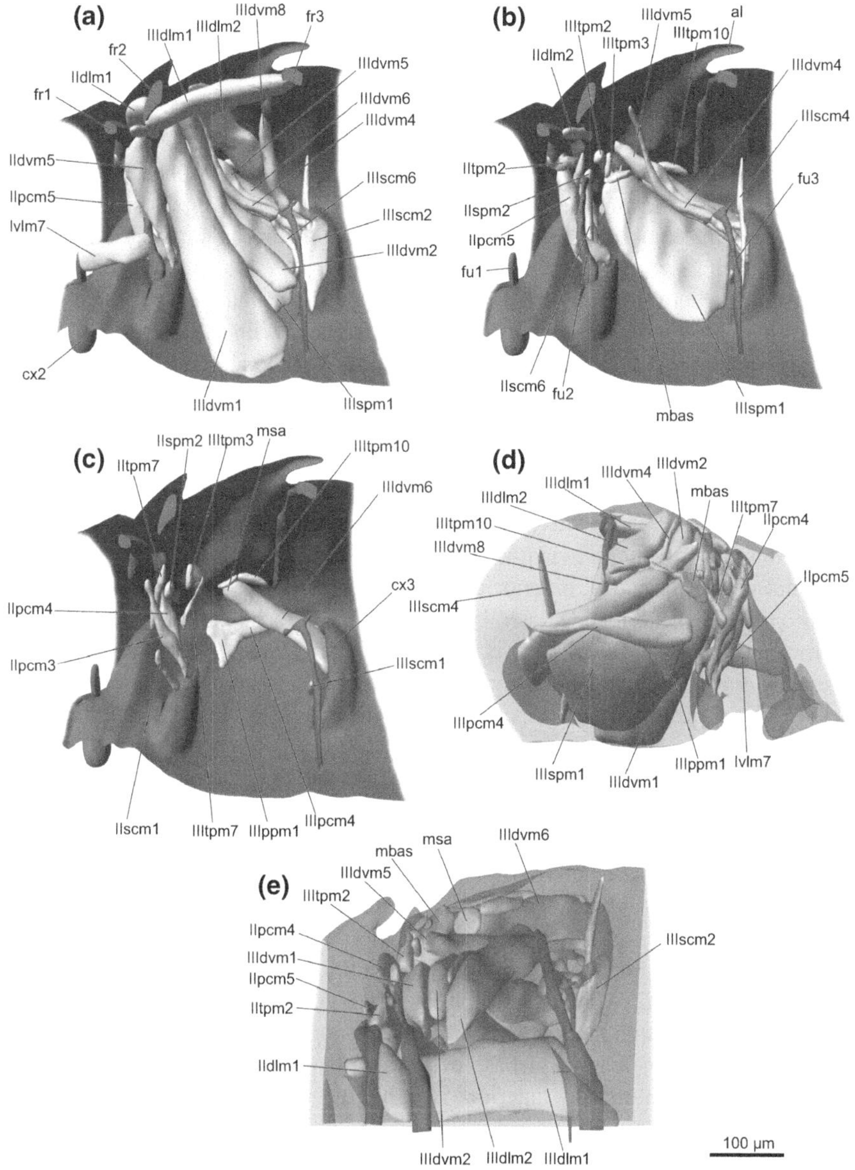

Fig. 6.15 Musculature of meso- and metathorax in adult *Sericoderus lateralis*, 3D: **a–c** lateral internal view; **d** lateral external view; **e** dorsal view; *al* alacrista, *cx2.3* meso- and metacoxae, *fr1.2.3* pro-, meso-, and metaphragma, *fu1.2.3* pro-, meso-, and metafurca, *mbas* muscle disc of basalare, *msa* disc of subalare, *scl2* mesoscutellum. Musculature see text

(M. tentorioparaglossalis) fused: O, tentorial arms; I, margin prementum. 0hy1 (M. frontooralis): O, frons; I, posterior margin of epipharynx. 0ci1 (M. clypeopalatalis): O, clypeal region; I, dorsal part of epipharynx. 0bu2 (M. frontobuccalis anterior): O, frons, I, dorsolateral part of pharynx. 0bu3 (M. frontobuccalis posterior): of several separate fibers, O, frons behind base of 0lb2; I, dorsal part of pharynx. 0ph2 (M. tentoriopharyngealis): O, tentorial bridge; I, ventral wall of pharynx. 0st1 (M. annularis stomodaei), transverse musculature, developed only on dorsal surface of pharynx. Internal musculature of antennae and mouthparts not studied because of extremely small size.

Musculature of thorax. Prothorax. Idlm1 (M. prophragma-occipitalis): O, medial part of notum; I, postocciput, together with Idlm2. Idlm2 (M. pronoto-occipitalis): O, anterior part of notum; I, medial part of postocciput, together with Idlm1. Idvm2 (M. cervico-occipitalis medialis): O, cervical region; I, notum, in front of Idvm6. Idvm6 (M. pronoto-cervicalis medialis): O, anterolateral part of notum; I, cervical region. Idlm5 (M. pronoto-phragmalis anterior): O, prophragma; I, medial part of notum. Idvm10 (M. profurca-phragmalis): O, profurca; I, lateral part of prophragma. Idvm13 (M. pronoto-trochantinalis anterior): O, lateral part of notum; I, trochanter, via fine tendon. Idvm17 (M. pronoto-coxalis posterior): O, lateral part of notum; I, anterior and lateral margin of base of coxa. Itpm3 (M. pronoto-pleuralis anterior): O, lateral part of notum; I, dorsal part of apodemal plate of trochantinopleura. Itpm6 (M. pronoto-intersegmentalis): O, posterior part of notum, lateral to Idlm5; I, intersegmental membrane, in front of mesepisternum. Ipcm4 (M. propleuro-coxalis superior): O, apodemal plate of trochantinopleura; I, anterior margin of base of coxa. Ipcm8 (M. propleuro-trochanteralis): O, apodemal plate of trochantinopleura; I, trochanter, via fine tendon. Ivlm1 (M. profurca-cervicalis): O, profurca; I, cervical region. Ivlm3 (M. profurca-tentorialis): O, profurca; I, posterior margin of gular plate. Ivlm7 (M. profurca-mesofurcalis): O, profurca; I, mesofurca. Iscm2 (M. profurca-coxalis posterior): O, profurca; I, posteromedial margin of base of coxa. Internal muscles of legs not studied because of extremely small size.

Mesothorax. IIdlm1 (M. prophragma-mesophragmalis): O, prophragma; I, mesophragma. IIdlm2 (M. mesonoto-phragmalis): O, medial part of prophragma; I, lateral part of mesophragma. IIdvm4 (M. mesonoto-coxalis anterior): O, mesophragma; I, posterior margin of base of coxa. IIdvm5 (M. mesonoto-coxalis posterior): O, episternum; I, anterior margin of base of coxa. IItpm2 (M. mesopleura-praealaris): O, pleural apodeme; I, anterolateral margin of notum. IItpm7 (M. mesanepisterno-axillaris): O, mesepisternum, I, third axillary plate. IItpm9 (M. mesepimero-axillaris tertius): fused with IItpm7 or absent. IIspm2 (M. mesofurca-pleuralis): O, apex of mesofurca; I, pleural ridge. IIpcm3 (M. mesanepisterno-coxalis anterior): O, basalare; I, anterior margin of base of coxa, together with IIdvm5. IIpcm5 (M. mesanepisterno-trochanteralis): O, basalare; I, trochanter, via fine tendon. IIscm1 (M. mesofurca-coxalis anterior): O, mesofurca; I, anterior margin of base of coxa. IIscm2 (M. mesofurca-coxalis posterior): O, mesofurca; I, posterolateral margin of base of coxa. IIscm6 (M.

mesofurca-trochanteralis): O, mesofurca; I, trochanter, via fine tendon. Internal muscles of legs not studied because of extremely small size.

Metathorax. IIIdlm1 (M. mesophragma-metaphragmalis): O, mesophragma; I, metaphragma. IIIdlm2 (M. metanoto-phragmalis): O, scutum; I, metaphragma. IIIdvm1 (M. metanoto-sternalis): O, anterior part of metanotum and lateral part of mesophragma; I, ventrite. IIIdvm2 (M. metanoto-trochantinalis anterior): O, scutellum; I, trochanter. IIIdvm4 (M. metanoto-coxalis anterior): O, notum; I, posterolateral margin of base of coxa. IIIdvm5 (M. metanoto-coxalis posterior): O, notum, lateral to IIIdvm2; I, lateral margin of base of coxa. IIIdvm6 (M. metacoxa-subalaris): O, posterior margin of base of coxa; I, subalare. IIIdvm8 (M. metafurca-phragmalis): O, apex of metendosternite; I, metaphragma. IIItpm2 (M. metapleura-praealaris): O, prealar sclerite; I, pleural ridge. IIItpm3 (M. metanoto-basalaris): O, lateral margin of notum; I, muscular disc of basalare. IIItpm7 (M. metanepisterno-axillaris) fused with IIItpm9. IIItpm9 (M. metepimero-axillaris tertius) fused with IIItpm7: O, anterior part of pleural ridge; I, third axillary plate. IIItpm10 (M. metepimero-subalaris): O, epimeron; I, subalare. IIIppm1 (M. metatransanapleuralis): O, pleural ridge; I, lateral margin of ventrite. IIIspm1 (M. metapleura-sternalis): O, ventrite, lateral to IIIdvm1; I, muscle disc of basalare. IIIpcm4 (M. metanepisterno-coxalis posterior): O, episternum; I, anterolateral margin of base of coxa. IIIscm1 (M. metafurca-coxalis anterior): O, metendosternite; I, anteromedial margin of base of coxa. IIIscm2 (M. metafurca-coxalis posterior): O, metendosternite; I, posterior margin of base of coxa. IIIscm4 (M. metafurca-coxalis lateralis): O, metendosternite; I, could not be determined precisely, lateral margin of base of coxa or posterior margin of pleurite. IIIscm6 (M. metafurca-trochanteralis): O, metendosternite; I, trochanter, via fine tendon. Internal muscles of legs not studied because of extremely small size.

Musculature of abdomen. Dorsal longitudinal muscles (Mm. dorsales), several parallel fibers: O, anterior phragma; I, posterior phragma. Ventral longitudinal muscles (Mm. ventrales), several parallel fibers: O, anterior margin of segment; I, posterior margin of segment. Dorsoventral muscles, several tergosternal muscles, homology unknown.

Reproductive system

Male reproductive system consisting of paired testes, spermiducts, and accessory glands, as well as copulatory apparatus (Fig. 6.16c, d).

Female reproductive system consisting of paired ovaries and oviducts, spermatheca, and accessory glands. Each ovary consisting of 2–4 teleotrophic ovarioles. Oviduct forming dilation, egg chamber, in which egg develops (Fig. 6.16a, b).

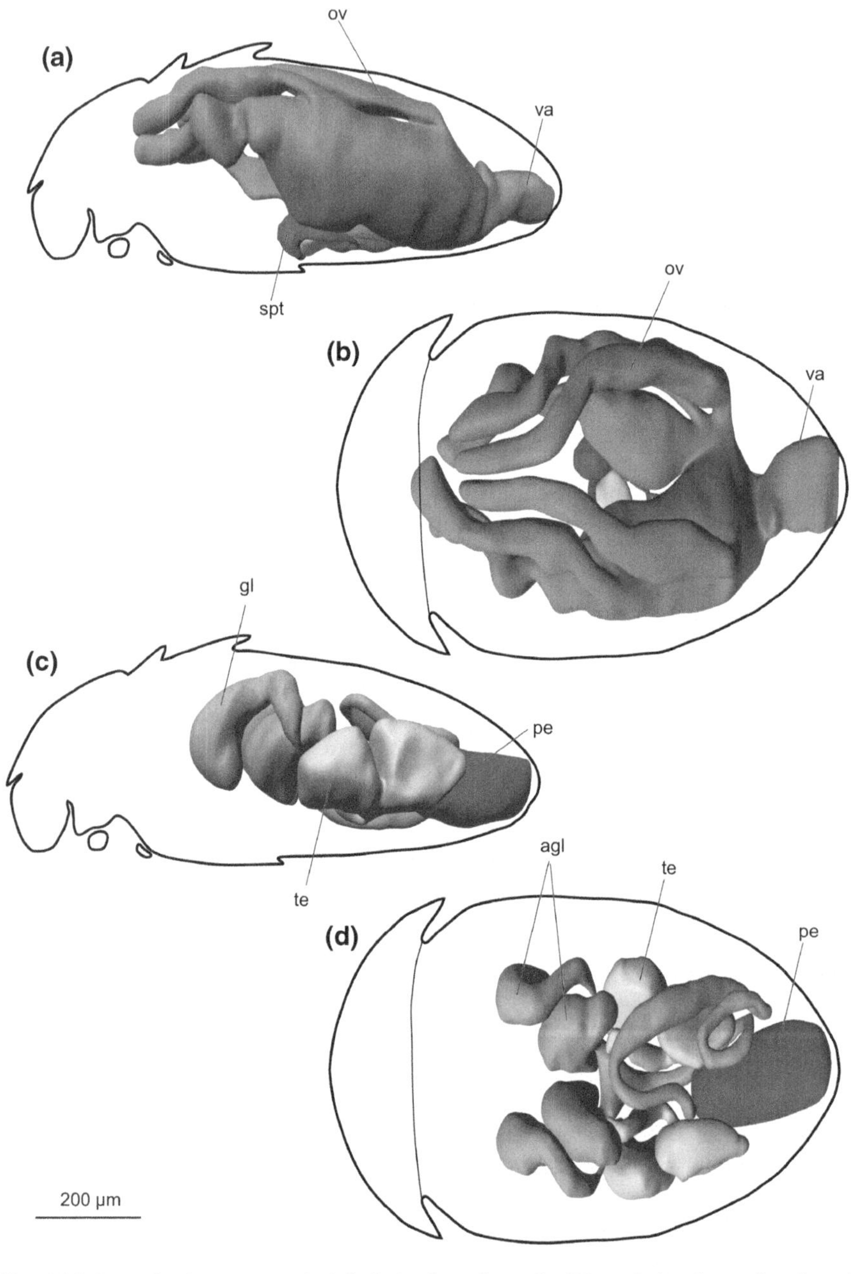

Fig. 6.16 Reproductive system of adult *Sericoderus lateralis*, 3D: **a**, **b** female; **c**, **d** male; *agl* accessory glands, *ov* ovariole, *pe* penis, *spt* spermatheca, *te* testis, *va* vagina

6.4 Morphology of Larvae

6.4.1 External Morphology

Body length of first instar larva in *Sericoderus lateralis* 540–660 μm (M = 597, n = 5), body length of last instar larva 1480–1950 (M = 1700, n = 22)

Morphology of head

Head prognathous, slightly flattened in dorsoventral direction (Fig. 6.17a, b). Cranium solid, without sutures. In some species of *Holopsis* in late instar larvae head protruding as long rostum (Fig. 6.18). Tentorium represented by only posterior arms and bridge, anterior and dorsal arms absent. Ocelli present, two on each side of head. Antennae well developed, 3-segmented, less often 2-segmented (*Sericoderus*), or strongly shortened (*Orthoperus*). Antennomere 2 bearing sensorium (sensory cone) and several trichoid sensilla, apical antennomere bearing 3–4 trichoid sensilla and 2–4 cuticular processes of various length. Labrum movably joined with cranium and appearing as simple plate, less often fused with cranium (*Rypobiini*). Mandibles well developed, symmetrical or almost symmetrical (Fig. 6.17c, d). Short, with wide base, well developed molar process, prostheca absent. Less often molar process reduced (Rypobiini). Incisor bearing 3–5 large teeth. Body of maxilla consisting of indistinctly divided cardo and stipes. Maxillary palp 3-segmented, less often 2-segmented (*Orthoperus*). Palpomere 1 without setae; palpomere 2 with trichoid sensilla and apically with sensillum ampullaceum; palpomere 3 apically with several basiconical sensilla (probably contact chemoreceptors). Lacinia and galea not differentiated, mala bearing several pointed digitiform processes apically. Labium membranous, weakly sclerotized. Mentum and submentum fused with cranium; prementum separated from mentum weakly by discernible suture. Labial palps 2-segmented (in *Orthoperus* 1-segmented), with ligula between them.

Morphology of thorax

Thorax consisting of three separate segments, prothorax slightly larger than other segments. Thoracic segments with only weakly sclerotized and weakly pigmented dorsal sclerites (absent in early instar larvae). Pair of spiracles present between pro- and mesothorax, as in other beetles. Profurca very small. Legs consisting of coxa, trochanter, femur, fused tibia and tarsus, and simple claw (Fig. 6.17e, f). Elongate empodium present under claw.

Structure of abdomen

Abdomen 10-segmented. All segments without sclerites. Segments 1–7 bearing rounded spiracles. Segments 1 and 8 bearing paired orifices of repugnatorial glands (Fig. 6.17h). Lateral surfaces of segments 1–8 with areas of spherical sensilla. Urogomphi absent. Pygopod well developed. Anal hooks absent.

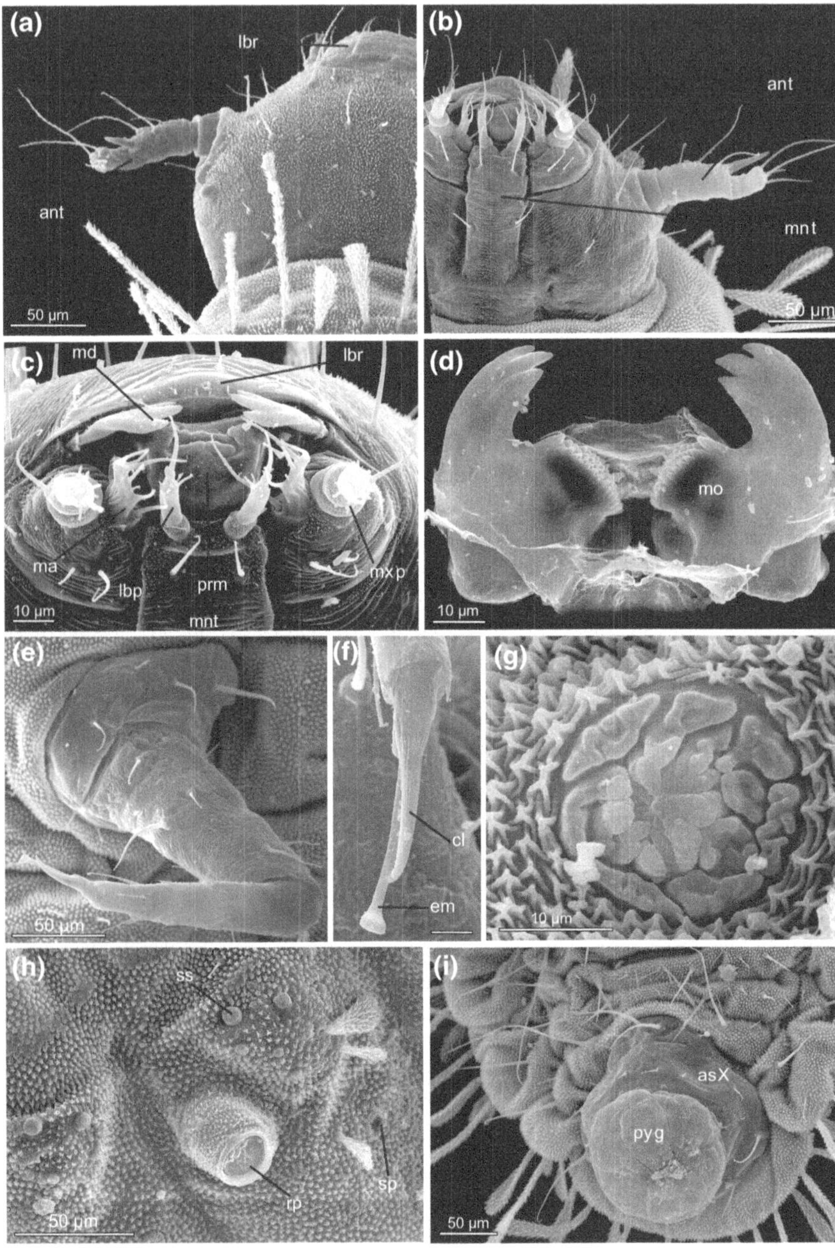

Fig. 6.17 External morphology of last instar larva in *Sericoderus lateralis*, SEM: **a** head, dorsal view; **b** head, ventral view; **c** mouthparts, frontal view; **d** mandibles; **e**, **f** hindleg; **g** spiracle of abdominal segment 2; **h** lateral part of abdominal segment 2; **i** apex of abdomen; *ant* antenna, *as* segment of abdomen, *cl* claw, *em* empodium, *lbp* labial palp, *lbr* labrum, *md* mandible, *mnt* mentum, *ma* mala, *mo* mola, *mxp* maxillary palp, *prm* prementum, *pyg* pygopod, *rg* repugnatorial gland, *sp* spiracle, *ss* spherical sensillum

Fig. 6.18 External morphology of larva in *Holopsis* sp., SEM: **a**, **b** first instar larva; **c**–**f** last instar larva; **a**, **b**, **e** lateral view; **c**, **f** ventral view; **d** dorsal view; **b** head; **f** mouthparts; *ant* antenna, *lbp* labial palp, *lbr* labrum, *md* mandible, *mxp* maxillary palp

6.4.2 Internal Morphology

General plan of internal structure: central part of body cavity occupied by intestine, considerable part of thorax occupied by musculature (Fig. 6.19).

Integument

Larval integument consisting of cuticle, hypoderm, and basal membrane. In first instar larva cuticle thickness 0.7–3.5 μm (M = 1.7; n = 80); in last instar larva cuticle thickness 1.7–13.4 μm (M = 5.1; n = 80), cuticle weakly sclerotized. Surface with stellate microsculpture.

Digestive and excretory systems

Digestive system of larvae, as that of adults, generally complies in Corylophidae with plan typical of insects (Fig. 6.20). Foregut divided into pharynx, oesophagus, and crop. Muscular stomach undeveloped. Pharynx V-shaped, with strongly sclerotized ventral wall. One pair of labial glands present. Oesophagus straight, short, with one thin layer of circular muscles. Midgut appearing as short straight tube. Wall consisting of unilayered epithelium, in anterior part of gut cells cylindrical, in posterior part cells flattened. Entire midgut with slender unilayered peritrophic membrane. Muscles of midgut not found. Hindgut divided into small intestine and rectum. Small intestine curved; rectum with rectal ampullae.

Boundary between mid- and hindgut bearing four Malpighian tubules forming several loops.

Circulatory system and fat body

Circulatory system strongly reduced, heart appearing as slender tube, other vessels absent. **Fat body** consisting of cells of various shapes and sizes, containing electron transparent inclusions.

Tracheal system

Tracheal system strongly simplified, with only longitudinal stems in thorax and abdomen and small number of weakly branching tracheae.

Nervous system

Larval central nervous system of Corylophidae different from that of adults in less strongly pronounced condensation and oligomerization of ganglia, especially in last instar larva (Fig. 6.21). Cerebrum shifted into thorax. Only rather small area of deutocerebrum in first instar larva situated in posterior part of cranium. Suboesophageal ganglion fused with prothoracic ganglion and situated in prothorax. Thoracic ganglia closely set, but separated; abdominal ganglia fused into one synganglion, partly shifted into metathorax. Central nervous system of last instar larvae symmetrical, but in first instar larva cerebrum strongly asymmetrical; right half of protocerebrum forming large distal process. Abdominal synganglion of first instar larva also asymmetrical, shifted left. Protocerebrum in first instar larva lobe-shaped, forming several large processes lying between gut and tufts of musculature. Ultrastructural organization of central nervous system and plan of structure of ganglia as in adult. Cerebrum displaying early stages of differentiation of optic lobes, peduncles of mushroom bodies, and central body complex.

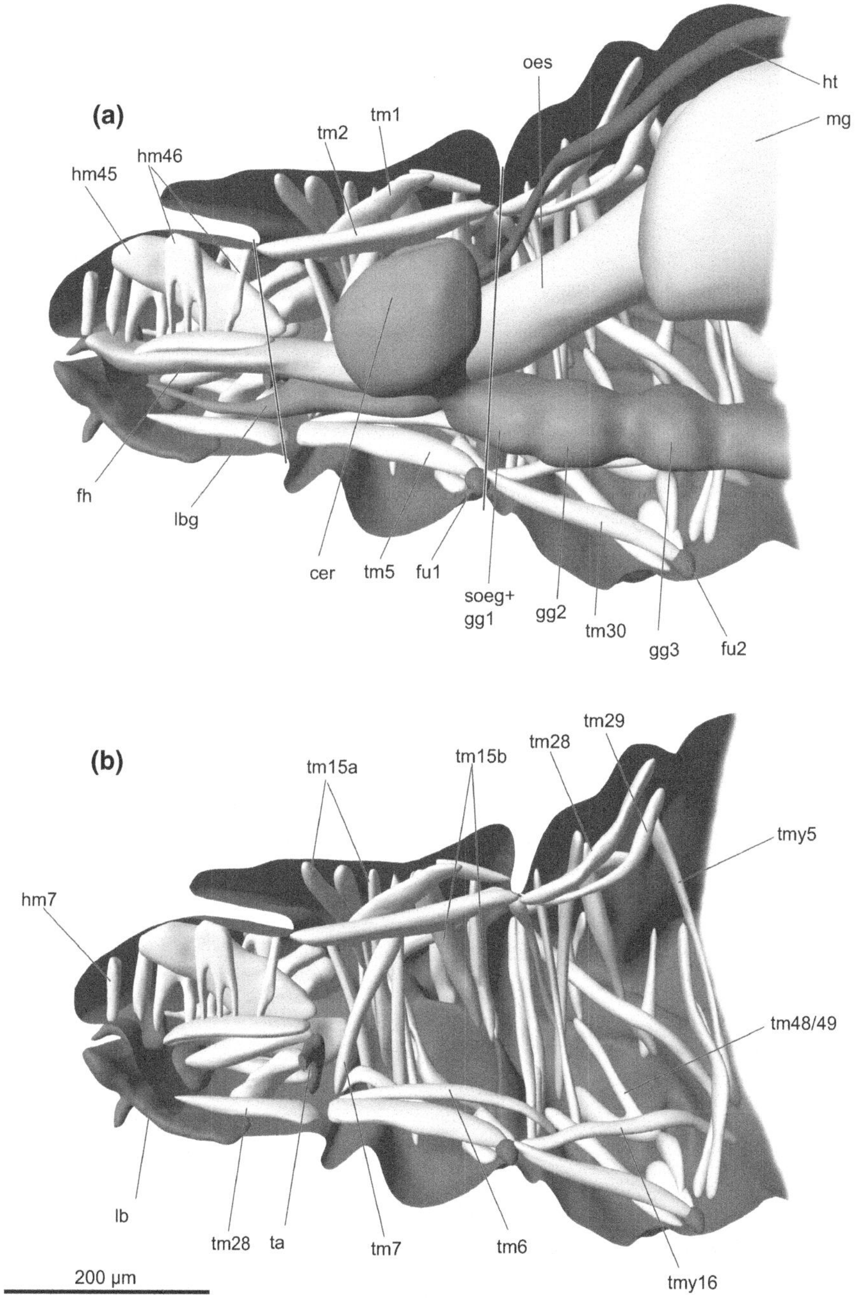

Fig. 6.19 Internal morphology of last instar larva in *Sericoderus lateralis*, 3D, lateral internal view; *cer* cerebrum, *fu1.2* pro- and mesofurca, *gg1.2.3* pro-, meso-, and metathoracic ganglia, *hm* musculature of head (see text), *ht* heart, *lbg* labial gland, *lbr* labrum, *mg* midgut, *oes* oesophagus, *ph* pharynx, *soeg* suboesophageal ganglion, *ta* tentorial arms, *tm* musculature of thorax (see text)

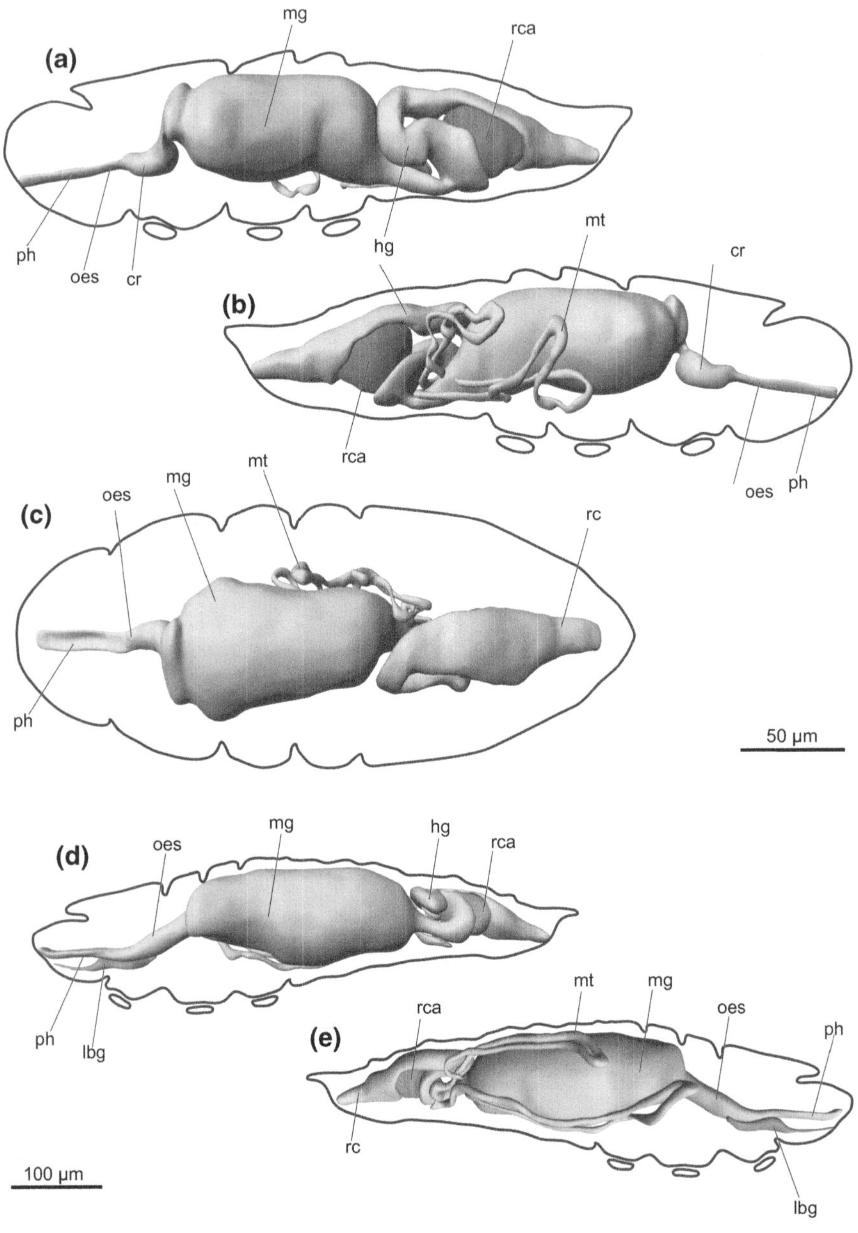

Fig. 6.20 Structure of digestive system in larva of *Sericoderus lateralis*, 3D: **a–c** first instar larva; **d**, **e** last instar larva; **a**, **b**, **d**, **e** lateral view; **c** dorsal view; *cr* crop, *hg* hindgut, *lbg* labial gland, *mg* midgut, *mt* Malpighian tubules, *oes* oesophagus, *ph* pharynx, *rc* rectum, *rca* rectal ampulla

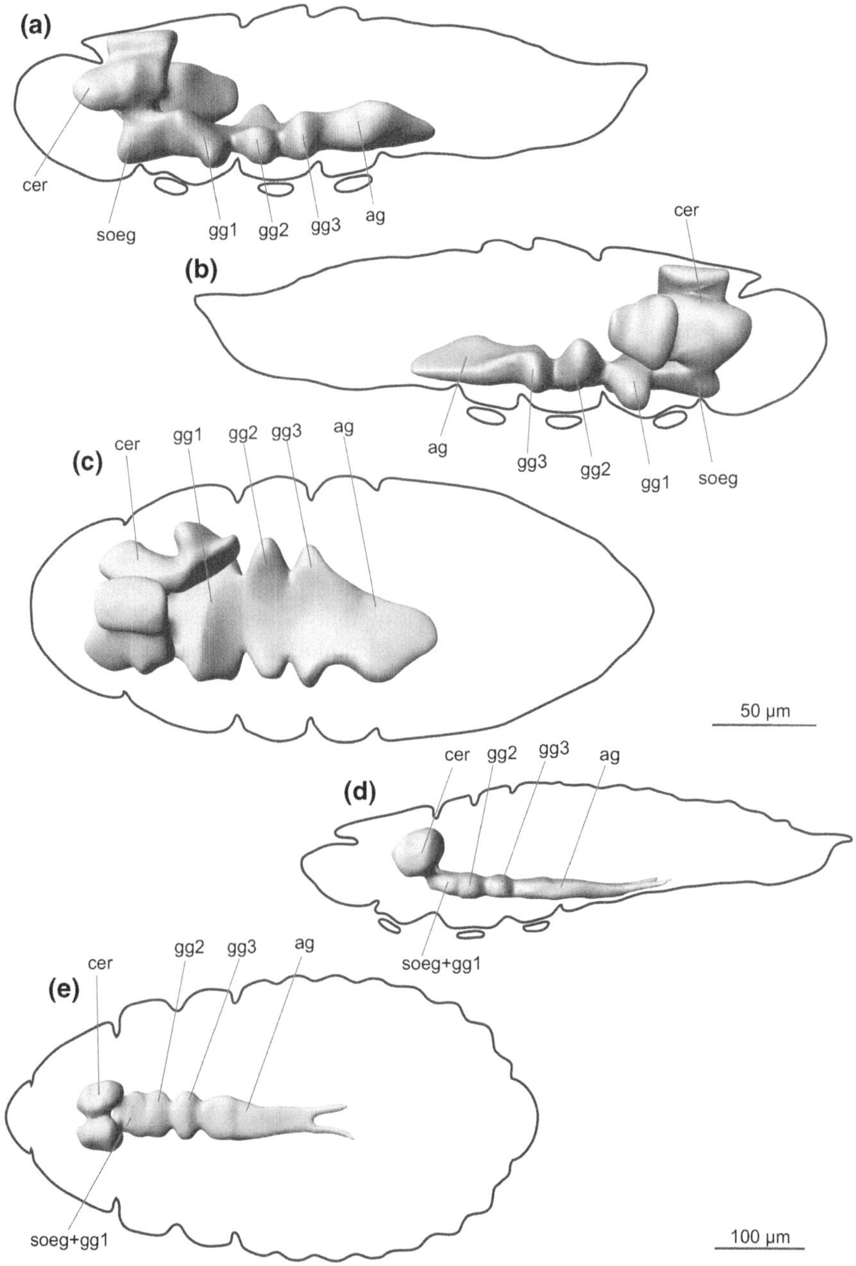

Fig. 6.21 Structure of central nervous system in larva of *Sericoderus lateralis*, 3D: **a**–**c** first instar larva; **d**, **e** last instar larva; **a**, **b**, **d** lateral view; **c**, **d** dorsal view; *ag* abdominal ganglion, *cer* cerebrum, *gg1.2.3* pro-, meso-, and metathoracic ganglia, *soeg* suboesophageal ganglion

Deutocerebrum symmetrical, in spite of asymmetry of protocerebrum. Tritocerebrum also symmetrical, represented by paired formations at sides of pharynx. Frontal ganglion strongly pronounced.

Cell size in nervous system of *S. lateralis* 2.29–2.71 μm (M = 2.49; n = 332). Cerebrum of first instar larva containing about 7000 cells.

Muscular system

Corylophid larvae, in spite of their small size, have all principal groups of musculature and almost all particular muscles typical of beetles. First and last instar larvae have been studied in *Sericoderus lateralis* (Fig. 6.22, Table 5.2).

Musculature of head. M1 (M. tentorioscapalis anterior), absent in first instar larva, present in last instar larva: O, lateral surface of pharynx; I, base of scape. M2 (M. tentorioscapalis posterior): O, posterior part of frons; I, base of scape. M4 (M. tentorioscapalis medialis): O, posterior part of frons; I, base of scape. M7 (M. labroepipharyngalis): O, central part of labrum; I, anterior part of epipharynx. M9 (M. frontoepipharyngalis): O, anterior part of frons, I, middle of epipharynx. M11 (M. craniomandibularis internus): largest muscle of head; O, lateral and posterior part of cranium; I, adductor of mandible. M12 (M. craniomandibularis externus): O, posterolateral part of cranium; I, abductor of mandible. M15 (M. craniocardinalis externus): O, ventrolateral part of cranium; I, lateral part of base of maxilla. M17 (M. tentoriocardinalis) and M18 (M. tentoriostipitalis) fused: O, tentorial bridge; I, ventral surface of base of maxilla. M19 (M. craniolacinialis): O, lateral part of cranium; I, distal part of maxilla. M28 (M. submentopraementalis): O, posterior part

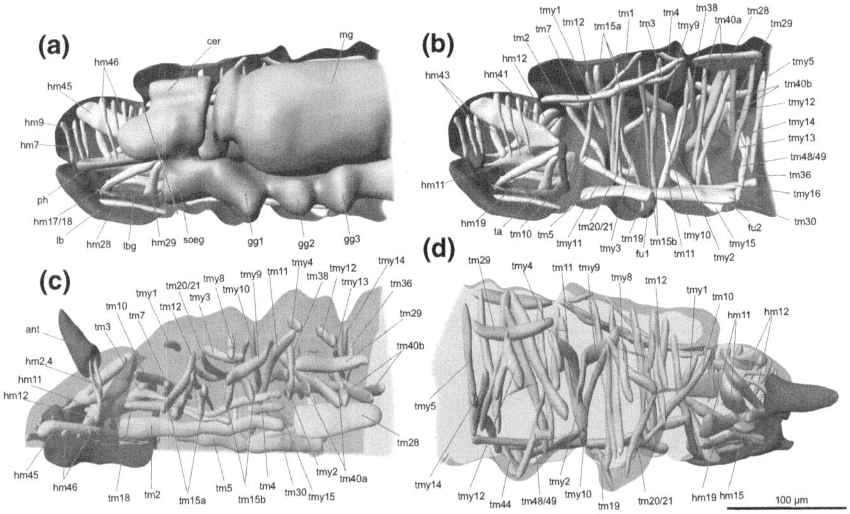

Fig. 6.22 Internal morphology of first instar larva in *Sericoderus lateralis*, 3D: **a**, **b** medial section; **c** dorsal view; **d** lateral view; *ant* antenna, *cer* cerebrum, *fu1.2* pro- and mesofurca, *hm* musculature of head (see text), *gg1.2.3* pro-, meso-, and metathoracic ganglia, *lb* labium, *lbg* labial gland, *mg* midgut, *soeg* suboesophageal ganglion, *ta* tentorial arms, *tm* musculature of thorax (see text)

of submentum; I, prementum. M29 (M. tentoriopraementalis inferior): O, posterior tentorial arms; I, lateral part of submentum. M41 (M. frontohypopharyngalis): O, frons; I, epipharynx. M43 (M. clypeopalatalis): O, frontal region, I, dorsolateral part of epipharynx. M45 (M. frontobuccalis anterior): O, frontal region; I, dorsal wall of pharynx. M46 (M. frontobuccalis posterior): O, frontal region behind M45; I, dorsal wall of pharynx. Transverse musculature of pharynx developed only on dorsal surface. Internal musculature of antennae and mouthparts not studied because of extremely small size.

Musculature of thorax. Prothorax. M1 (M. pronoti primus): O, pronotum; I, dorsal part of cervical membrane. M2 (M. pronoti secundus): O, medial part of prophragma; I, dorsal part of postoccipital region. M3 (M. pronoti tertius): O, prophragma lateral to M2; I, dorsal part of cervical membrane. M4 (M. pronoti quartus): O, prophragma; I, posterior part of pronotum. M5 (M. prosterni primus): O, profurca; I, ventral part of postoccipital suture. M6 (M. prosterni secundus) absent in first instar larvae, present in last instar larvae: O, fold between pro- and metathorax; I, ventral part of posterior margin of cranium. M7 (M. dorsoventralis primus): O, pronotum; I, ventrolateral part of postoccipital suture. M9 (M. dorsoventralis tertius): absent or fused with M7. M10 (M. dorsoventralis quartus): O, sternite prothorax; I, dorsolateral margin of postoccipital suture. M11 (M. dorsoventralis quintus): O, lateral part of prophragma; I, profurca. M12 (M. notopleuralis): O, lateral part of pronotum; I, pleural part of prothorax. M15a (M. notocoxalis anterior): O, pronotum; I, anterior margin of base of coxa. M15b (M. notocoxalis posterior): O, posterior part of pronotum; I, posterior margin of base of coxa. M19 (M. furcacoxalis) of two fibers: O, profurca; I, central margin and wall of coxa. M 20 (M. pleurotrochanteralis): O, pleural region, I, trochanter together with M 21 (M. coxotrochanteralis medialis). Several muscles of unclear homology: MY1 (Speyer 1922: mI10c?; Grebennikov and Beutel 2002: y; Polilov and Beutel 2009: MY1): O, anterior part of pronotum, I, anterior part of sternal region. MY2 (Speyer 1922: mI12c?; Polilov and Beutel 2009: MY2?): O, prophragma; I, ventrally between pro- and mesothorax. MY3 (Speyer 1922: 10g?; Polilov and Beutel 2009: MY3): O, posterior part of pronotum; I, posterior part of pleural region. MY8: O, lateral part of pronotum; I, sternal region. MY9 (Speyer 1922: mI12b?): O, lateral part of pronotum; I, ventrolateral surface between pro- and metathorax. MY10 (Speyer 1922: mI12a?): O, lateral part of pronotum; I, ventral surface between pro- and mesothorax MY2. MY11 (Speyer 1922: mI10a?): O, profurca; I, ventrolateral part of postoccipital suture. Internal muscles of legs not studied because of extremely small size.

Mesothorax. Musculature identical to prothoracic, except for absence of cervical musculature. All muscles of unclear homology found in prothorax present also in mesothorax: MY4 (in mesothorax), MY1 (in prothorax): MY5, MY2. MY6, MY3, MY12, MY8, MY13, MY9, MY14, MY10. MY15, MY11. Another muscle of unclear homology, MY16 (Speyer 1922: mII14b?): O, ventral surface between pro- and mesothorax; I, profurca.

Metathorax. Musculature identical to that of mesothorax.

Musculature of abdomen. Segments I–IX. Dorsal longitudinal muscles (dlm, Mm. dorsales abdominis) consisting of several fibers: O, anterior phragma; I, posterior phragma. Ventral longitudinal muscles (vlm, Mm. ventrales abdominis) consisting of several fibers, O, anterior phragma; I, posterior phragma. Dorsoventral muscles (dvm): O, lateral part of tergite; I, lateral part of sternite.

References

Amourig, L. (1960). Histologie de l'apparil digestif et des tubes de Malpighi de la larva de *Coccinella septempunctata* L. *Bulletin de la Société zoologique de France, 85*, 21–35.

Amourig, L. (1963). Ganglions cerebroides de la larva de Coccinella septempunctata L. Anatomie compare des corps pedoncules et du corps central de la larve des Coccinelles et du premier stade jeune de Gryllus domesticus. *Bulletin de la Société zoologique de France, 132*, 1–30.

Boving, A. G., & Craighead, F. C. (1931). *An illustrated synopsis of the principal larvae forms of the Coleoptera*. New York: The Brooklyn Entomological Society.

Bowestead, S. (1999). *A revision of the Corylophidae (Coleoptera) of the west Palaearctic region*. Geneva: Muséum d'histoire Naturelle.

Crowson, R. A. (1981). *The biology of the Coleoptera*. London: Academic Press.

De Marzo, L. (2000). Larve di coleotteri in detriti vegetali di orgine agricola: lineamenti morfologici e presenza stagionale (Polyphaga: 20 familglie). *Entomologica Bari, 2000*(34), 65–131.

Grebennikov, V. V., & Beutel, R. G. (2002). Morphology of the minute larva of *Ptinella tenella*, with special reference to effects of miniaturisation and the systematic position of Ptiliidae (Coleoptera: Staphylinoidea). *Arthropod Structure & Development, 31*(2), 157–172.

Hunt, T., Bergsten, J., Levkanicova, Z., et al. (2007). A comprehensive phylogeny of beetles reveals the evolutionary origins of a super-radiation. *Science, 318*, 1913–1916.

Ibrahim, M. M. (1948). The morphology and anatomy of *Coccinella undecimpunctata aegyptiaca* Reiche. *Bulletin de la Société entomologique d'Égypte, 32*, 305–316.

Landis, B. J. (1936). Alimentary canal and Malpigian tubules of *Ceratomegilla fuscilabris* Muls. *Annals of the Entomological Society of America, 29*, 15–27.

Lawrence, J. F., & Newton, A. F. (1995). Families and subfamilies of Coleoptera (with selected genera, notes, references and data family-group names) In J. F. Lawrence, & A. F. Newton (Eds.), *Biology, phylogeny and classification of Coleoptera: Papers celebrating the 80th birthday of R.A. Crowson* (pp. 779–1006). Warzawa: Muzeum i Instytut Zoologii PAN.

Lawrence, J. F., Ślipiński, A., Seago, A. E., Thayer, M. K., Newton, A. F., & Marvaldi, A. E. (2011). Phylogeny of the Coleoptera based on morphological characters of adults and larvae. *Annals of Zoology, 61*(1), 1–217.

Matthews, M. A. (1899). *A monograph of the coleopterous families Corylophidae and Sphaeriidae*. Janson and Son: London, O.E.

Paulian, R. (1950). Les Corylophidae d'Afrique. *Mémoires de l'Institut français d'Afrique noire, 12*, 1–126.

Polilov, A. A. (2011). Thoracic musculature of *Sericoderus lateralis* (Coleoptera, Corylophidae): miniaturization effects and flight muscle degeneration related to development of reproductive system. *Entomological Review, 91*(6), 735–742.

Polilov, A. A., & Beutel, R. G. (2009). Miniaturization effects in larvae and adults of *Mikado* sp. (Coleoptera: Ptiliidae), one of the smallest free-living insects. *Arthropod Structure & Development, 38*(3), 247–270.

Polilov, A. A., & Beutel, R. G. (2010). Developmental stages of the hooded beetle *Sericoderus lateralis* (Coleoptera: Corylophidae) with comments on the phylogenetic position and effects of miniaturization. *Arthropod Structure & Development, 39*(1), 52–69.

Praghan, S. (1938). Neuromuscular study of the mouthparts of *Coccinella septempunctata* L. with a comparison of the mouthparts in carnivorous and herbivorous Coccinellids. *Records of the Indian Museum, 40*, 341–358.

Praghan, S. (1939). The alimentary canal and pro-epithelian regeneration in *Coccinella septempunctata* L., with comparison of the mouthparts in carnivorous and herbivorous Coccinellids. *Quarterly Journal of Microscopical Science, 81*, 451–478.

Praghan, S. (1942). Reasociation of Malpighian tubules in Coccinellid beetles. *Indian Journal of Entomology, 4*, 11–21.

Praghan, S. (1953). The genitalia and their role in copulation of *Epilachnia indica* with a discussion on the morphology of genitalia in the family. *Proceedings of the Academy of Sciences University of Allahabad, 5*, 250–263.

Priore, R. (1966). Anatomia ed istologia della Rodolia cardinalis Muls Coleoptera Coccinellidae. *Bollettino del Laboratorio di Entomologia Agraria Filippo Silvestri, 24*, 247–316.

Robertson, J. A., Ślipiński, A., Hiatt, K., Miller, K. B., Whiting, M. F., & McHugh, J. V. (2013). Molecules, morphology and minute hooded beetles: A phylogenetic study with implications for the evolution and classification of Corylophidae (Coleoptera: Cucujoidea). *Systematic Entomology, 38*(1), 209–232.

Savoiskaya, G. I. (1983). Личинки кокцинеллид (Coleoptera, Coccinellidae) фауны СССР (Lady beetle larvae (Copeoptera, Coccinellidae) of the USSR). In: Opredeliteli po faune SSSR, izdavaemye Zoologicheskim institutom AN SSSR (Vol. 137). Leningrad: Nauka.

Seago, A. E., Giorgi, J. A., Li, J., & Ślipiński, A. (2011). Phylogeny, classification and evolution of ladybird beetles (Coleoptera: Coccinellidae) based on simultaneous analysis of molecular and morphological data. *Molecular Phylogenetics and Evolution, 60*(1), 137–151.

Ślipiński, A., Tomaszewska, W., & Lawrence, J. F. (2009). Phylogeny and classification of Corylophidae (Coleoptera: Cucujoidea) with descriptions of new genera and larvae. *Systematic Entomology, 34*, 409–433.

Speyer, W. (1922). Die Muskulatur der Larve von *Dytiscus marginalis* L. Ein Beitrag zur Kenntnis des Insektenko¨rpers. *Zeitschrift für wissenschaftliche Zoologie, 119*, 423–492.

Verhoeff, K. (1895). Beitrage zur verglaichenden Morphologie des Abdomens der Coccinellidae. *Wiegmann's Archiv für Naturgeschichte, 61*, 1–80.

Yavorskaya, M. I., & Polilov, A. A. (2016). Morphology of the head of *Sericoderus lateralis* (Coleoptera, Corylophidae) with comments on effects of the miniaturization. *Entomological Review, 96*(4), 395–406.

Yavorskaya, M. I., Leschen, R. A. B., Polilov, A. A., & Beutel, R. G. (2014). Unique rostrate larvae and basidiomycophagy in the beetle family Corylophidae. *Arthropod Structure & Development, 43*, 153–162.

Chapter 7
Structure of the Principal Groups of Microinsects. V. Fairy Flies (Hymenoptera: Mymaridae)

7.1 Introduction

Fairy flies (Mymaridae) are small insects distributed worldwide; their larvae develop in eggs of other insects. Mymarids are important for discussions of miniaturization-related problems, because the smallest known insect, *Dicopomorpha echmepterygis*, belongs to this family (Mockford 1997; Polilov 2015). Mymarids have attracted the attention of many researchers also as agents of the biological method of plant protection. Nevertheless, questions of external morphology and especially anatomy of adult mymarids have been insufficiently treated in the literature until recently. External morphology has been studied in detail in several mymarid genera (Debauche 1948). The structure of many mymarids was briefly described in several large-scale taxonomic studies (Nikolskaya 1952; Noyes and Valentine 1989a, b; Huber 1997). The external morphology of larvae was studied much more thoroughly (Balduf 1928; Bakkendorf 1934; Jackson 1961; Boivin et al. 1993; van Baaren et al. 1997; Boivin 2010). The internal morphology of adults was poorly studied; the only available publication were studies on such particular aspects as the structure of the reproductive system (King and Copland 1969; Jackson 1969; Sahad 1982; Viggiani 1988; Chiappini and Mazzoni 2000; Boivin 2010), general plan of the digestive system (Sahad 1982), and structure of the cerebrum (Makarova and Polilov 2013). The very strongly simplified internal morphology of mymarid larvae has been described in several studies (Balduf 1928; Bakkendorf 1934; Ivanova-Kazas 1954, 1961; Jackson 1961; Sahad 1984). Specialized work on the study of peculiar miniaturization-related external and internal features of Mymaridae started rather recently (Polilov 2007), but to date they have been studied in rather much detail (Huber and Noyes 2013; Makarova and Polilov 2013; Polilov 2016).

According to modern data on the taxonomy and phylogeny of Hymenoptera, Mymaridae, and Trichogrammatidae belong to the superfamily Chalcidoidea (Gibson 1986; Gibson et al. 1999; Munro 2011; Heraty et al. 2011). Most

A.A. Polilov, *At the Size Limit—Effects of Miniaturization in Insects*,
DOI 10.1007/978-3-319-39499-2_7

chalcidoids are small, and many experts believe that mymarids are the basal lineage of Chalcidoidea (Munro 2011; Heraty et al. 2011). The earliest known chalcidoids had a body length of 1.3–2.4 mm (Rasnitsyn et al. 2004; Rasnitsyn and Brothers 2007), and therefore many relatively large chalcidoids (2–3 mm long) are more similar in size to the early forms than at least the smaller representatives of Mymaridae and Trichogrammatidae (0.14–0.4 mm). Hence, they can be used in comparative analysis. The comparative morphology of different families of Chalcidoidea has been described in several studies (Nikolskaya 1952; Bradley 1955; Rasnitsyn 1980; Gibson 1985, 1986, 1997, 1999; Noyes and Valentinem 1989b; Vilhelmsen et al. 2010). The morphology of the larvae of different representatives of Chalcidoidea has been treated in studies on development and biology (Hanna 1934; Ivanova-Kazas 1954, 1961; Cutler 1955; Saakyan-Baranova and Muzafarov 1972). The external and internal morphology of Chalcidae has been studied especially thoroughly in *Euchalcidia caryobori* (Hanna 1934, 1935). Morphology has also been described in detail in such families of Chalcidoidea as Eurytomidae (James 1926) and Callimomidae (Bucher 1948).

7.2 Brief Description of the Structure of Mymaridae

Small hymenopteras, adults 0.14–4 mm long (usually 0.4–1 mm). Body compact, abdomen sometimes elongate (Fig. 7.1). Coloration monotonous, from yellow to black, without metallic sheen. Integument usually without complex microsculpture or dense pubescence. Head hypognathous. Compound eyes lateral, consisting of 60–200 ommatidia; three ocelli present. In some representatives, eyes and ocelli lost. Antennae 8–13-segmented, attached between eyes. Transverse medial suture, not found in other Chalcidoidea, running dorsal to base of antennae. Mesosoma consisting of fused pro-, meso-, metathorax, and abdominal segment 1. Mesothorax larger than any other segment of mesosoma; metathorax represented by metanotum, with other sclerites of metathorax fused with abdominal segment 1 to form propodeum. Both pairs of wings usually present, sometimes hindwing strongly reduced, and in some cases all wings absent. Wing consisting of narrow wing blade with strongly depleted venation, on perimeter of wing with fringe of long setae. Membrane of hindwing not reaching wing base. Legs divided into coxa, trochanter, femur, tibia, and tarsus, typical of insects. Trochanter in most representatives of this family 2-segmented. Tarsi 4–5-segmented, in some representatives 1-segmented or fused with tibia (male *Dicopomorpha*). Number visible segments of metasoma 5–7, because abdominal segment 1 included in mesosoma, segment 2 forming petiole, and apical segments modified; number of visible tergites and sternites often different in males and females of one species. Petiole sometimes rather long (*Mymar*) or very short (*Anaphes*). External male genitalia represented by phallobase bearing parameres, and simple aedeagus. Females with ovipositor consisting of external and internal valves and pair of stylets. External sexual dimorphism strongly pronounced. Males and females different in size, proportions, number of visible

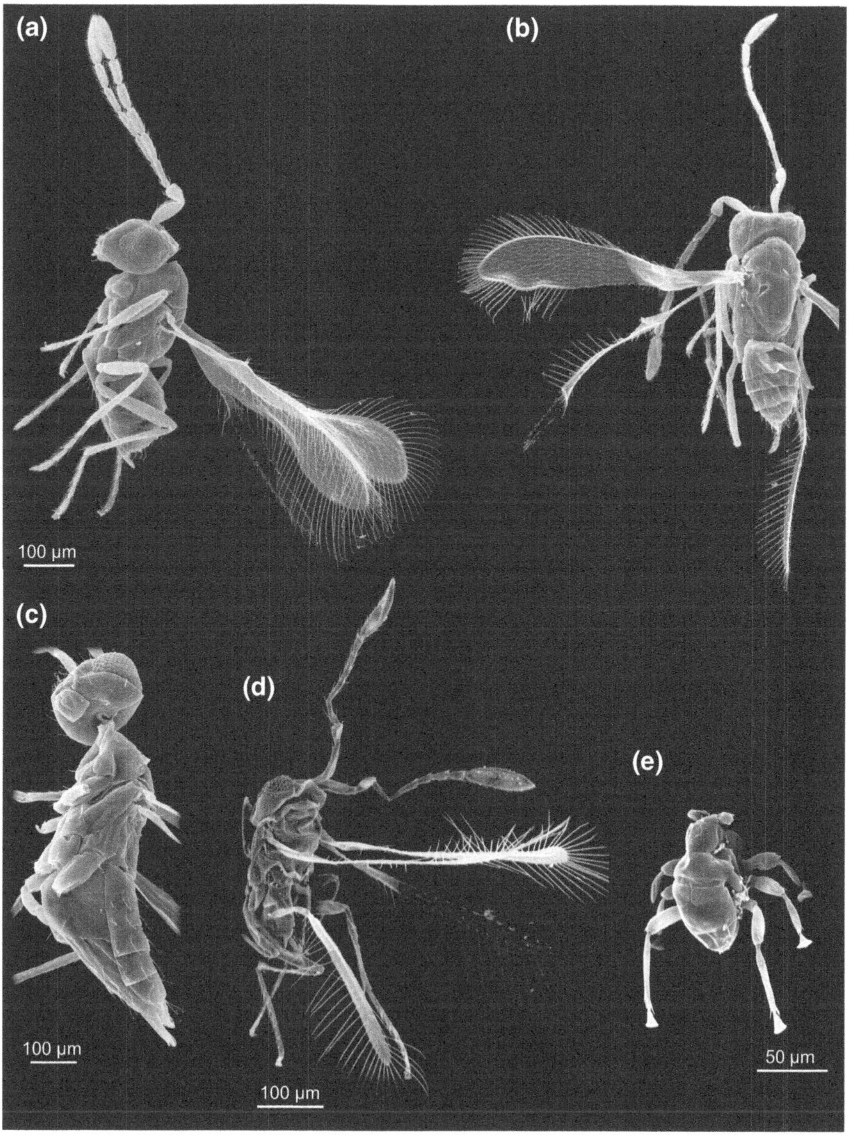

Fig. 7.1 External morphology of Mymaridae, SEM: **a**, **b** *Anaphes flavipes*; **c** *Anagrus* sp.; **d**, **e** *Dicopomorpha echmepterygis*; **d** female, **e** male; **a**, **c**, **d** lateral view; **b**, **e** dorsal view

sclerites of abdomen, number of antennomeres, and armature of legs. In some cases, e.g., in *Dicopomorpha*, males considerably smaller than females and with strongly simplified structure: lacking wings, eyes, ocelli, and mouthparts and having 1-segmented antennae (Fig. 7.2).

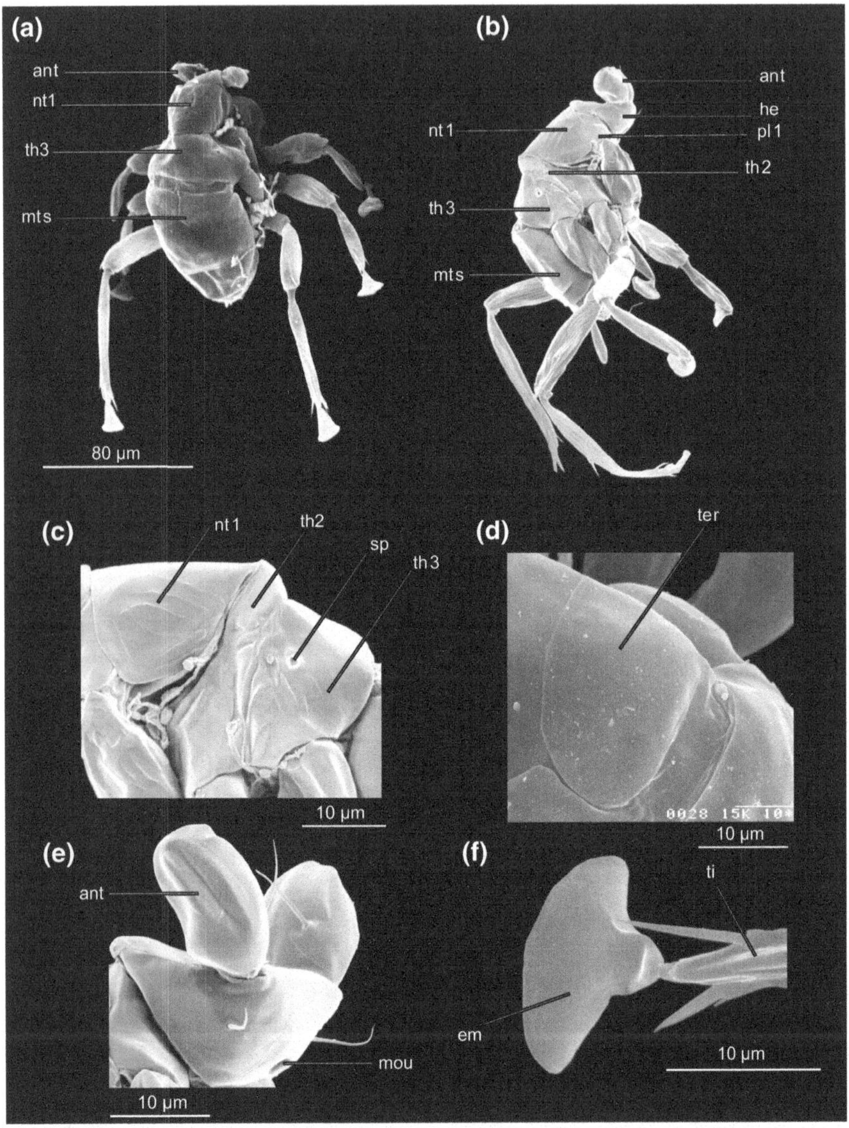

Fig. 7.2 External morphology of adult *Dicopomorpha echmepterygis*, male, SEM: **a** dorsal view; **b** lateral view; **c** mesosoma, lateral view; **d** base of metasoma, dorsal view; **e** head, lateral view; **f** pretarsus; **b**, **c**, **e** from Huber and Landry (1999), modified; *ant* antenna, *em* empodium, *he* head, *mts* metasoma, *mou* oral orifice, *nt1* pronotum, *ppe* prepectus, *sp* spiracle, *ter* tergite, *th2.3* meso- and metathorax, *ti* tibia

The larvae of Mymaridae develop in eggs of other insects; they are strongly desembryonized and simplified because of their parasitoid mode of life. First, instar larvae are mymariform, sacciform, or elongate; last instar larvae are sacciform.

7.3 Morphology of Adult Mymaridae

7.3.1 External Morphology

Body length of adult *Anaphes flavipes* 440–480 (M = 465, n = 10); body length of male *Dicopomorpha echmepterygis* 142–163 (M = 152, n = 2), but the smallest measured specimen of this species had a body length of 139 μm (Mockford 1997).

Structure of Head

Head hypognathous (Fig. 7.3a). One typical morphological feature of Mymaridae: presence of H-shaped system of sutures (trabeculae), formed by medial suture running dorsal to bases of antennae and sutures running from medial suture to mouthparts (Fig. 7.3d). In cross-section these sutures appearing as helically twisted areas of cuticle (Fig. 7.4a, b). Other sutures of cranium not pronounced, except for hypostomal suture and poorly discernible postoccipital suture. Clypeus not separated from cranium. Tentorium, as in other chalcidoids, with anterior and posterior arms; dorsal arms reduced (Fig. 7.4c).

Compound eyes lateral, consisting of 60–200 ommatidia, considerably fewer than in other hymenopterans. Number of ocelli usually 3. In males of *Dicopomorpha echmepterygis* eyes and ocelli absent.

Antennae in females 8–11-segmented with 1–2-segmented club, in males 10–13-segmented, filiform. Each antenna in *Anaphes* bearing about 560 sensilla of seven types (van Baaren et al. 1999). In males of *Dicopomorpha echmepterygis* antennae 1-segmented, receptor apparatus represented by one trichoid sensilla on each antenna.

Mouthparts of structure typical of chalcidoids, consisting of well-developed mandibles, maxillae, and labium (Figs. 7.3, 7.4d, e and 7.5a, b). Labrum weakly developed and represented by rather small triangular membranous plate. Mandibles bearing 2–3 large teath on molar surface. Maxillae connected with labium by membranous septum into labiomaxillary complex. Maxillae consisting of cardo, stipes, fused galea and lacinia, and maxillary palp. Maxillary palp 1-segmented, strongly reduced, in some representatives (*Anagrus*, *Dicopomorpha*) reduced to one llarge sensillum. Galea bearing large setae and spines, lacinia fused with galea and appearing as brush of setae. Labium consisting of mentum, bearing 1-segmented palps (in *Anagrus* reduced to only one sensillum), ligula, and paraglossae. In males of *Dicopomorpha echmepterygis* mouthparts absent.

Structure of Mesosoma

Prothorax consisting of semicircular pronotum and propectus, formed by prosternite and propleurites. Pleurites with lobes sunk into body. Anterior part of propectus bearing paired cervical processes; head attached to these processes. Profurca, as in most other chalcidoids, V-shaped with flattened arms.

Mesothorax more developed than other segments of metasoma (Figs. 7.6a–c and 7.7). Notum consisting of two parts, divided by scutoscutellar suture. Anterior

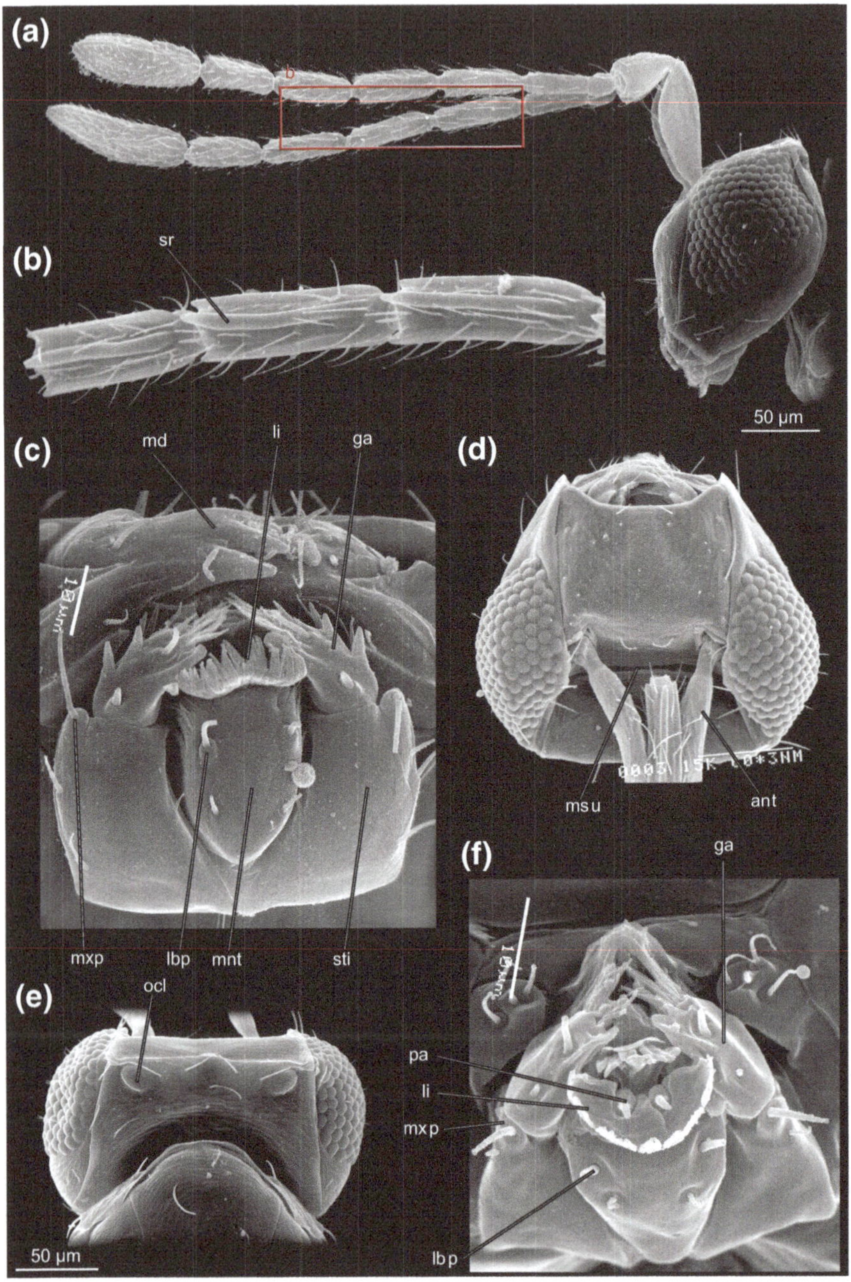

Fig. 7.3 External morphology of Mymaridae, SEM: **a–e** *Anaphes flavipes*; **e** *Anagrus* sp., **d**, **e** head; **b** antenna; **c**, **f** mouthparts; **a** lateral view, **b**, **c** ventral view, **d** frontal view, **e** dorsal view; *ga* galea, *lbp* labial palp, *li* ligula, *md* mandible, *mnt* mentum, *msu* medial suture, *mxp* maxillary palp, *pa* paraglossae, *ocl* ocellus, *sr* sensory ridge, *sti* stipes

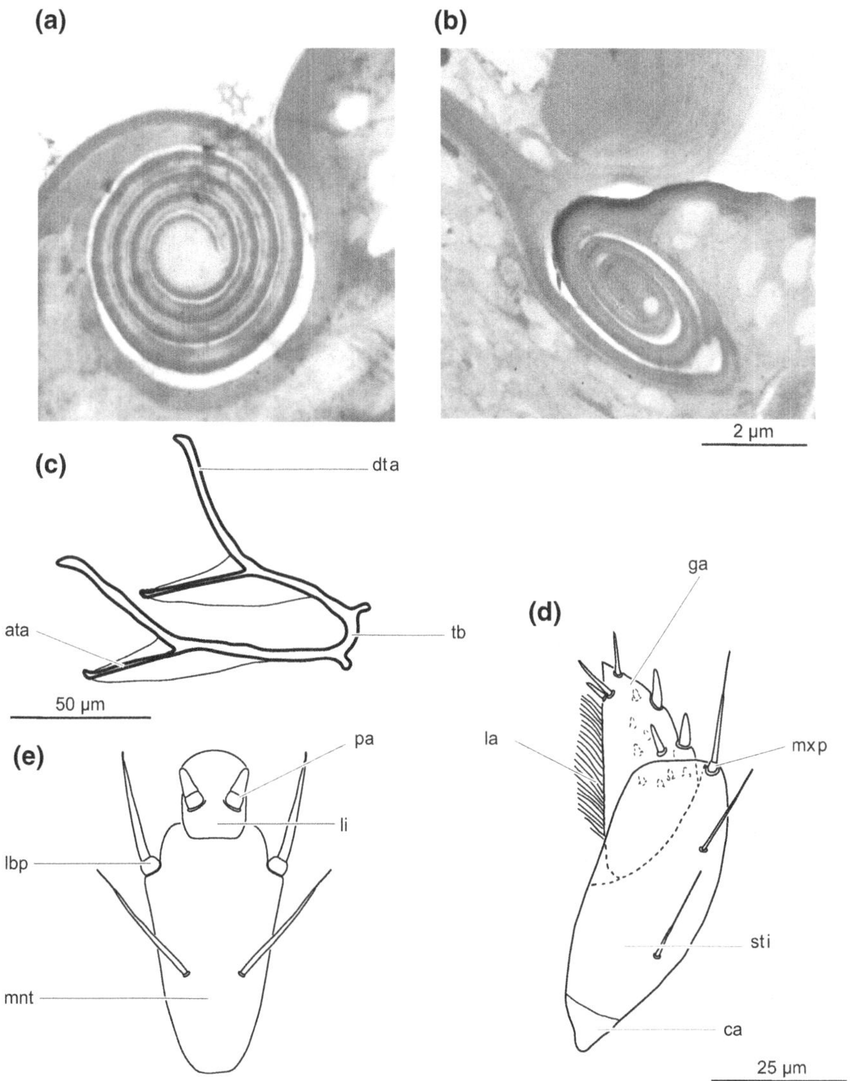

Fig. 7.4 *Anaphes flavipes*, SEM: **a** medial suture, TEM; **b** longitudinal suture of trabecula, TEM; **c** tentorium; **d** maxilla; **e** labium; *ata* anterior tentorial arms, *ca* cardo, *dta* dorsal tentorial arms, *lc* lacinia, *li* ligula, *mnt* mentum, *mxp* maxillary palp, *pa* paraglossae, *sti* stipes, *tb* tentorial bridge

part divided by longitudinal parapsidal striae into scutum and scapulae (lateral lobes of scutum). Posterior part divided into scutellum and axillary sclerites. In contrast to most chalcidoids, scutellum in Mymaridae divided by transverse suture into anterior part and posterior part. Lateral part of mesothorax not divided into episternum and epimeron, in contrast to many chalcidoids. Prepectus (secondary formation found in all chalcidoids) present in Mymaridae between pro- and mesothorax. Tegulae

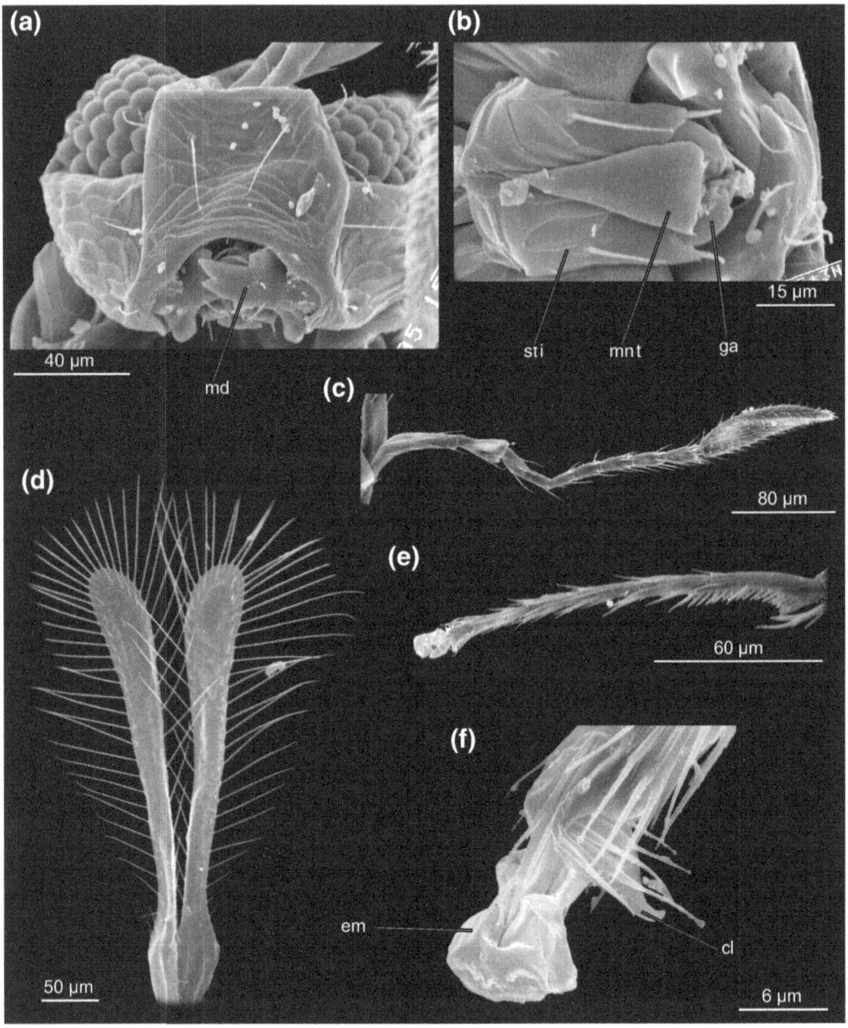

Fig. 7.5 External morphology of adult *Dicopomorpha echmepterygis*, female, SEM: **a** head, frontal view; **b** mouthparts, ventral view; **c** antenna; **d** wings; **e**, **f** tarsus and pretarsus; *cl* claw, *em* empodium, *ga* galea, *md* mandible, *mnt* mentum, *sti* stipes

well-developed. Mesofurca typical of Chalcidoidea, lateral arms well developed. Pair of spiracles present between pro- and mesothorax.

Metathorax represented by narrow semicircle of metanotum; other sclerites of metathorax fused with abdominal segment 1, forming propodeum. Propodeum bearing pair of abdominal spiracles. Metapleurites fused with propodeum or, in some representatives, separated from propodeum by weakly pronounced suture. Metafurca represented by two widely set arms with flattened apically.

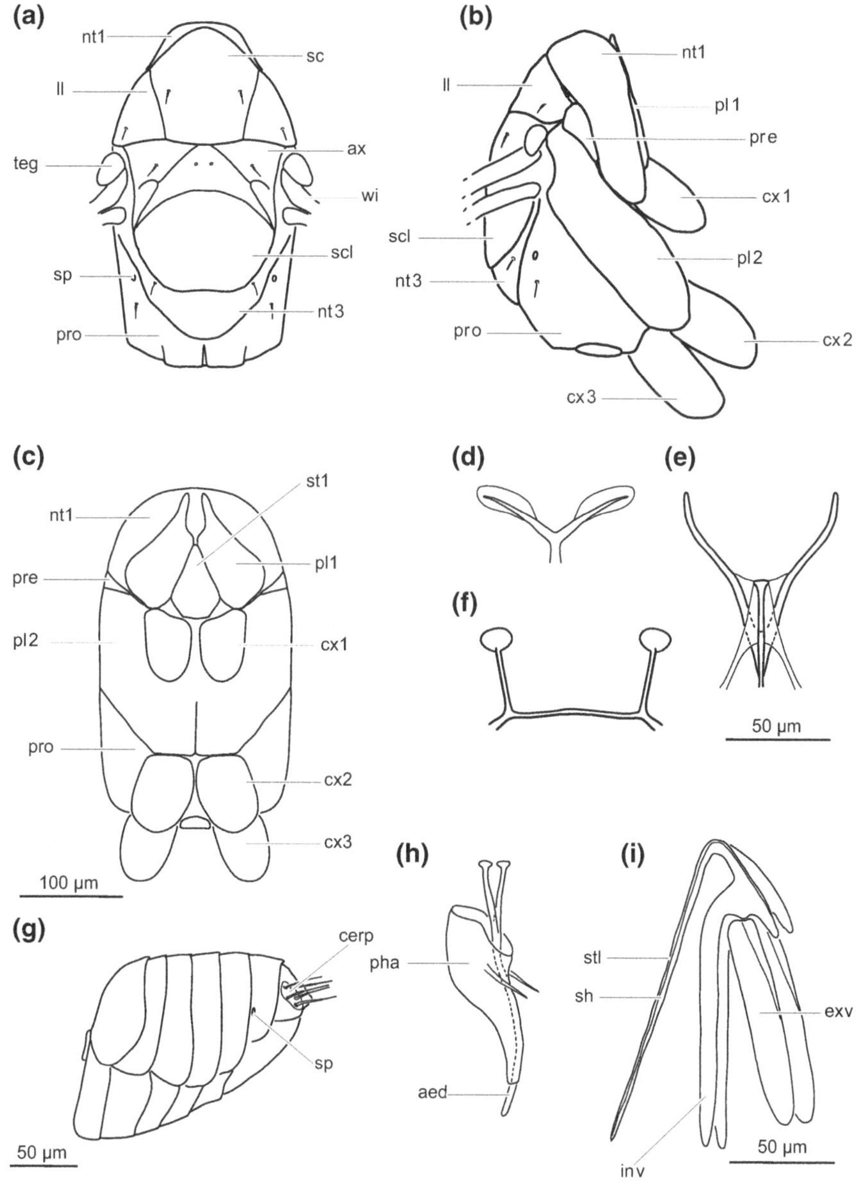

Fig. 7.6 Structures of *Anaphes flavipes*, schematic: **a–c** mesosoma; **d**, **e**, **f** pro-, meso-, and metafurca; **g** metasoma; **h** external male genitalia, **i** ovipositor; **a** dorsal view; **b**, **g–i** lateral view; **c** ventral view; *aed* aedeagus, *ax* axillary sclerite, *cx1.2.3* pro-, meso-, and metacoxae, *exv* external valves of ovipositor, *inv* internal valves of ovipositor, *ll* scapulae, *nt1.3* pro- and metanotum, *pha* phallobase, *pl1.2* pleurite of pro- and mesothorax; *pre* prepectus, *pro* propodeum, *sc* scutum, *scl* scutellum, *sh* sheath of ovipositor, *sp* spiracle, *st1* prosternite, *stl* stylet of ovipositor, *teg* tegula, *wi* wing

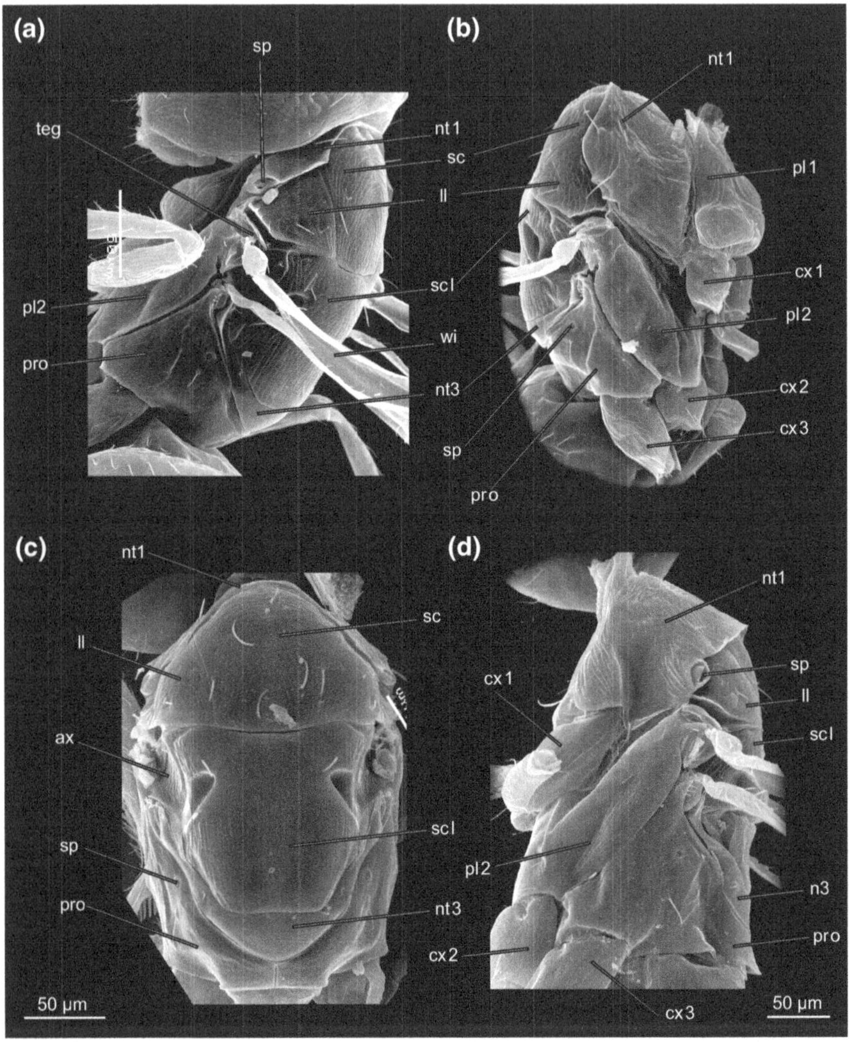

Fig. 7.7 Mesosoma in adult Mymaridae, SEM: **a–c** *Anaphes flavipes*; **d** *Anagrus* sp.; **a**, **b**, **d** lateral view; **c** dorsal view; *ax* axillary sclerite, *cx1.2.3* pro-, meso-, and metacoxae, *ll* scapulae, *nt1.3* pro- and metanotum, *pl1.2* pleurite pro- and metathorax, *pre* prepectus, *pro* propodeum, *sc* scutum, *scl* scutellum, *sp* spiracle, *teg* tegula, *wi* wing

Mesosoma of males in *Dicopomorpha echmepterygis* with soft weakly sclerotized integument containing poorly discernible sclerites of rather unclear homology with sclerites of other Mymaridae. Each thoracic segment practically consisting of one sclerite covering its dorsal and lateral surfaces: pre- and metathoracic sclerites wide, whereas metathoracic sclerite narrow. Propodeum fused with metathorax. Petiole not pronounced.

Wings consisting of narrow wing blade with strongly depleted venation, on perimeter of wing with fringe of long setae (Fig. 7.8a, b, d). Plate of hindwing not reaching base, and in some representatives (*Mymar*) reduced and hindwing appearing as seta. In some representatives wings reduced or absent. Forewing with three preserved veins: submarginal, marginal, and stigmal, often fused into one vein near anterior margin of wing and resulting from fusion of subcostal and radius; homology of particular parts is discussed in earlier studies (Burks 1938; Brandley 1955). In contrast to other chalcidoids, postmarginal, medial, and cubital veins and parastigma absent. Wing blade usually covered with small setae. Hindwing with only one submarginal vein, forming petiole of wing. In some mymarids wings strongly reduced; some species include both forms with normally developed wings and forms with incompletely developed wings (Noyes and Valentine 1989a). In males of *Dicopomorpha echmepterygis* wings absent.

Legs consisting of coxa, 1–2-segmented trochanter, femur, tibia, 4–5-segmented tarsus. Tibiae apically bearing well developed, often branched, spurs. Protarsomere 1 with dense row of setae. Apical tarsomere bearing two claws and well developed arolium (Fig. 7.8i, f). Tarsus in males of *Dicopomorpha echmepterygis* fused with tibia, claws reduced, arolium modified to form funnel-shaped sucker (Fig. 7.2f).

Structure of Metasoma

Abdomen in Mymaridae, as in other hymenopterans, 10-segmented. Segment 1 included in propodeum, segment 1 forming petiole, distal segments usually modified and sunk into metasoma (Fig. 7.9). Petiole sometimes rather long (*Mymar*), sometimes short, or secondarily lost (male *Dicopomorpha echmepterygis*). Males usually with seven visible tergites and six visible sternites, females usually with six or seven visible tergites and five or six visible sternites; apical sternites divided longitudinally into two sclerites. Tergite 6 bearing spiracles. Tergite 7 (epipygium) bearing cercal plates with group of long setae. In males of *Dicopomorpha echmepterygis* metasoma consisting of four weakly discernible segments.

Ovipositor consisting of outer ovipositor plates (derivates of tergite 9), inner ovipositor plates (Val3), sheath (fused Val2), and stylets (Val1) (Fig. 7.6i). Spermatheca rather small, rounded.

External male genitalia represented by simple aedeagus and phallobase encircling aedeagus (Fig. 7.6h). In some genera parameres also present (Chiappini and Mazzoni 2000).

7.3.2 *Internal Morphology*

General plan of internal structure: most of head occupied by supraoesophageal and suboesophageal ganglia; considerable part of mesosoma occupied by musculature; reproductive system occupying most of metasoma (Fig. 7.10).

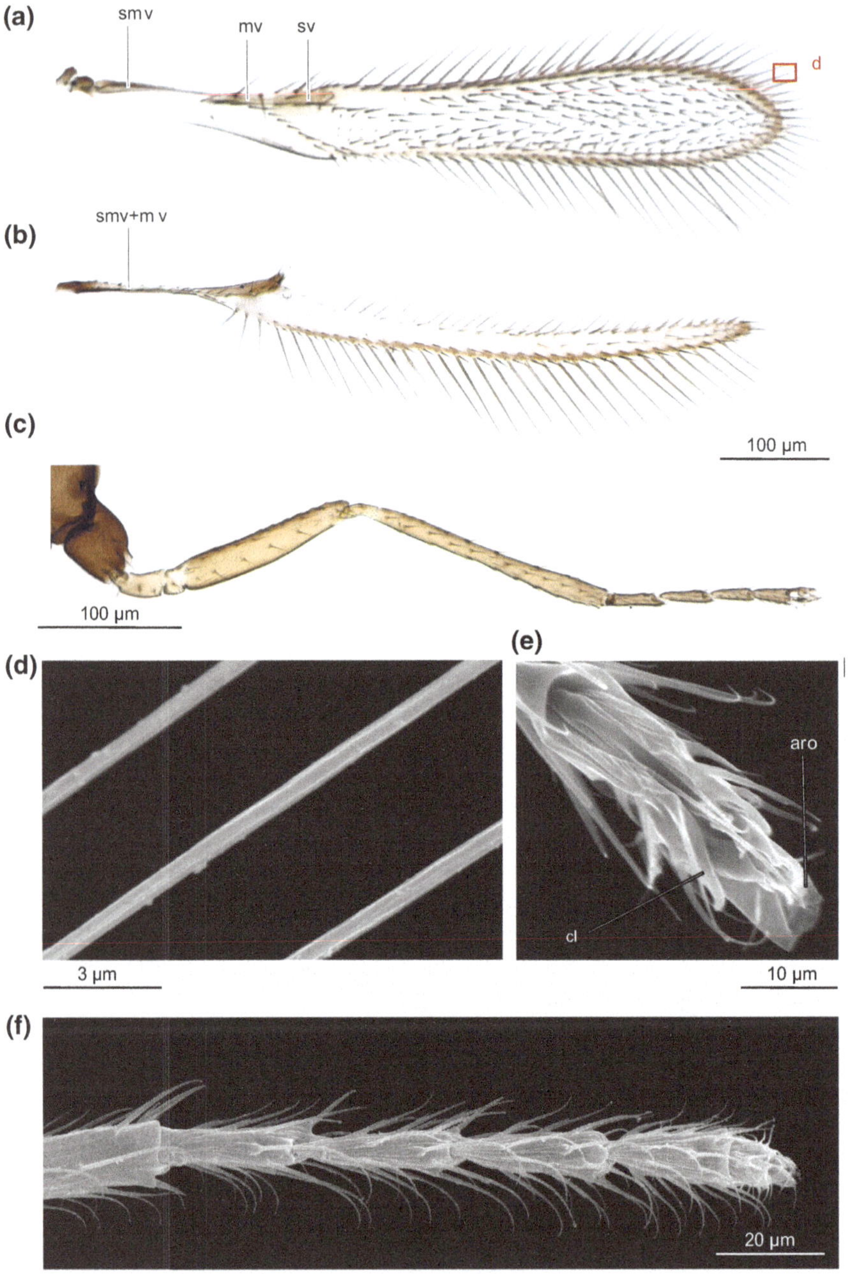

Fig. 7.8 Adult *Anaphes flavipes*: **a**, **d** forewing; **b** hindwing; **c** leg; **e** pretarsus, **f** tarsus; *aro* arolium, *cl* claw, *mv* marginal vein, *smv* submarginal vein, *sv* stigmal vein

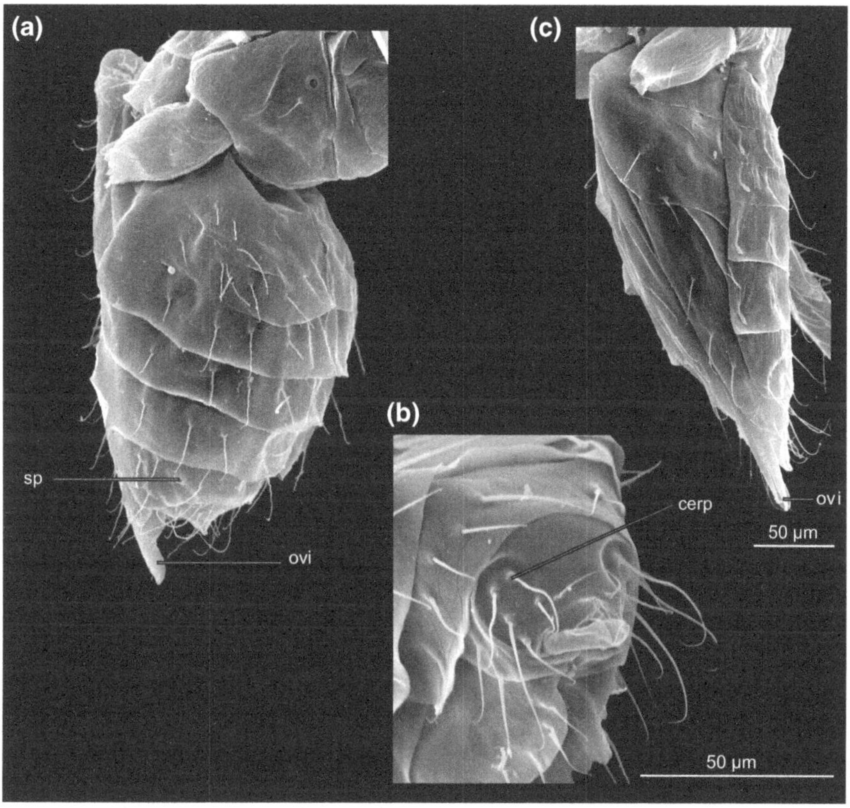

Fig. 7.9 Metasoma in Mymaridae, SEM: **a**, **b** *Anaphes flavipes*; **c** *Anagrus* sp.; **a**, **c** lateral view; **b** posterior view; *cerp* cercal plate, *ovi* ovipositor, *sp* spiracle

Integument

Integument consisting of cuticle, hypoderm, and basal membrane. Cuticle thickness in *Anaphes flavipes* 1.1–3.7 µm (M = 2.2, n = 80). Cuticle consisting of epicuticle and exo- and endocuticle. In many areas of integument exo- and endocuticle not discernible (Fig. 7.11a, b). Procuticle homogeneous. Hypoderm represented by flattened cells 0.5–1.5 µm thick.

Digestive and Excretory Systems

Digestive system complying with general plan typical of insects. Digestive canal divided into fore-, mid-, and hindgut (Fig. 7.12a, b). Fore- and hindgut with thin cuticular lining. Midgut without cuticular lining. Intestinal canal somewhat longer than body, forming loop in metasoma. Of digestive glands, only labial salivary glands known.

Foregut divided into pharynx, oesophagus, and crop. Oesophagus straight, running through entire mesosoma. Muscles of oesophagus absent. Crop situated in metasoma.

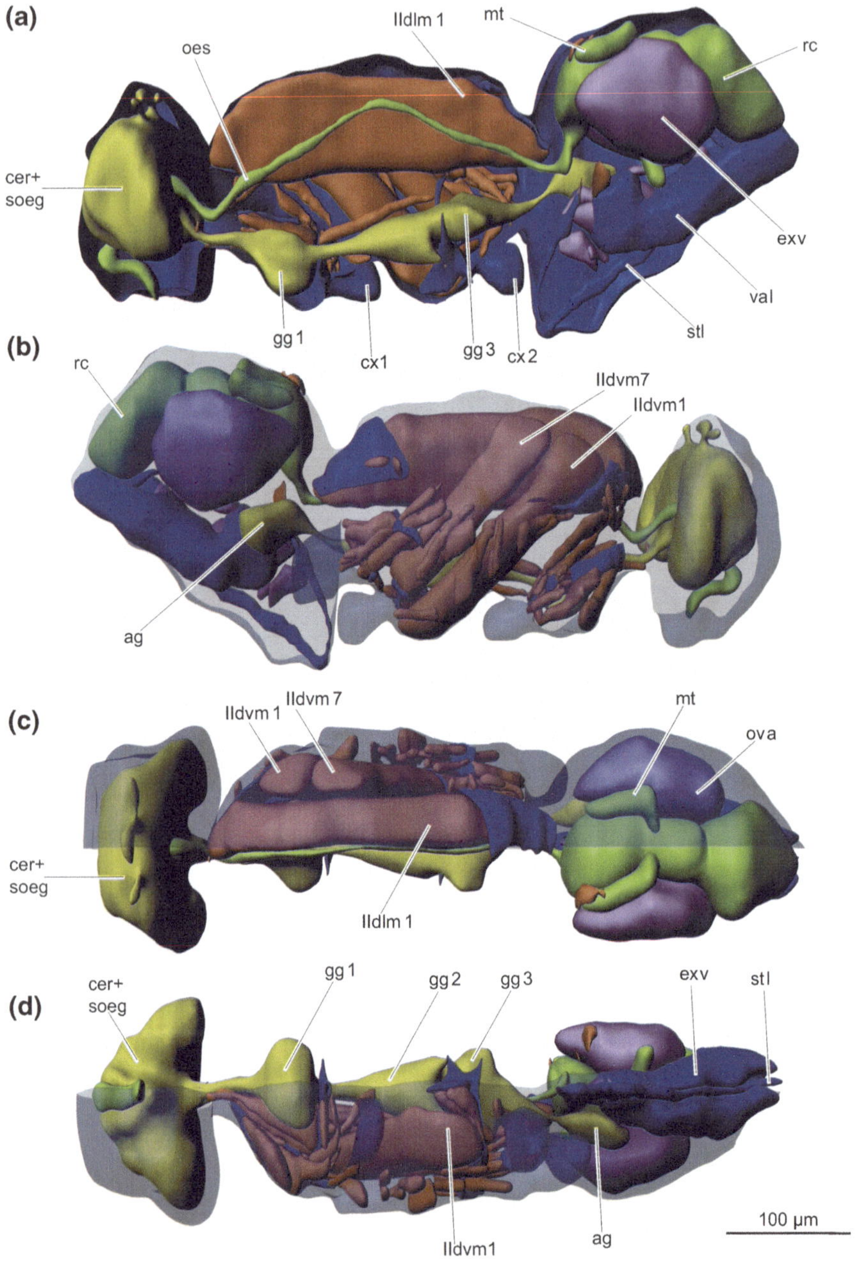

Fig. 7.10 Internal morphology of *Anaphes flavipes*, 3D: **a** lateral internal view; **b** lateral external view; **c** dorsal view; **d** ventral view; *ag* abdominal ganglion, *cer* cerebrum, *exv* external valves of ovipositor, *gg1.2.3* pro-, meso-, and metathoracic ganglia, *mg* midgut, *mt* Malpighian tubules, *oes* oesophagus, *ova* ovary, *rc* rectum, *soeg* suboesophageal ganglion, *stl* stylet of ovipositor. Colors: *blue* cuticle, *green* digestive system, *yellow* central nervous system, *brown* musculature, *purple* reproductive system. Musculature see text

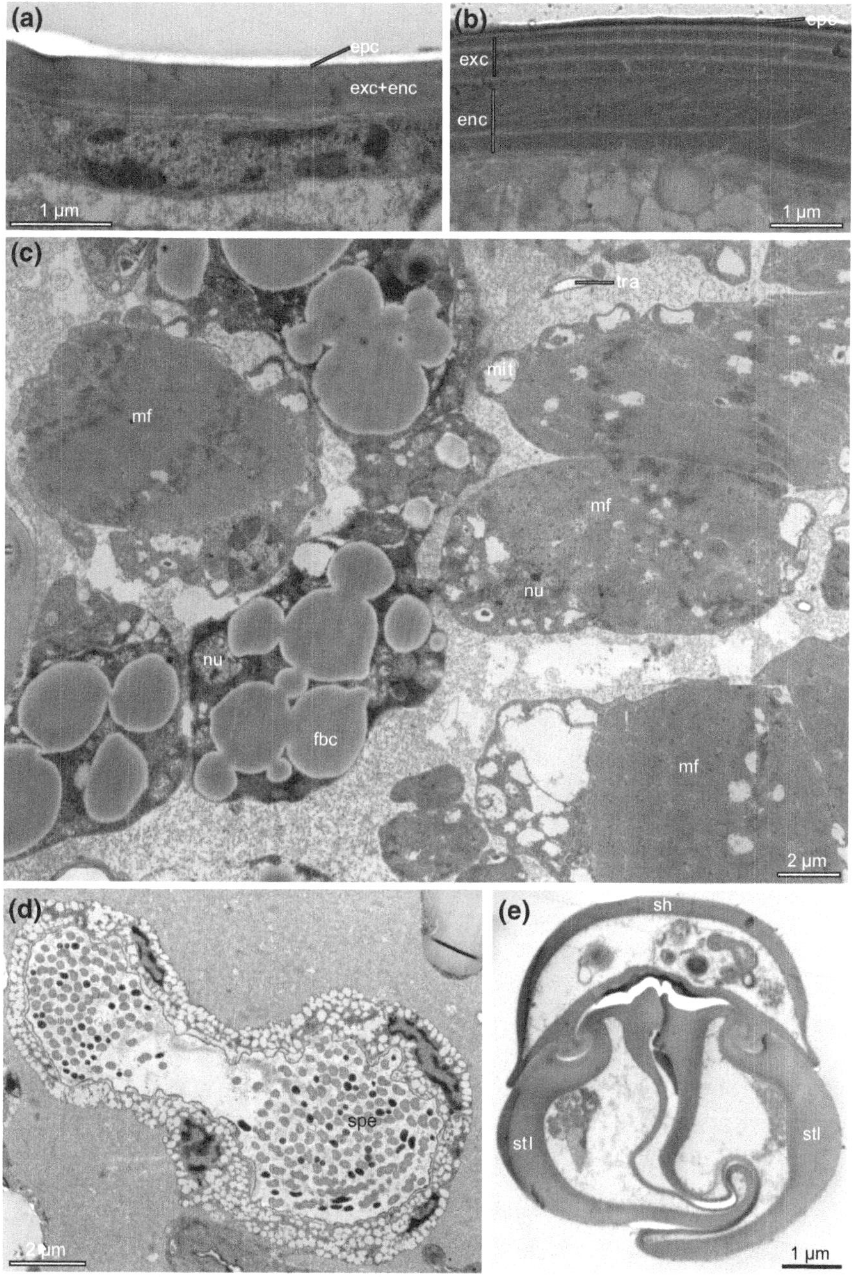

Fig. 7.11 Ultrastructure of *Anaphes flavipes*, TEM: **a**, **b** integument, mesothorax; **c** fat body and musculature, mesothorax; **d** spermatheca; **e** ovipositor; *enc* endocuticle, *epc* epicuticle, *exc* exocuticle, *fbc* cell of fat body, *mf* muscle fiber, *mit* mitochondrion, *nu* nucleus, *sh* sheath of ovipositor, *spe* spermatozoon, *stl* stylet of ovipositor, *tra* trachea

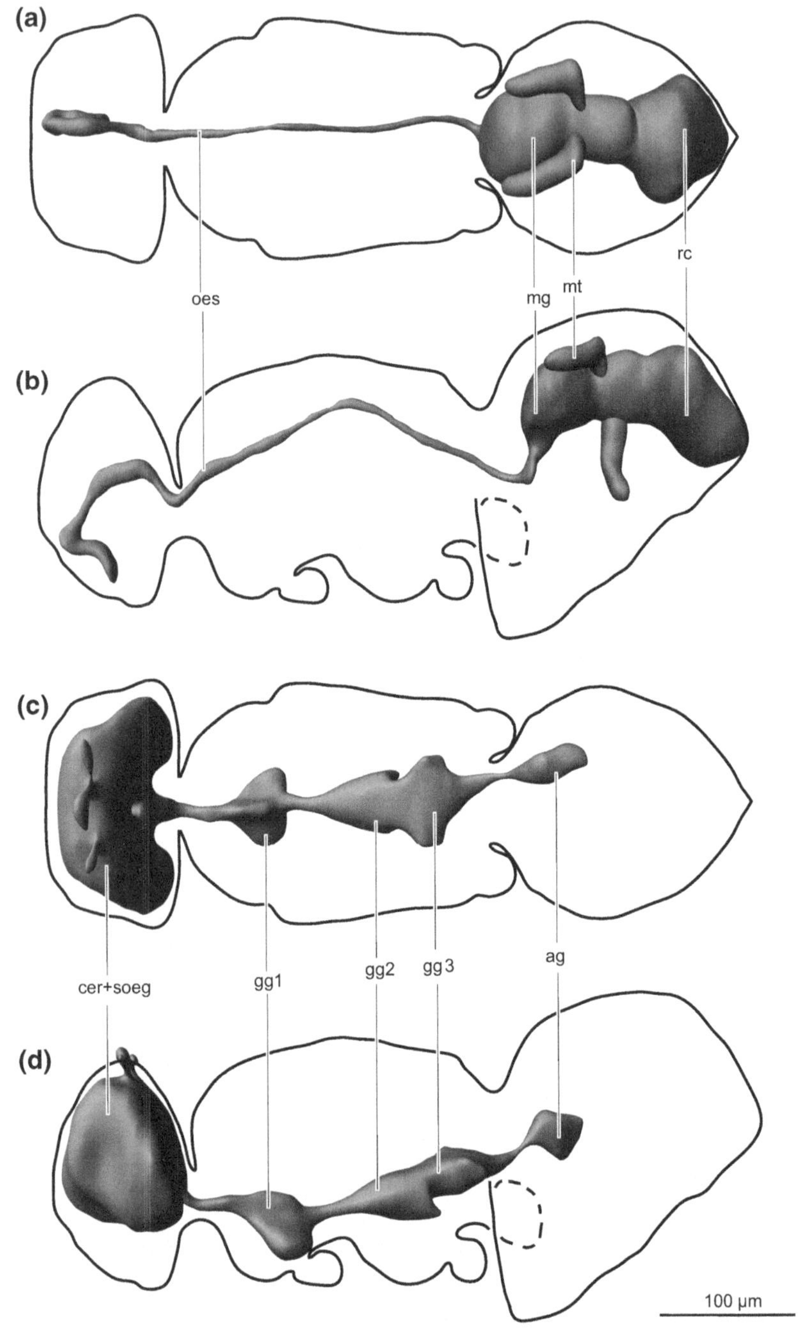

Fig. 7.12 Internal morphology of *Anaphes flavipes*, 3D: **a**, **b** intestine and Malpighian tubules; **c**, **d** central nervous system; **a**, **c** dorsal view; **b**, **d** lateral view; *ag* abdominal ganglion, *cer* cerebrum, *gg1.2.3* pro-, meso-, and metathoracic ganglia, *mg* midgut, *mt* Malpighian tubules, *oes* oesophagus, *rc* rectum, *soeg* suboesophageal ganglion

Midgut short, wide, in females situated dorsomedially in posterior half of metasoma, in males shifted right. Walls of midgut formed by strongly flattened cells. Muscles of midgut not found, distinguishing mymarids from other chalcidoids (in which layer of circular muscles present). Peritrophic membrane not found.

Hindgut short, divided into small intestine and rectum.

Boundary between mid- and hindgut bearing three Malpighian tubules, shaped as short tubules (in larger chalcidoids six Malpighian tubules present).

Circulatory System and Fat Body

Circulatory system strongly simplified and represented by rather small formation shaped as tube in metasoma, not traceable into mesosoma and probably representing reduced heart. Other vessels not found, probably because of diffusion sufficing for transport of molecules in organisms of such small sizes. Fat body occupying almost all cavities between organs in metasoma and to a smaller degree in mesosoma (Fig. 7.11c).

Tracheal System

Tracheal system strongly simplified. Only few weakly branching tracheae present, connected to mesosomal and metasomal spiracles. Transverse stems and air sacs absent, probably because of diffusion sufficing for oxygen transport in organisms of such small sizes. Tracheae with structure typical of insects, consisting of hypoderm and intima, intima with helical thickenings—taenidia.

Nervous System

In contrast to Coleoptera, in Hymenoptera, because of their very mobile head and narrow neck, head ganglia not shifted into thorax (Fig. 7.12c, d). Cerebrum and suboesophageal ganglion fused into one head synganglion occupying almost all cranium. Ganglia of thoracic neural chain separated. Mesothoracic and metathoracic ganglia closely set. Abdominal ganglia fused into one synganglion, abdominal synganglion in males situated medially, in females shifted right from body axis.

Ensheathment of standard structure, typical of ganglionic sheaths of insects. Extraneural sheath strongly pronounced, in contrast to that of above-described smallest coleopterans, in *Anaphes flavipes* represented by thick, almost continuous layer of cells of fat body, 2–4 μm thick. Thickness of neural lamella 0.3–0.5 μm. Thickness of perineurium 0.2–1 μm. As in smallest coleopterans, cells of nervous system displaying markedly reduced volume of cytoplasm. Nuclei of cells in nervous system rounded, occupying about 70–80 % of area of neurons. Chromatin strongly compacted. In spite of strongly reduced volume of cytoplasm, cellular organelles distinguishable in cell bodies of neurons at cellular level. Flattened, densely packed cisternae of granular endopasmic reticulum discernible, with accumulations of ribosomes between them. Large mitochondria, Golgi apparatus, lysosome clearly discernible. Cytoplasm of large axons displaying mitochondria, neurofilaments, microtubules, and some cisternae of agranular endoplasmic reticulum (Fig. 7.13). Neuropil with distinguishable surface, cortical, neuropil glia. Neuropil containing numerous synapses of longitudinal, transverse, and divergent

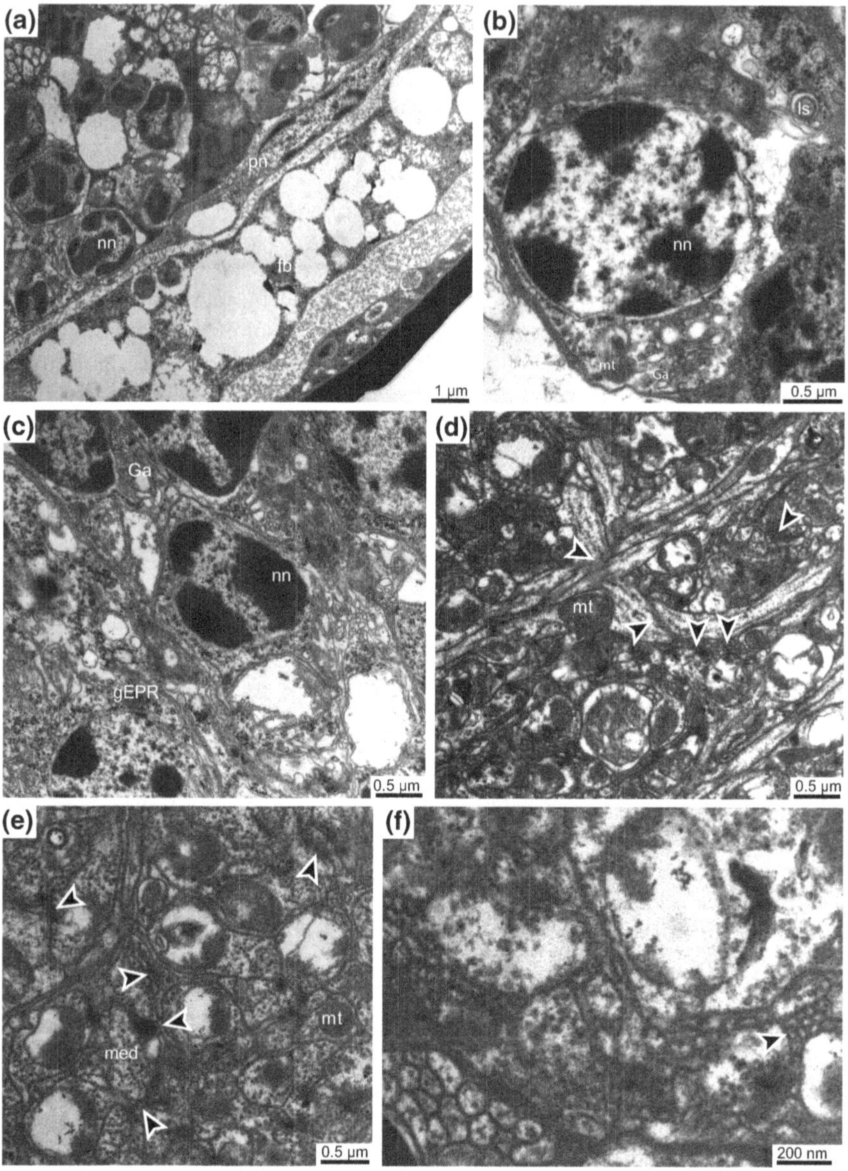

Fig. 7.13 Ultrastructure of cerebrum in *Anaphes flavipes*, TEM: *fb* extraneural layer of ensheathment, *Ga* Golgi apparatus, *gEPR* granular endoplasmic reticulum, *ls* lysosomes, *med* granules of mediator, *mt* mitochondrion, *nn* nucleus, *pn* perineurium, *syn* synaptic contact. **d**, **e** *arrows* indicate synaptic contacts; **f** *arrow* indicates group of fibers of smallest diameter

types. Vast majority of synaptic contacts represented by chemical synapses of asymmetrical type. Minimum diameter of projections in *A. flavipes* about 0.03 μm.

Cell size in cerebrum of *A. flavipes* 1.31–3.14 μm (M = 2.03, n = 494). Cerebrum of *A. flavipes* containing about 17,000 cells.

Cerebrum appearing generally more compacted than in Coleoptera. All neuropilar structures of cerebrum closely set.

Most volume of cerebrum occupied by protocerebrum. Central body complex occupying medial position and represented by two main compartments of neuropils: large central body and protocerebral bridge (Fig. 7.14). Central body differentiated

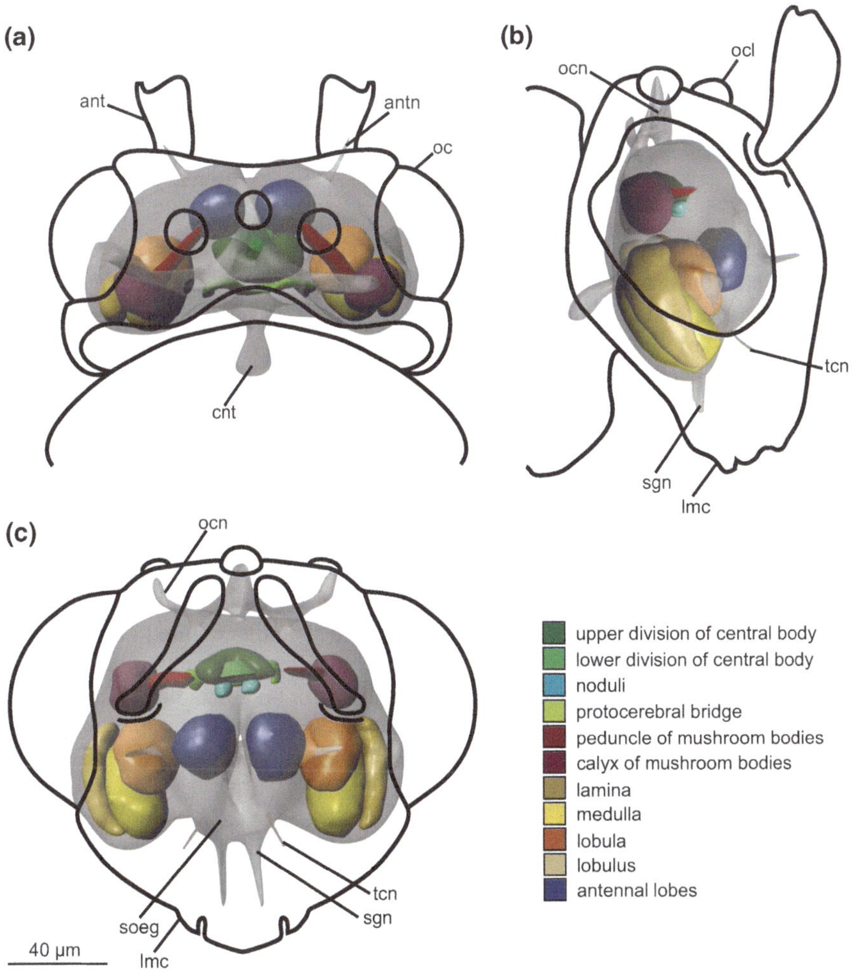

Fig. 7.14 3D reconstructions of cerebrum in *Anaphes flavipes* (modified from Makarova and Polilov 2013): **a** dorsal view; **b** lateral view; *in* frontal view; *ant* antennae, *antn* antennal nerves, *cnt* connectives linking suboesophageal and prothoracic ganglia, *oc* compound eyes, *ocl* ocelli, *ocn* ocellar nerves, *lmc* mouthparts, *sgn* nerves of suboesophageal ganglion, innervating mouthparts, *soeg* suboesophageal ganglion, *tcn* paired nerves of tritocerebrum

into dorsal region, consisting of dorsal arch and fan-shaped body, ventral region (ellipsoid body), and noduli.

Mushroom bodies situated on both sides of central body. Mushroom bodies strongly differentiated into calyx and peduncle. Calyx of double structure.

Optic lobes represented by three neuropils. Lamina elongate, sickle-shaped. Lobular complex differentiated into lobula and lobular plate. Optic neuropils closely set and tightly adjoining each other. Optic lobes S-shaped, running from dorsomedial part of cerebrum to eye. Parietal part of protocerebrum with strongly pronounced ocellar ganglia producing fibers running to ocelli.

Deutocerebrum with discernible large antennal lobes of rounded shape and glomerular structure. Diameter if glomerules about 4–5 μm. Antennal lobes containing about 160 glomerules.

Tritocerebrum smaller than any other part of cerebrum and similar in structure to that of smallest coleopterans, described above.

Elements of stomatogastric system not found, probably fused with central nervous system.

Muscular System

Structure of muscular system in mymarids, in spite of their small size, complying with general plan typical of insects. Musculature has been studied in *Anaphes flavipes* (Polilov 2016); musculature of the mesosoma was studied in *Gonatocerus morrilli* (Howard 1908) (Vilhelmsen et al. 2010) (Table 7.1).

Musculature of head (Fig. 7.15). 0an1 (M. tentorioscapalis anterior): O, anterior tentorial arms; I, anterior margin of base of scape. 0an2 (M. tentorioscapalis posterior): O, anterior tentorial arms, I, posterior margin of base of scape. 0an3 (M. tentorioscapalis lateralis): O, anterior tentorial arms; I, lateral margin of base of scape. Two muscles of unclear homology, possibly, modified subunits 0an4 (M. tentorioscapalis medialis): Mh1. O, anterior part of gena; I, lateral margin of base of scape. Mh2. O, anterior part of frons; I, anterior margin of base of scape. 0lb2 (M. frontoepipharyngalis): O, lateral part of frons; I, lateral part of epipharynx. 0md1 (M. craniomandibularis internus) consisting of two subunits: O, first, lateral part of cranium, second, gular zone, I, adductor of mandible. 0md3 (M. craniomandibularis externus): O, lateral part of cranium; I, abductor of mandible. 0md4 (M. hypopharyngomandibularis): O, anterior tentorial arms; I, adductor of mandible. 0mx1 (M. craniocardinalis): O, posterior part of cranium; I, ventrolateral part of cardo. 0mx3 (M. tentoriocardinalis): O, anterior tentorial arms; I, cardo. 0mx4 (M. tentoriostipitalis anterior): O, anterior tentorial arms; I, base of stipes. 0la5 (M. tentoriopraementalis): O, anterior tentorial arms; I, posterior margin prementum. 0la6 (M. tentorioparaglossalis): O, anterior tentorial arms; I, anterior margin prementum. 0hy1 (M. frontooralis): O, frons; I, posterior margin of epipharynx. 0hy3 (M. craniohypopharyngealis): O, anterior tentorial arms; I, anterior margin of epipharynx. 0ci1 (M. clypeopalatalis): O, clypeus; I, dorsal part of epipharynx. 0bu2 (M. frontobuccalis anterior): O, frons; I, dorsolateral part of pharynx. 0bu3

Table 7.1 Musculature of adult Chalcidoidea and Ichneumonoidea

	Head muscles																			
	0an1	0an2	0an3	0an4	0lb2	0md1	0md3	0md4	0mx1	0mx3	0mx4	0la5	0la6	0hy1	0hy3	0ci1	0bu2	0bu3	0ph1	0st1
Megaphragma	+	+	+	+	–	+	+	+	+	+	+	+	+	+	+	+	+	+	–	+
Trichogramma	+	+	+	+	–	+	+	+	+	+	+	+	+	+	+	+	+	+	+	+
Anaphes	+	+	+	+	+	+	+	+	+	+	+	+	+	+	+	+	+	+	+	+
Hemiptarsenus	+	+	+	+	+	+	+	+	+	+	+	+	+	+	+	+	+	+	+	+

	Mesosoma muscles											
	Idlm1	Idlm2	Idlm5	Idvm2	Idvm5	Idvm6	Idvm7	Idvm9	Idvm18	Itpm2	Itpm3	Itpm4–5
Megaphragma	–	–	+	+	a	a	+	+	+	+	+	+
Trichogramma	–	–	+	+	a	a	+	+	+	+	+	+
Anaphes	–	–	+	+	a	a	+	+	+	+	+	+
Gonatocerus	?	?	+	+	a	a	+	+	+	+	+	+
Hemiptarsenus	–	+	+	+	a	a	+	+	+	+	+	+
Megastigmus	?	?	+	+	a	a	+	+	+	+	+	+
Spalangia	?	?	+	+	a	a	+	+	+	+	+	+
Doryctes	+	?	?	+	a	a	+	+	+	–	+	+
Urosigalphus	+	?	?	+	a	a	+	+	+	+	+	+

	Mesosoma muscles											
	Ipcm2	Ipcm3	Ipcm4	Ipcm8	Ivlm1	Ivlm3	Ivlm7	Iscm1	Iscm2	Iscm3	Iscm5	Iscm6
Megaphragma	–	+	+	+	+	+	+	+	+	+	+	–
Trichogramma	–	+	+	+	+	+	+	+	+	+	+	–
Anaphes	–	+	+	+	+	+	+	+	+	+	+	–
Gonatocerus	+	+	+	–	+	+	+	?	+	+	+	–
Hemiptarsenus	+	+	+	+	+	+	+	+	+	+	+	–

(continued)

Table 7.1 (continued)

	Mesosoma muscles											
	Ipcm2	Ipcm3	Ipcm4	Ipcm8	Ivlm1	Ivlm3	Ivlm7	Iscm1	Iscm2	Iscm3	Iscm5	Iscm6
Megastigmus	+	+	+	+	+	+	+	?	+	+	+	–
Spalangia	+	+	+	+	+	+	+	?	+	+	+	–
Doryctes	+	+	+	+	+	+	–	?	+	+	+	+
Urosigalphus	+	+	+	+	+	+	–	?	+	+	+	+

	IIdlm1	IIdvm1	IIdvm6	IIdvm7	IIdvm8	IItpm2	IItpm4	IItpm6	IItpm7	IItpm9	IItpm10	IItpm11
Megaphragma	+	–	–	+	+	+	b	b	–	+	–	–
Trichogramma	+	–	–	+	+	+	b	b	–	+	+	+
Anaphes	+	+	–	+	+	+	b	b	–	+	+	+
Gonatocerus	+	+	–	+	+	+	b	b	+	+	?	+
Hemiptarsenus	+	+	–	+	+	+	+	+	–	+	+	+
Megastigmus	+	+	–	+	+	+	b	b	+	+	?	+
Spalangia	+	+	+	+	+	+	b	b	+	+	?	+
Doryctes	+	+	+	–	+	?	b	b	+	+	?	+
Urosigalphus	+	+	+	–	+	?	b	b	+	+	?	+

	IIppm1	IIppm2	IIspm1	IIspm2	IIpcm4	IIvlm3	IIscm1	IIscm2	IIscm3	IIscm6	V54	V55
Megaphragma	–	+	–	+	+	+?	+	+	+	–	–	–
Trichogramma	–	+	+	+	+	+	+	+	+	–	–	–
Anaphes	–	+	+	+	+	+	+	+	+	–	–	–
Gonatocerus	–	+	+	+	+	+	?	+	+	–	–	+
Hemiptarsenus	–	+	+	+	+	+	+	+	+	–	–	–
Megastigmus	–	+	+	+	+	+	?	+	+	–	–	+

(continued)

Table 7.1 (continued)

	IIppm1	IIppm2	IIspm1	IIspm2	IIpcm4	IIvlm3	IIscm1	IIscm2	IIscm3	IIscm6	V54	V55
Spalangia	−	+	+	+	+	+	?	+	+	−	−	+
Doryctes	+	+	+	+	+	+	?	+	+	+	−	+
Urosigalphus	+	+	+	+	+	+	?	+	+	+	+	+

	IIIdlm1	IIIdvm2	IIIdvm6	IIItpm3	IIItpm5	IIItpm6	IIItpm7	IIItpm9	IIItpm11
Megaphragma	+	+	−	−	c	c	+	+	+
Trichogramma	+	+	−	−	c	c	+	+	+
Anaphes	+	+	−	−	c	c	+	+	+
Gonatocerus	+	+	−	−	+	+	+	+	+
Hemiptarsenus	+	+	−	−	+	+	+	+	+
Megastigmus	+	+	−	−	+	+	+	+	+
Spalangia	+	+	−	−	+	+	+	+	+
Doryctes	+	−	+	+	+	+	+	+	+
Urosigalphus	+	−	+	+	+	+	+	+	+

	IIIppm1	IIIspm1	IIIpcm3	IIIpcm4	IIIpcm6	IIIvlm2	IIIscm1	IIIscm2	IIIscm5	IIIscm6	V118	mx
Megaphragma	−	+	+	+	+	−	+	−	−	−	+	+
Trichogramma	−	+	+	+	+	+	+	+	+	+	+	−
Anaphes	−	+	+	+	+	+	+	+	+	+	+	−
Gonatocerus	−	+	?	+	+	+	+	+	?	+	+	−
Hemiptarsenus	−	+	+	+	+	+	+	+	+	+	+	−
Megastigmus	−	+	?	+	+	+	+	+	?	+	+	−
Spalangia	−	+	?	+	+	+	+	+	?	+	+	−
Doryctes	+	+	?	+	+	+	+	+	?	+	+	−
Urosigalphus	+	+	?	+	+	+	+	+	?	+	+	−

Partly from Vilhelmsen et al. 2010

+ present; − absent; *a*, *b* fused muscles, ? unknown

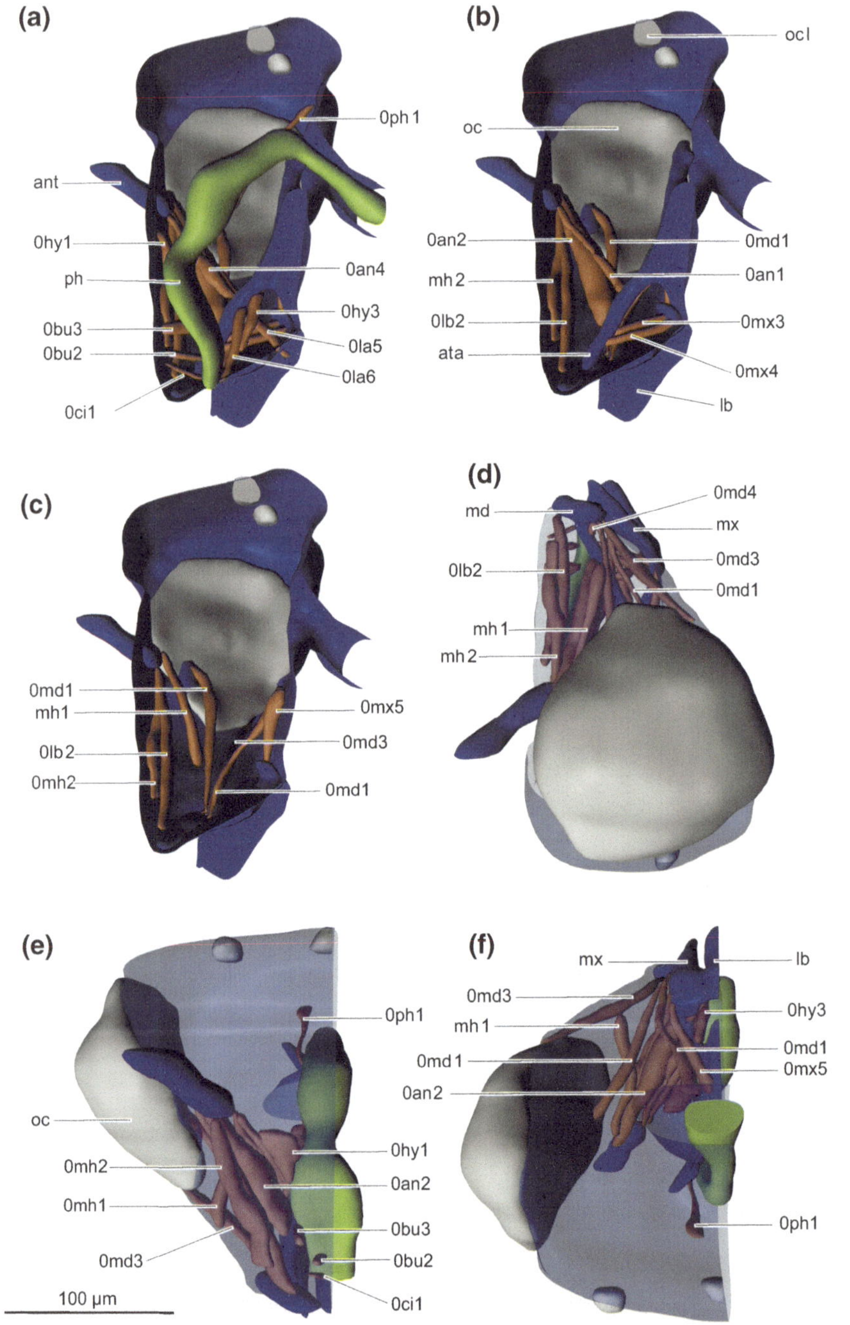

Fig. 7.15 Musculature of head in *Anaphes flavipes*, 3D: **a–c** lateral internal view; **d** lateral external view; **e** dorsal view; **f** ventral view; *ant* antenna, *ata* anterior tentorial arms, *lb* labium, *md* mandible, *mx* maxilla, *oc* eye, *ocl* ocellus, *ph* pharynx. Musculature see text

(M. frontobuccalis posterior): O, frons; I, dorsal part of pharynx. 0ph1 (M. verticopharyngealis): O, occiput; I, ventral wall of pharynx. 0st1 (M. annularis stomodaei), transverse musculature, developed only on dorsal surface of pharynx. Internal musculature of antennae and mouthparts not studied because of extremely small size.

Musculature of mesosoma (Fig. 7.16). **Prothorax**. Idlm1 (M. prophragma-occipitalis) and Idlm2 (M. pronoto-occipitalis) absent in *Anaphes*, no reliable data available about *Gonatocerus*. Idlm5 (M. pronoto-phragmalis anterior): O, prophragma; I, medial part of pronotum. Idvm2 (M. cervico-occipitalis medialis): O, cervical region; I, occipital region. Idvm5 (M. pronoto-cervicalis anterior) and Idvm6 (M. pronoto-cervicalis medialis) fused: O, anterolateral part of notum; I, cervical region. Idvm9 (M. profurca-occipitalis): O, profurca; I, occipital region. Idvm18 (M. pronoto-coxalis lateralis): O, lateral part of notum; I, lateral margin of base of coxa. Itpm2 (M. propleuro-occipitalis): O, cervical region; I, anterior part of pleurite of prothorax. Itpm3 (M. pronoto-pleuralis anterior): O, lateral part of notum; I, pleurite of prothorax. Itpm4-5 (M. pronoto-apodemalis): O, lateral part of pronotum; I, pleurite of prothorax. Ipcm4 (M. propleuro-coxalis superior): O, pleural apodeme of prothorax; I, anterior margin of base of coxa. Ipcm8 (M. propleuro-trochanteralis) present in *Anaphes*: O, pleural apodeme of prothorax; I, trochanter, via fine tendon, in *Gonatocerus* absent. Ivlm1 (M. profurca-cervicalis): O, profurca; I, cervical region. Ivlm3 (M. profurca-tentorialis): O, profurca; I, postoccipital region. Ivlm7 (M. profurca-mesofurcalis): O, profurca; I, mesofurca. Iscm1 (M. profurca-coxalis anterior): O, profurca; I, anterior margin of base of coxa. Iscm2 (M. profurca-coxalis posterior): O, profurca; I, posterior margin of base of coxa. Iscm3 (M. profurca-coxalis medialis): O, profurca; I, medial margin of base of coxa. Iscm4 (M. profurca-coxalis lateralis): O, profurca; I, lateral margin of base of coxa. Iscm5 (M. prospina-coxalis): O, fold between pro- and metathorax; I, posterior margin of base of coxa. Internal muscles of legs not studied because of extremely small size.

Mesothorax. IIdlm1 (M. prophragma-mesophragmalis), largest muscle: O, prophragma, I, mesophragma. IIdvm1 (M. mesonoto-sternalis): O, mesonotum; I, mesopleurite. IIdvm5 (M. mesonoto-coxalis posterior): O, mesopleurite; I, anterior margin of base of coxa. IIdvm7 (M. mesonoto-trochanteralis): O, mesonotum; I, apodeme of trochanter. IIdvm8 (M. mesofurca-phragmalis): O, mesofurca; I, mesophragma. IItpm2 (M. mesopleura-praealaris): O, mesopleurite; I, prealar zone. IItpm4 (M. mesonoto-pleuralis anterior): O, mesopleurite; I, third axillary plate. IItpm6 (M. mesonoto-pleuralis posterior) in *Anaphes* fused with IItpm4. In *Gonatocerus*: O, mesopleurite; I, mesonotum. IItpm7 (M. mesanepisterno-axillaris) in *Anaphes* not found, in *Gonatocerus* present: O, mesepisternum; I, axillary plate. IItpm9 (M. mesepimero-axillaris tertius): O, mesopleurite; I, third axillary plate. IItpm11 (M. mesopleura-subalaris): O, mesopleurite; I, subalare. IIppm2 (M. mesobasalare-intersegmentalis): O, intersegmental juncture; I, basalare. IIspm1 (M. mesopleura-sternalis): O, mesopleurite; I, basalare. IIspm2 (M. mesofurca-pleuralis): O, apex of mesofurca, I, mesopleurite. IIvlm3 (M. mesofurca-metafurcalis): O, mesofurca; I, metafurca. IIscm1 (M. mesofurca-coxalis anterior): O, mesofurca; I, anterior margin

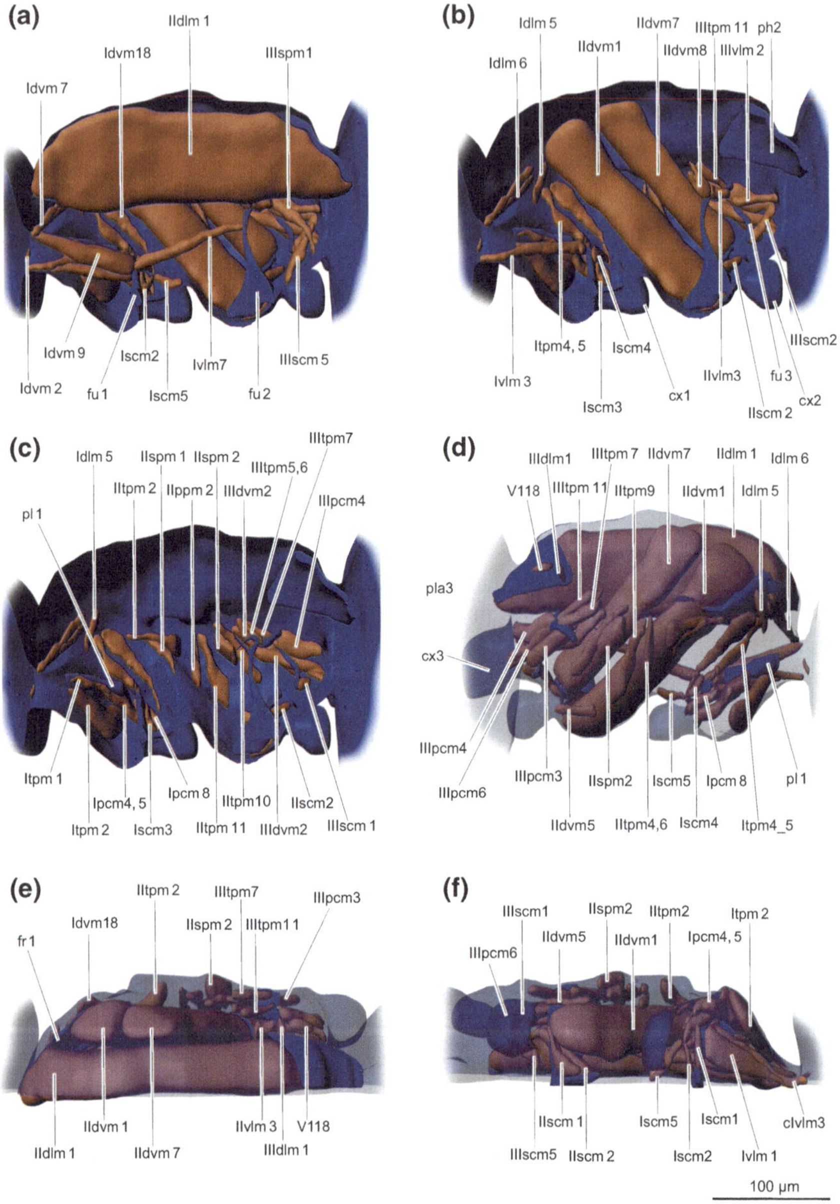

Fig. 7.16 Musculature of mesosoma in *Anaphes flavipes*, 3D: **a–c** lateral internal view; **d** lateral external view; **e** dorsal view; **f** ventral view; *pl1* propleurite, *pla3* pleural apodeme of metathorax, *fr1.2.3* pro-, meso-, and metaphragma, *fu1.2.3* pro-, meso-, and metafurca, *cx1.2.3* pro-, meso-, and metacoxae. Musculature see text

of base of coxa. IIscm2 (M. mesofurca-coxalis posterior): O, mesofurca; I, postero-lateral margin of base of coxa. IIscm3 (M. mesofurca-coxalis medialis): O, mesofurca; I, medial margin of base of coxa. Muscle V55, described only in hymenopterans (Vilhemsen et al. 2010: no. 55, t2p–t3, M. posterior mesonoto-metanotal) in *Anaphes* not found, in *Gonatocerus* present. Internal muscles of legs not studied because of extremely small size.

Metathorax and propodeum. IIIdlm1 (M. mesophragma-metaphragmalis): O, mesophragma; I, metaphragma. IIIdvm1 (M. metanoto-sternalis): O, anterior part of metanotum and lateral part of mesophragma; I, ventral part of pleurite. IIIdvm2 (M. metanoto-trochantinalis anterior): O, metanotum; I, trochanter. IIItpm5 (M. metanoto-pleuralis medialis) and M92 IIItpm6 (M. metanoto-pleuralis posterior) in *Anaphes* fused, in *Gonatocerus* two sparate muscles with identical O and I: O, pleural apodeme of metathorax; I, lateral margin of metanotum. IIItpm7 (M. metanepisterno-axillaris): O, pleural apodeme; I, third axillary plate. IIItpm9 (M. metepimero-axillaris tertius): O, pleural apodeme of metathorax; I, third axillary plate. IIItpm11 (M. metapleura-subalaris): O, pleural apodeme; I, subalare. IIIspm1 (M. metapleura-sternalis): O, ventral part of propodeum; I, basalare. IIIspm2 (M. metafurca-pleuralis) in *Anaphes* absent, no reliable data available about *Gonatocerus*. IIIpcm4 (M. metanepisterno-coxalis posterior): O, pleural apodeme; I, anterolateral margin of base of coxa. IIIpcm6 (M. metapleura-trochanteralis): O, pleural apodeme; I, trochanter. IIIvlm2 (M. metafurca-abdominosternalis): O, metafurca; I, second sternite of abdomen. IIIscm1 (M. metafurca-coxalis anterior): O, metafurca; I, anterior margin of base of coxa. IIIscm2 (M. metafurca-coxalis posterior): O, metafurca; I, posterior margin of base of coxa. IIIscm5 (M. metaspina-coxalis): O, fold between meso- and metasoma; I, posterior margin of base of coxa. IIIscm6 (M. metafurca-trochanteralis): O, metafurca; I, trochanter, via fine tendon. Muscle V118, described only in hymenopterans (Vilhemsen et al. 2010: no. 118, ph3–T2. M. metaphragma-second abdominal tergal): O, metaphragma; I, tergite of abdominal segment 2. Internal muscles of legs not studied because of extremely small size.

Musculature of metasoma. Dorsal longitudinal muscles (Mm. dorsales), several parallel fibers: O, anterior phragma; I, posterior phragma. Ventral longitudinal muscles (Mm. ventrales), several parallel fibers: O, anterior margin of segment, I, posterior margin of segment. Dorsoventral muscles, several of urotergosternal muscles, homology unknown: O, lateral part of tergite; I, lateral part of sternite.

Reproductive System

In males consisting of paired testes, spermiducts, accessory glands, and copulatory apparatus (Fig. 7.17c, d). Right testis considerably smaller than left testis.

In females consisting of paired ovaries, oviducts, spermatheca: accessory glands and of ovipositor (Fig. 7.17a, b). Each ovary consisting of three polytrophic ovarioles.

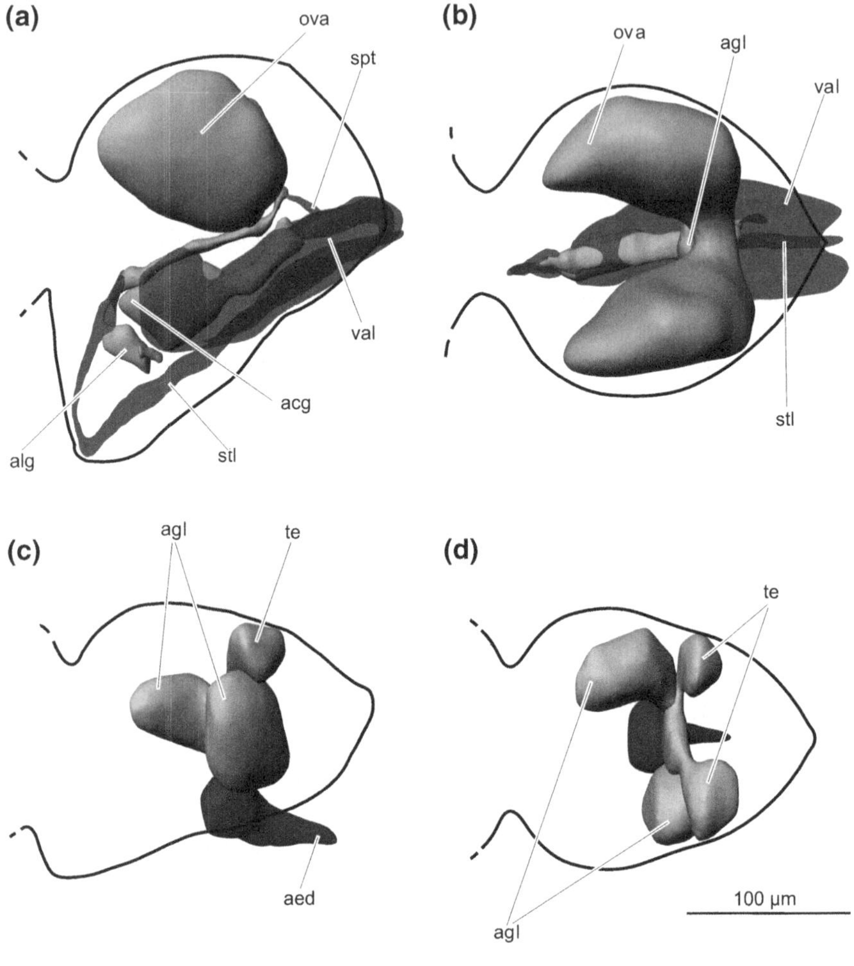

Fig. 7.17 Reproductive system of *Anaphes flavipes*, 3D: **a**, **b** female; **c**, **d** male; **a**, **c** lateral view; **b**, **d** dorsal view; *acg* acid gland, *aed* aedeagus, *agl* accessory glands, *alg* alkaline gland, *ova* ovary, *spt* spermatheca, *stl* stylet, *te* testis, *val* valves of ovipositor

7.4 Morphology of Larvae

Larvae of Mymaridae lack legs, mouthparts (sometimes except for mandibles or primordial of mandibles), antennae, or eyes. The length of first instar larva in *Anaphes* is 145–195 µm; the length of last instar larva is 814 µm (Boivin et al. 1993; van Baaren et al. 1997). Four morphological types of first instar larvae are recognized (Bakkendorf 1934; Jackson 1961): sacciform, typical of *Anagrus* (without pronounced segmentation or with weakly pronounced segmentation; head distinguished; tufts of setae absent); mymariform, typical of *Anaphes*, *Polynema*, and *Enasius* (body C-shaped, segmented; head fully formed, with frontal process; with long caudal process and tufts of setae); simplified mymariform, typical of *Ooctonus*, *Gonatocerus*, *Alaptus*, and *Erythmelus* (body straight, segmented; frontal process and caudal process rather short; tufts of setae absent); and elongate, typical of *Caraphractus* (body elongate, cylindrical, without visible segmentation). Larvae of later instars are sacciform or cylindrical, some occasionally described as 'histriobdellid' (*Anagrus*, Jackson 1961), without distinct cranium and without segmentation. Inner organs include only closed intestine (Balduf 1928; Bakkendorf 1934; Jackson 1961), salivary glands, present in most (Bakkendorf 1934) or absent (*Caraphractus*, Jackson 1961), and primordia of central nervous system (Ivanova-Kazas 1961).

References

Bakkendorf, O. (1934). Biological investigations on some Danish hymenopterous egg-parasites, especially in homopterous and heteropterous eggs, with taxonomic remarks and descriptions of new species. *Entomologiske Meddelelser, 19*, 1–134.

Balduf, W. V. (1928). Observations on the buffalo tree-hopper *Ceresa bubalus* Fabr. (Membracidae, Homoptera) and the bionomics of an egg parasite, *Polynema striaticorne* Girault (Mymaridae, Hymenoptera). *Annals of the Entomological Society of America, 21*(3), 419–435.

Boivin, G., Picard, C., & Auclair, J. L. (1993). Preimaginal development of *Anaphes* n.sp. (Hymenoptera: Mymaridae), an egg parasitoid of the carrot weevil (Coleoptera: Curculionidae). *Biological Control, 3*(3), 176–181.

Boivin, G. (2010). Reproduction and immature development of egg parasitoids. In F. L. Consoli, J. R. P. Parra, & R. A. Zucchi (Eds.), *Egg parasitoids in agroecosystems with emphasis on trichogramma* (pp. 1–23). Houten: Springer.

Bradley, J. C. (1955). The wing-venation of Chalcidoidea and of some allied Hymenoptera. *Memoires de la Societe Entomologique de Belgique, 27*, 127–137.

Bucher, G. (1948). The anatomy of *Monodontomerus dentipes* Boh., an entomophagus chalcid. *Canadian Journal of Research, 26*, 230–281.

Burks, B. D. (1938). A study of chalcidoid wing. *Annals of the Entomological Society of America, 31*, 157–160.

Chiappini, E., & Mazzoni, E. (2000). Differing morphology and ultrastructure of the male copulatory apparatus in species-groups of *Anagrus* Haliday (Hymenoptera: Mymaridae). *Journal of Natural History, 34*(8), 1661–1676.

Cutler, J. R. (1955). The morphology of the head of the final instar larva of *Nasonia vitripennis* Walker (Hymenoptera: Chalcidoidea). *Proceedings of the Royal Entomological Society of London Series A General Entomology, 30*(4–6), 73–81.

Debauche, H. R. (1948). Étude sur les Mymarommidae et les Mymaridae de la Belgique (Hymenoptera: Chalcidoidea). *Mémoires du Musée royal d'histoire naturelle de Belgique, 108*, 1–248.

Gibson, G. A. P. (1985). Some pro-and mesothoracic structures important for phylogenetic analysis of Hymenoptera, with a review of terms used for the structures. *Canadian Entomologist, 117*, 1395–1443.

Gibson, G. A. P. (1986). Evidence for monophyly and relationships of Chalcidoidea, Mymaridae, and Mymarommatidae (Hymenoptera: Terebrantes). *Canadian Entomologist, 118*, 205–240.

Gibson, G. A. P. (1997). Chapter 2: Morphology and Terminology. In G. A. P. Gibson, J. T. Huber, & J. B. Woolley (Eds.), *Annotated Keys to the Genera of Nearctic Chalcidoidea (Hymenoptera)* (pp. 16–44). Ottawa, Canada: National Research Council of Canada Research Press.

Gibson, G. A. P. (1999). Sister-group relationship of the Platigastroidea and Chalcidoidea (Hymenoptera)—An alternate hypothesis to Rasnitsyn (1988). *Zoologica Scripta, 28*, 125–138.

Gibson, G. A. P., Heraty, J. M., & Woolley, J. B. (1999). Phylogenetics and classification of Chalcidoidea and Mymarommatoidea—A review of current concepts (Hymenoptera, Apocrita). *Zoologica Scripta, 28*, 87–124.

Hanna, A. D. (1934). The male and female genitalia and the biology of *Euchalcidia caryobori* Hanna (Hymenoptera, Chalcidinae). *Transactions of the Royal Entomological Society of London, 82*, 107–136.

Hanna, A. D. (1935). The morphology and anatomy of *Euchalcidia caryobori* Hanna (Hymenoptera-Chalcidinae). *Bulletin de la Société Royale Entomologique d'Égypte, 19*, 326–361.

Heraty, J., Ronquist, F., Carpenter, J. M., et al. (2011). Evolution of the hymenopteran megaradiation. *Molecular Phylogenetics and Evolution, 60*, 73–88.

Huber, J. T. (1997). Review of the genus *Stephanodes* (Hymenoptera: Mymaridae). *Proceedings of the Entomological Society of Ontario, 128*, 27–63.

Huber, J. T., & Landry, J.-F. (1999). Cutio nanissimus incredibilis. Nouv'ailes: *Bulletin de nouvelles de l'Association des entomologistes amateurs du Québec 9*(3), 11.

Huber, J. T., & Noyes, J. (2013). A new genus and species of fairyfly, *Tinkerbella nana* (Hymenoptera, Mymaridae), with comments on its sister genus *Kikiki*, and discussion on small size limits in arthropods. *Journal of Hymenoptera Research, 32*, 17–44.

Ivanova-Kazas, O. M. (1954). Вопросы эволюции эмбрионального развития у перепончатокрылых (Hymenoptera) (On the evolution of embryonic development in Hymenoptera). *Trudy Vsesoyuznogo ehntomologicheskogo obshchestva, 44*, 301–335.

Ivanova-Kazas, O. M. (1961). *Очерки по сравнительной эмбриологии перепончатокрылых (Essays on the comparative embryology of Hymenoptera)*. Leningrad: Leningrad Univ. Press.

Jackson, D. J. (1961). Observations on the biology of *Caraphractus cinctus* Walker (Hymenoptera, Mymaridae), a parasitoid of the eggs of Dytiscidae. II. Immature stages and seasonal history with a review of mymarid larvae. *Parasitology, 51*, 269–294.

Jackson, D. J. (1969). Observation on the female reproductive organs and the poison apparatus of *Caraphractus cinctus* Walker (Hymenoptera: Mymaridae). *Zoological Journal of the Linnean Society, 48*, 59–81.

James, H. C. (1926). The anatomy of a British phytophagous chalcidoid of the genus *Harmolita* (Isosoma). *Proceedings of the Zoological Society of London, 96*(1), 75–182.

King, P. E., & Copland, M. J. W. (1969). The structure of the female reproductive system in the Mymaridae (Chalcidoidea: Hymenoptera. *Journal of Natural History, 3*, 349–365.

Makarova, A. A., & Polilov, A. A. (2013). Peculiarities of the brain organization and fine structure in small insects related to miniaturization. 2. The smallest Hymenoptera (Mymaridae, Trichogrammatidae). *Entomological Review, 93*(6), 714–724.

Mockford, E. L. (1997). A new species of *Dicopomorpha* (Hymenoptera: Mymaridae) with diminutive, apterous males. *Annals of the Entomological Society of America, 90*(2), 115–120.

Munro, J. B., Heraty, J. M., Burks, R. A., et al. (2011). A Molecular Phylogeny of the Chalcidoidea (Hymenoptera). *PLoS ONE, 6*(11), e27023.

Nikolskaya, M. N. (1952). Хальциды фауны СССР (Chalcidoidea) (Chalcidoids of the USSR). In *Opredeliteli po faune SSSR, izdavaemye Zoologicheskim institutom AN SSSR* (Vol. 44), Izdatelsvo Akademii nauk SSSR, Leningrad.

Noyes, J. S., & Valentine, E. W. (1989a). Mymaridae (Insecta: Hymenoptera)—Introduction, and review of genera. *Fauna of New Zealand, 17*, 1–95.

Noyes, J. S., & Valentine, E. W. (1989b). Chalcidoidea (Insecta: Hymenoptera)—Introduction, and review of genera in smaller families. *Fauna of New Zealand, 18*, 1–91.

Polilov, A. A. (2007). Морфологические особенности Mymaridae связанные с миниатюризацией (Miniaturization-related structural features of Mymaridae). *Studies on Hymenopterous insects: Collection of scientific papers* (pp. 50–64). Moscow: KMK.

Polilov, A. A. (2015). Small is beautiful: Features of the smallest insects and limits to miniaturization. *Annual Review of Entomology, 60*, 103–121.

Polilov, A. A. (2016). Features of the structure of hymenoptera associated with Miniaturization. 1. Anatomy of fairyfly *Anaphes flavipes* (Hymenoptera, Mymaridae). *Entomological Review, 94*(4), 419–431.

Rasnitsyn, A. P. (1980). Происхождение и эволюция перепончатокрылых насекомых (Origin and evolution of Hymenoptera). *Trudy Paleontologicheskogo instituta Akademii nauk SSSR, 174*, 1–192.

Rasnitsyn, A. P., & Brothers, D. J. (2007). Two new hymenopteran fossils from the mid-Cretaceous of southern Africa (Hymenoptera: Jurapriidae, Evaniidae). *African Invertebrates, 48*(1), 193–202.

Rasnitsyn, A. P., Basibuyuk, H. H., & Quicke, D. L. J. (2004). A basal chalcidoid (Insecta: Hymenoptera) from the earliest Cretaceous or latest Jurassic of Mongolia. *Insect Systematics & Evolution, 35*, 123–135.

Saakyan-Baranova, A. A., & Muzafarov, S. S. (1972). Строение, биология и взаимоотношения северного кермеса *Kermococcus quercus* L. (Homoptera, Kermococcidae) и его паразитов (Hymenoptera, Chalcidoidea): II. Гиперпаразиты и отношения северного кермеса и его паразитов (Morphology, biology, and relationship of the kermes *Kermococcus quercus* L. (Homoptera, Kermococcidae) and its parasites (Hymenoptera, Chalcidoidea): II. Hyperparasites and relationship of *Kermococcus quercus* and its parasites). *Entomologiceskoe obozrenie 51*(4), 697–715.

Sahad, K. A. (1982). Biology and morphology of *Gonatocerus* sp. (Hymenoptera, Mymaridae), an egg parsasitoid of the green rice leafhopper, *Nephotettix cincticeps* Uhler (Homoptera, Deltocephalidae). II. Morphology. *Kontyu, 50*(3), 467–476.

Sahad, K. A. (1984). Biology of *Anagrus optabilis* (Perkins) (Hymenoptera, Mymaridae), and egg parasitoid of delphacid planthoppers. *Esakia, 22*, 129–144.

van Baaren, J., Boivin, G., Le, Lannic J., & Nénon, J.-P. (1997). The male and female first instar larvae of *Anaphes victus* and A. listronoti (Hymenoptera, Mymaridae). *Zoomorphology, 117* (3), 189–197.

van Baaren, J., Boivin, G., & Le, Lannic J. (1999). Nénon J-P (1999), Comparison of antennal sensilla of Anaphes victus and A. listronoti (Hymenoptera, Mymaridae), egg parasitoids of Curculionidae. *Zoomorphology, 119*(1), 1–8.

Viggiani, G. (1988). A preliminary classification of the Mymaridae (Hymenoptera: Chalcidoidea) based on external male genitalic characters. *Bollettino del Laboratorio di Entomologia Agraria Filippo Silvestri, Portici, 45*, 141–148.

Vilhelmsen, L., Miko, I., & Krogmann, L. (2010). Beyond the wasp-waist: structural diversity and phylogenetic significance of the mesosoma in apocritan wasps (Insecta: Hymenoptera). *Zoological Journal of the Linnean Society, 159*(1), 22–194.

Chapter 8
Structure of the Principal Groups of Microinsects. VI. Trichogrammatid Wasps (Hymenoptera: Trichogrammatidae)

8.1 Introduction

Trichogrammatids are small insects distributed worldwide; their larvae develop in eggs of other insects. Trichogrammatids of the genus *Megaphragma* include some of the smallest insects and some of the smallest metazoans. Unique almost anucleate nervous system was recently described in *M. mymaripenne* (Polilov 2012), making this species especially interesting for studies of miniaturization in insects and miniaturization of the nervous system in animals in general. Since some trichogrammatids are widely used in biological methods of pest control, their biology and taxonomy are intensively studied (Nagarkatti and Nagaraja 1977). Much fewer studies treat the morphology of Trichogrammatidae. The external morphology of adults of many genera has been described in rather much detail (Chumakova 1966; Sorokina 1993; Pinto 2006). A separate large set of studies treats the sensillar armature of Trichogrammatidae (Voegelt et al. 1975; Schmidt and Smith 1985; Olson and Andow 1993; Amornsak et al. 1998; Ruschioni et al. 2012; Zhang et al. 2012). Only one detailed study treats the internal morphology of adult trichogrammatids, the species *Trichogramma evanescens* (Polilov 2016). Other available publications on the structure of trichogrammatids include studies on the anatomy of different *Trichogramma* species (Flanders 1937; Chumakova 1968; Boivin 2010) and of *Prestwichia aquatica* (Ivanova-Kazas 1952), descriptions of the structure of the cerebrum in *Megaphragma mymaripenne* (Polilov 2012) and *Trichogramma evanescens* (Makarova and Polilov 2013) and structure of the eyes (Fischer et al. 2011; Makarova et al. 2015), and ultrastructure of spermatozoa (Lino-Neto et al. 2000; Lino-Neto and Dolder 2001). Many studies are available on the development and external and internal morphology of trichogrammatid larvae (Flanders 1937; Ivanova-Kazas 1952, 1954, 1961; Chumakova 1966; Kochetova 1969; Dahlan and Gordh, 1996; Hawlitzky and Boulay 1982; Volkoff et al. 1995; Jarjees et al. 1998; Cônsoli et al. 1999; Wu et al. 2000; Jarjees and Merritt 2002, 2004; Boivin 2010).

A.A. Polilov, *At the Size Limit—Effects of Miniaturization in Insects*,
DOI 10.1007/978-3-319-39499-2_8

Literature on the structure of related groups of Hymenoptera is reviewed in the previous chapter of this book (Chap. 7).

8.2 Brief Description of the Structure of Trichogrammatidae

Small hymenopterans 0.17–2 mm long (usually 0.4–0.8 mm long). Body compact (Fig. 8.1). Coloration monotonous, from yellow to black, without metallic sheen. Integument without complex microsculpture or dense pubescence. Head hypognathous. Compound eyes lateral, consisting of 30–200 ommatidia; number of ocelli 3. In some representatives eyes and ocelli absent. Antennae 5–9-segmented, clavate, attached between eyes. Mesosoma consisting of fused pro-, meso-, metathorax, and abdominal segment 1. Two pairs of wings usually present; hindwings sometimes strongly reduced; in some cases wings absent. Wing venation strongly depleted; perimeter of wings with fringe of long setae. Legs divided into coxa, trochanter, femur, tibia, and tarsus, typical of insects. Trochanter in most representatives of this family 2-segmented. Tarsi 3-segmented. Number of visible segments of metasoma 5–7, because of abdominal segment 1 included in mesosoma, segment 2 forming petiole, and distal segments modified; number of visible tergites and sternites often different in males and females of one species. Petiole usually weakly pronounced. External male genitalia represented by phallobase, bearing parameres, and simple aedeagus. Females with ovipositor, often long, consisting of external and internal sheaths and pair of stylets.

Males and females different in size, proportions, number of visible sclerites of abdomen, number of antennomeres, and armature of antennae and legs. Some species include both winged and and wingless forms.

8.3 Morphology of Adult Trichogrammatidae

8.3.1 External Morphology

Body length of adult 370–420 (M = 390, n = 10) in *Trichogramma evanescens*, 220–255 (M = 235, n = 10) in *Megaphragma mymaripenne*. In smallest representative of family, *Megaphragma caribea*, body length only 170 μm (Delvare 1993).

Structure of head

Head hypognathous (Fig. 8.2). Cranium with only one weakly discernible postoccipital suture and several folds of unclear homology around base of antennae. Hypostome and clypeus not distinguished. Tentorium, as in other chalcidoids, with anterior and posterior arms, dorsal arms reduced. Occipital area with numerous folds; because of these folds, volume of cranium decreasing from pupae to adults.

Fig. 8.1 External morphology of Trichogrammatidae, SEM: **a**, **c**, **e** *Trichogramma evanescens*; **b**, **d**, **f** *Megaphragma mymaripenne*; **a**, **d** dorsal view, **b**, **c** lateral view; **e**, **f** ventral view

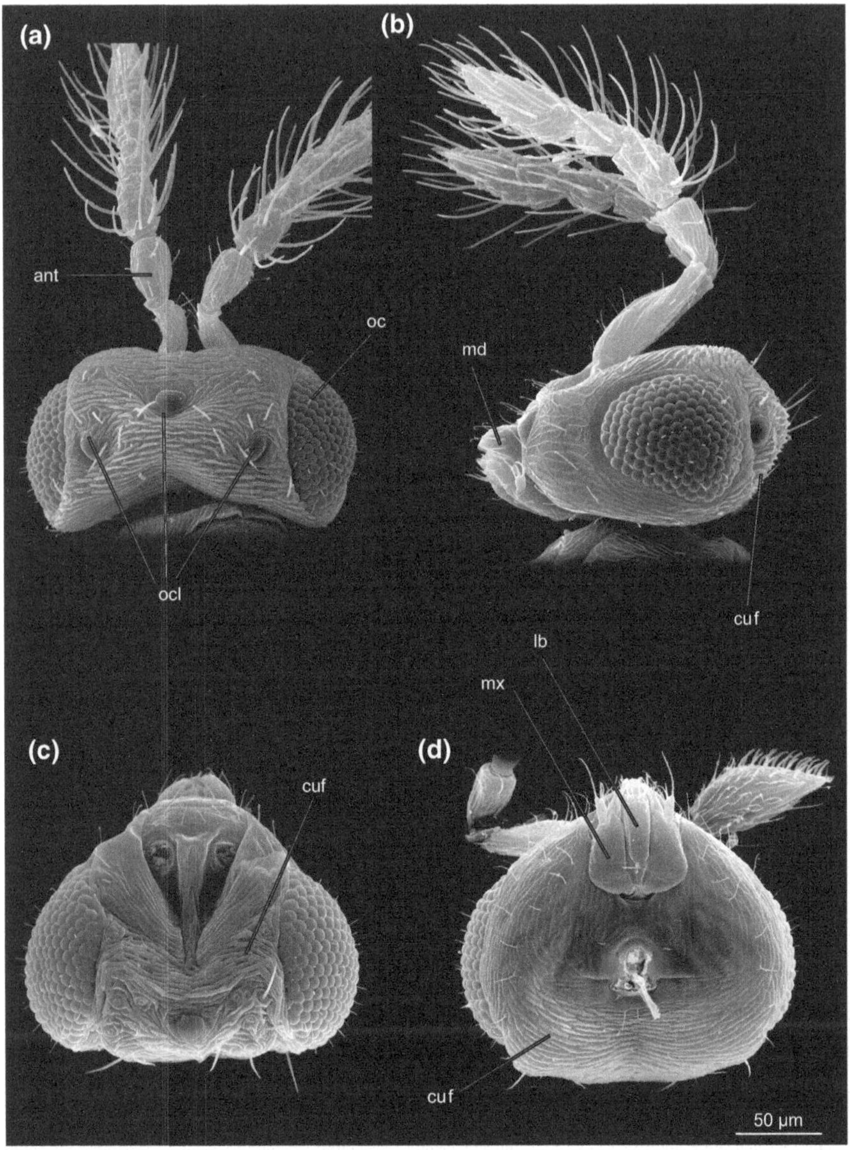

Fig. 8.2 Structure of head in *Trichogramma evanescens*, SEM: **a** dorsal view; **b** lateral view; **c** frontal view; **d** posterior view; **a**, **b** male; **c**, **d** female; *ant* antenna, *cuf* cuticular folds, *lb* labium, *md* mandible, *mx* maxilla, *oc* eye, *ocl* ocellus

Compound eyes lateral, consisting of 30–200 ommatidia, much fewer than in most other hymenopterans. Number of ocelli usually 3.

Antennae 5–9-segmented with 1–3-segmented club, consisting of scape, pedicel, anellus, funicle, and club. In males of some species segments of flagellum fused with club. Each antenna of *Trichogramma* bearing about 260 sensilla of 14 types (Amornsak et al. 1998). Antenna of *Megaphragma mymaripenne* bearing 38 sensilla of six types: 3 chaetoid on scape; 5 chaetoid on pedicel; 1 chaetoid and 5 trichoid (type 1) on flagellum; 14 trichoid (type 1), 2 trichoid (type 2), and 1 clavate on segment 1 of club; and 2 trichoid (type 1), 4 placoid, and 1 styloconic on segment 2 of club.

Mouthparts of structure typical of chalcidoids, consisting of well developed mandibles, maxillae, and labium (Figs. 8.3 and 8.4). Labrum weakly developed and represented by rather small triangular membranous plate. Mandibles with undulate margin and spines on internal surface. Maxillae connected with labium by membranous septum into labiomaxillary complex. Maxillae consisting of cardo, stipes, fused galea and lacinia, and maxillary palp. Maxillary palp 1-segmented, strongly reduced. Galea bearing large setae and spines, lacinia fused with galea and appearing as brush of setae. Labium consisting of mentum, bearing 1-segmented palps, ligulae, and paraglossae.

Structure of mesosoma

Prothorax narrow, consisting of semicircular pronotum and propectus, formed by sternite and pleurites of prothorax (Fig. 8.5a–c). Anterior part of propectus bearing paired cervical processes; head attached to these processes. Profurca Y-shaped with flattened arms. Pleurite bearing well developed apodeme.

Mesothorax more developed than other segments of metasoma (Figs. 8.5a–c and 8.6a–c). Mesonotum consisting of two parts, divided by scuto-scutellar suture. Anterior part divided by longitudinal parapsidal striae into scutum and scapulae. Posterior part divided into scutellum and axillary sclerites. Lateral part of mesothorax divided into episternum and epimeron by weakly discernible suture. Prepectus (secondary formation found in all chalcidoids) present in Trichogrammatidae between pro- and mesothorax. Mesofurca typical of Chalcidoidea, lateral arms well developed. Posterior margin of mesonotum forming mesophargma reaching middle of abdomen (*Trichogramma*) or almost reaching apex of abdomen (*Megaphragma*). The part of the mesophragma reaching into the metasoma is termed as postphragma by some authors (Sorokina 1993). Pair of spiracles present between pro- and mesothorax.

Metathorax represented by narrow semicircle of metanotum; other sclerites of metathorax fused with abdominal segment 1, forming propodeum. Propodeum bearing pair of abdominal spiracles. Metepisterna separated from propodeum by weakly pronounced suture. Epimera fused with propodeum. Pleural apodeme well developed, shaped as high longitudinal ridge with flattened top. Metafurca represented by two short, widely set arms (*Trichogramma*) or absent (*Megaphragma*).

Wings with strongly depleted venation (Fig. 8.7). In some representatives wings reduced or absent. Three veins preserved in forewing: submarginal (Pinto 2006: subcostal and premarginal), marginal, and stigmal (Sorokina 1993: radial), usually fused into one arch near anterior wing margin. These veins formed by fusion of

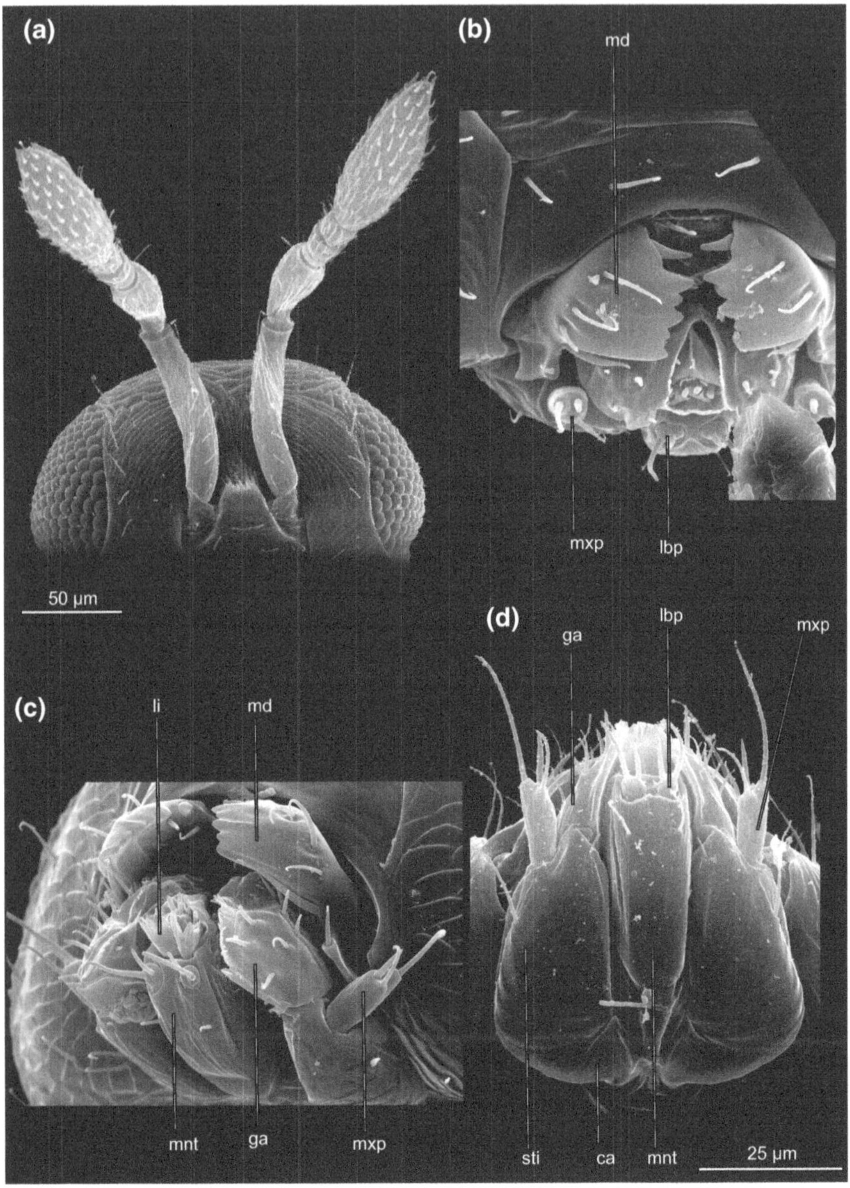

Fig. 8.3 *Trichogramma evanescens*, SEM: **a** male antennae; **b**–**d** mouthparts; **b** dorsal view; **c** lateral view; **d** ventral view; *ca* cardo, *ga* galea, *lbp* labial palp, *li* ligula, *md* mandible, *mnt* mentum, *mxp* maxillary palp, *sti* stipes

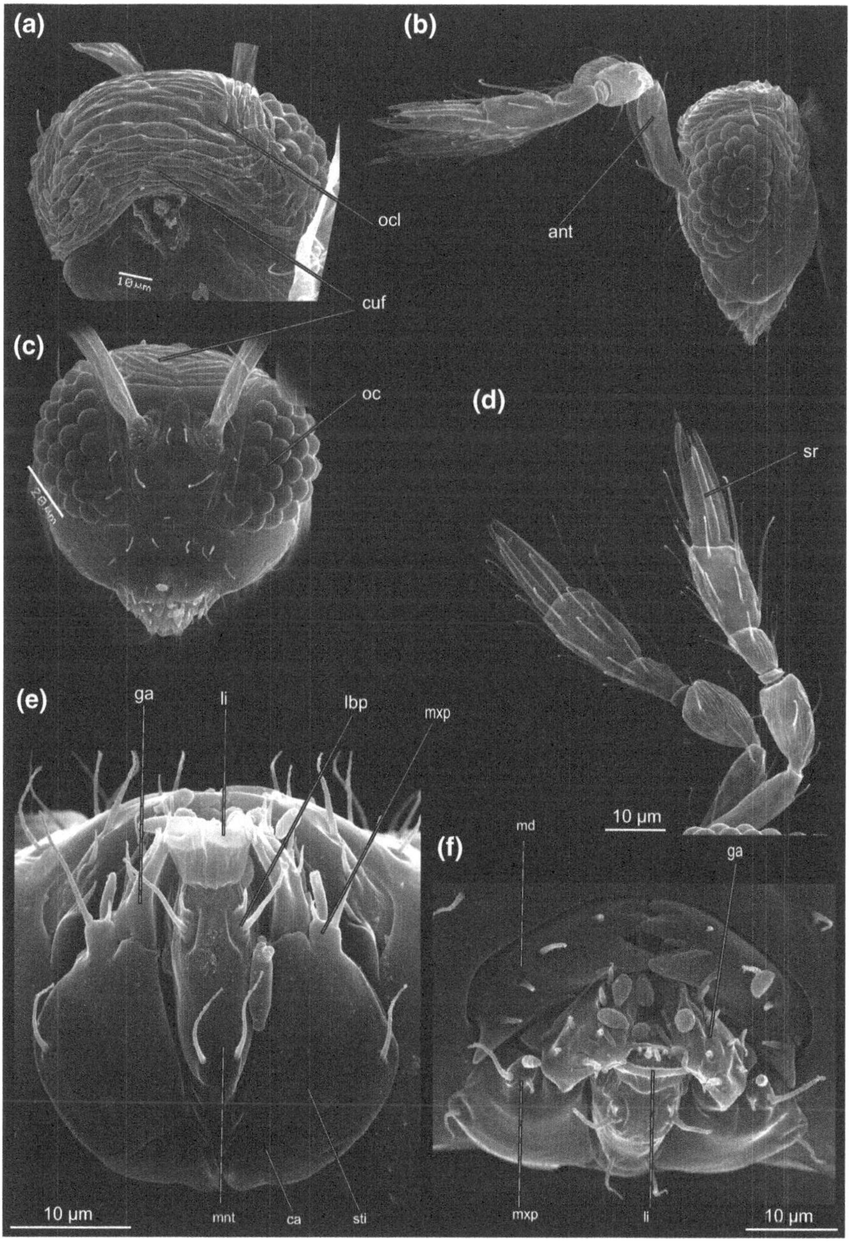

Fig. 8.4 Structure of head in *Megaphragma mymaripenne*, SEM: **a–c** head; **a** dorsal view; **b** lateral view; **c** frontal view; **d** antennae; **e** mouthparts, posterior view; **f** mouthparts, ventral view; *ant* antenna, *ca* cardo, *cuf* cuticular folds, *ga* galea, *lbp* labial palp, *li* ligula, *md* mandible, *mnt* mentum, *mxp* maxillary palp, *oc* eye, *ocl* ocellus, *sr* sensory ridge, *sti* stipes

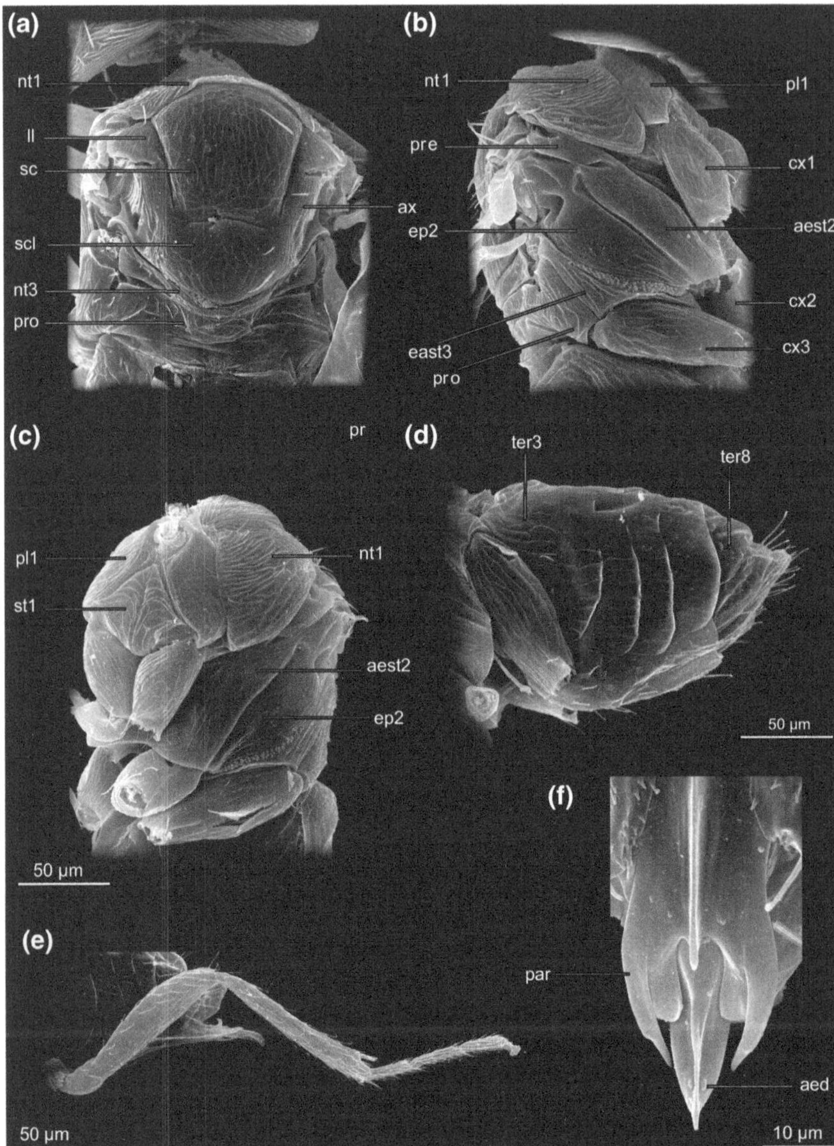

Fig. 8.5 Structure meso- and of metasoma in *Trichogramma evanescens*, SEM: **a**–**c** mesosoma; **d** metasoma; **e** leg; **f** male copulatory apparatus; **a** dorsal view; **b**, **d**, **e** lateral view; **c**, **f** ventral view; *aed* aedeagus, *aest2* mesepisternum, *aest3* metepisternum, *ax* axillary sclerite, *cx1.2.3* pro-, meso-, and metacoxae, *ep2* mesepimeron, *ll* scapulae, *nt1.3* pro- and metanotum, *par* parameres, *pl1* propleurite, *pre* prepectus, *pro* propodeum, *sc* scutum, *scl* scutellum, *st1* prosternite, *ter* tergite

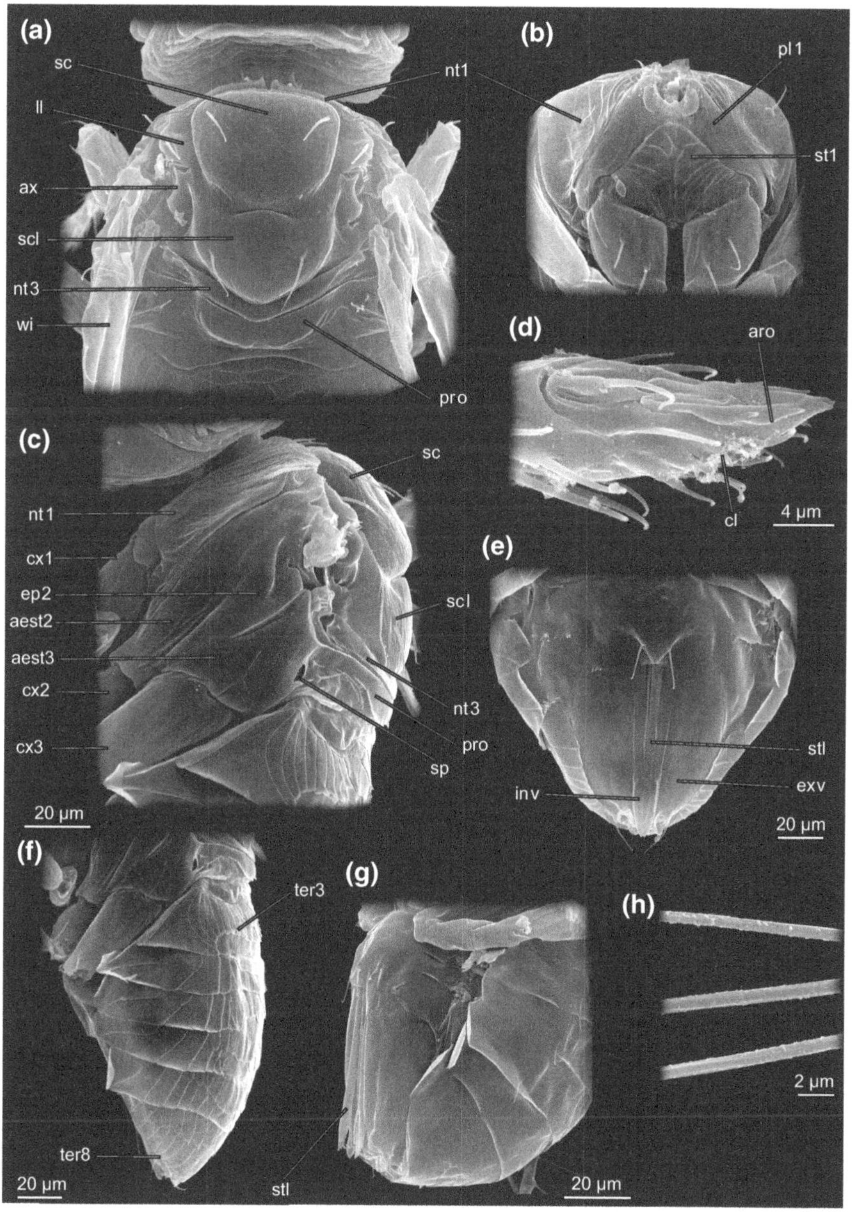

Fig. 8.6 Structure of meso- and of metasoma in *Megaphragma*, SEM: **a–f**, **h** *M. mymaripenne*; **g** *M. amalphitanum*; **a–c** mesosoma; **d** apex tarsi; **e–g** metasoma; **h** wing setae; **a**, **h** dorsal view; **b**, **d** ventral view; **c**, **d**, **f**, **g** lateral view; *aed* aedeagus, *aest2* mesepisternum, *aest3* metepisternum, *aro* arolium, *ax* axillary sclerite, *cl* claw, *cx1.2.3* pro-, meso-, and metacoxae, *ep2* mesepimeron, *exv* external valves of ovipositor, *inv* internal valves of ovipositor, *ll* scapulae, *nt1.3* pro- and metanotum, *par* parameres, *pl1* pleurite of prothorax, *pre* prepectus, *pro* propodeum, *prp* propectus, *sc* scutum, *scl* scutellum, *sp* spiracle, *stl* stylet of ovipositor, *ter* tergite, *wi* wing

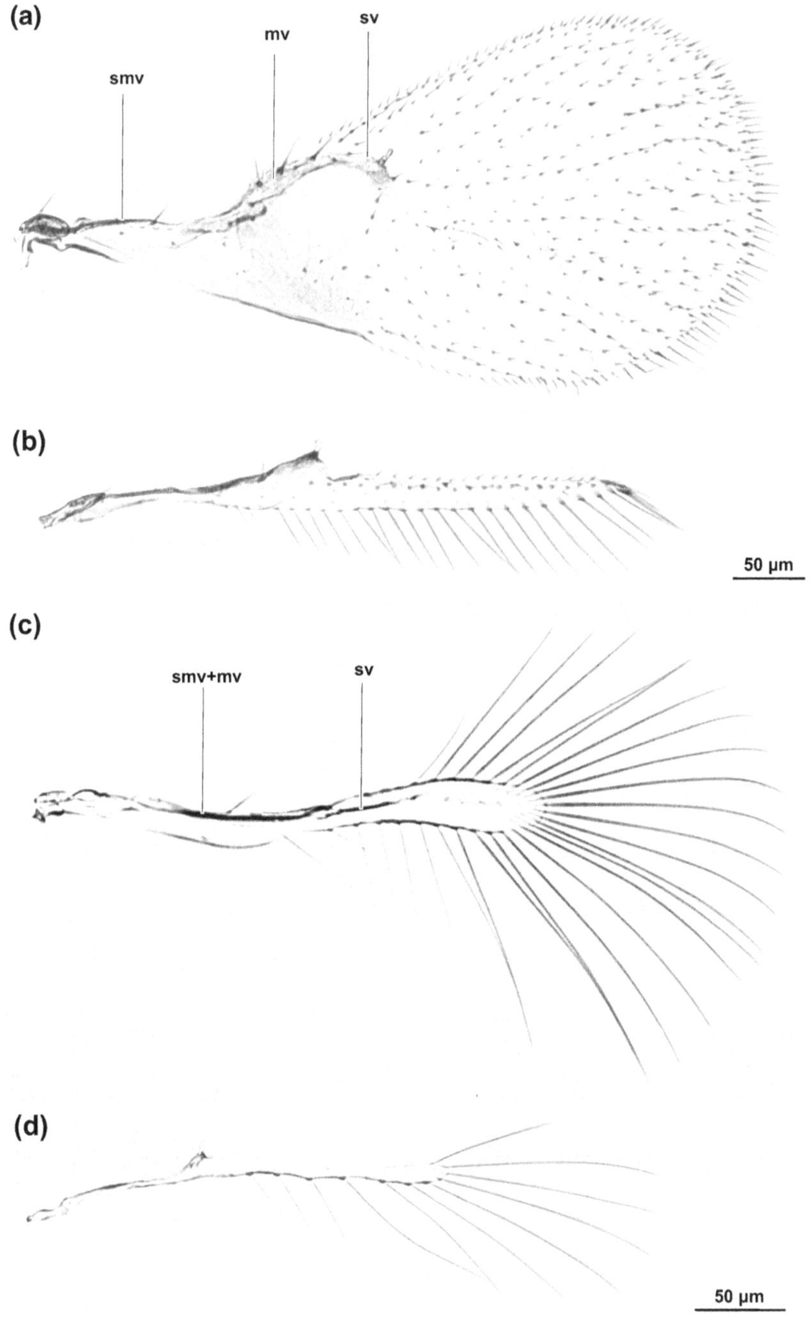

Fig. 8.7 Wings of Trichogrammatidae: **a**, **b** *Trichogramma evanescens*; **c**, **d** *Megaphragma mymaripenne*; **a**, **c** forewing, **b**, **d** hindwing

subcostal and radius; homology of particular parts is discussed in earlier studies (Burks 1938; Brandley 1955). Hindwings narrower than forewing. In contrast to other chalcidoids, postmarginal, medial, and cubital veins and parastigma absent. Forewings of *Trichogramma* rather wide, covered with small hairs, with longer hairs forming fringe on perimeter of wing. Hindwings of *Trichogramma* bearing fringe of long setae along posterior margin. Hindwing with only one short vein of unknown homology. Forewings of *Megaphragma* represented by narrow wing blade with fringe of long setae on perimeter.

Legs slender, ambulatorial, consisting of coxa, 1–2-segmented trochanter, femur, tibia, and 3-segmented tarsus. Tibiae of legs apically bearing well developed, often branched spurs. Apical tarsomere bearing two claws and well developed arolium.

Structure of metasoma

Abdomen in Trichogrammatidae 10-segmented, as in other hymenopterans. Segment 1 included in propodeum, segment 1 forming petiole, distal segments usually modified and sunk into metasoma. In trichogrammatids petiole not pronounced, mesosoma and metasoma broadly joined. Metasoma consisting of six of seven visible tergites and six visible sternites (*Trichogramma*), or sternites weakly discernible (*Megaphragma*) (Figs. 8.5d and 8.6e–g).

Ovipositor consisting of outer ovipositor plates (derivates of tergite 9), inner ovipositor plates (Val3), sheath (fused Val2), and stylets (Val1). Spermatheca rather small, rounded or pear-shaped.

External male genitalia represented by simple aedeagus, phallobase, and parameres.

8.3.2 Internal Morphology

General plan of internal structure: most of head occupied by supraoesophageal and suboesophageal ganglia; considerable part of metasoma occupied by musculature; very large muscle (IIdlm1) occupying much of meso- and metasoma; reproductive system occupying most of metasoma (Figs. 8.8 and 8.9).

Integument

Integument consisting of cuticle, hypoderm, and basal membrane. Cuticle thickness 0.8–3 μm (M = 1.7, n = 80) in *Trichogramma*, 0.7–2.4 (M = 1.2, n = 80) in *Megaphragma*. Cuticle consisting of epicuticle and procuticle. Procuticle homogeneous (Fig. 8.10a–c). Hypoderm represented by strongly flattened cells up to 1.5 μm thick. Many areas of hypoderm, especially in head, with numerous electron transparent vacuoles.

Digestive and excretory systems

Digestive system complying with general plan typical of insects. Digestive canal divided into fore-, mid-, and hindgut (Figs. 8.11a, b and 8.12a, b). Fore- and

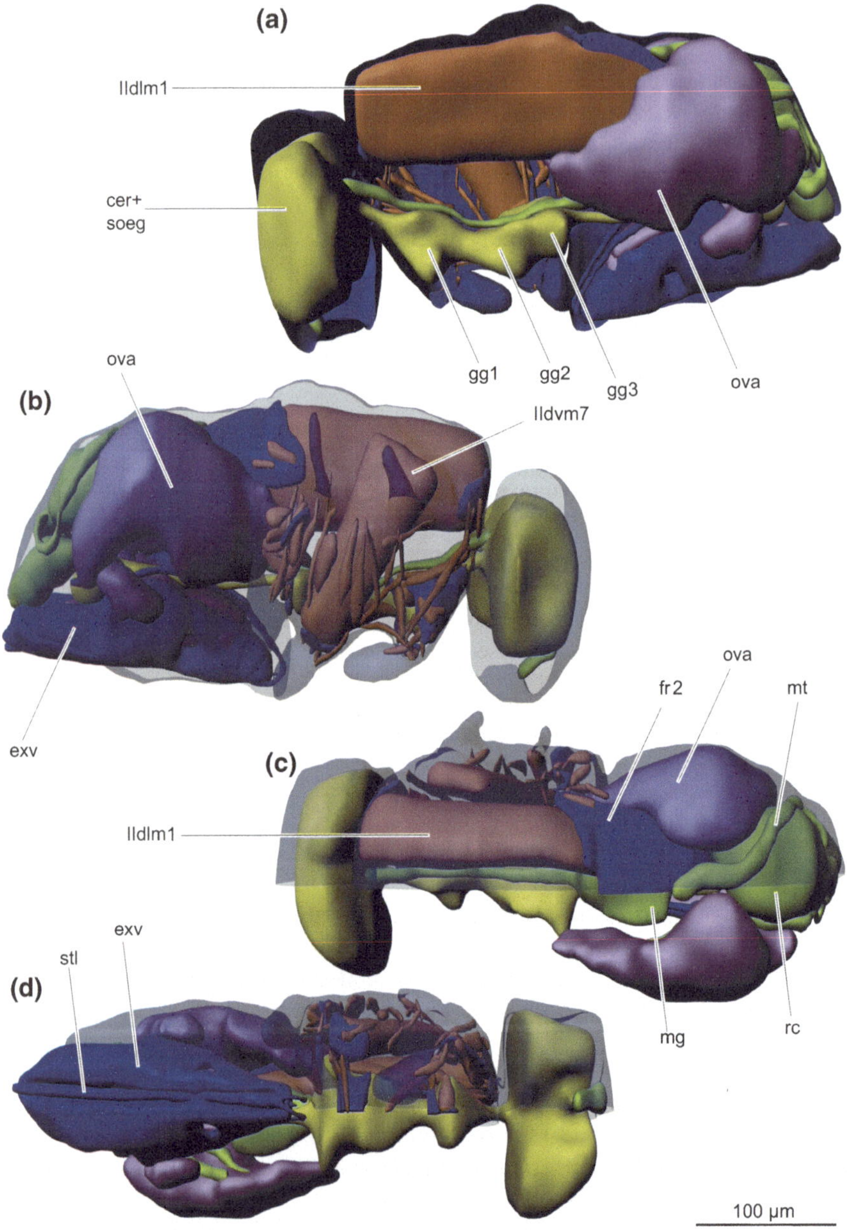

Fig. 8.8 Internal morphology of *Trichogramma evanescens*, 3D: **a** lateral internal view; **b** lateral external view; **c** dorsal view; **d** ventral view; *cer* cerebrum, *gg1.2.3* pro-, meso-, and metathoracic ganglia, *exv* external valves of ovipositor, *fr2* mesophragma; *mg* midgut, *mt* Malpighian tubules, *ova* ovary, *rc* rectum, *soeg* suboesophageal ganglion, *stl* stylet of ovipositor. Colors: *blue* cuticle, *green* digestive system, *yellow* central nervous system, *brown* musculature, *purple* reproductive system. Musculature see text

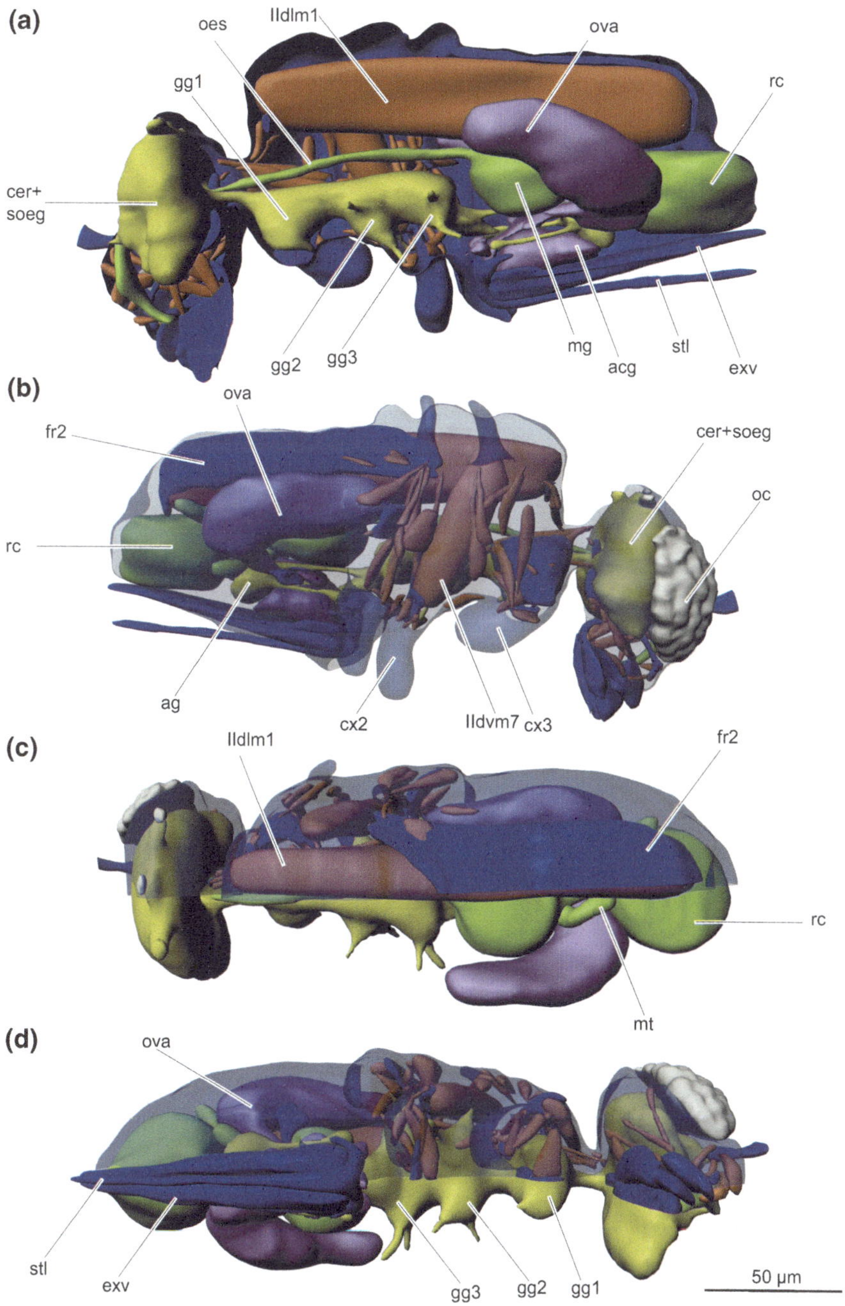

Fig. 8.9 Internal morphology of *Megaphragma mymaripenne*, 3D: **a** lateral internal view; **b** lateral external view; **c** dorsal view; **d** ventral view; *acg* acid gland, *ag* abdominal ganglion, *cer* cerebrum, *cx1.2* meso- and metacoxae, *exv* external valves of ovipositor, *gg1.2.3* pro-, meso-, and metathoracic ganglia, *fr2* mesophragma; *mg* midgut, *mt* Malpighian tubules, *oc* eye, *oes* oesophagus, *ova* ovary, *rc* rectum, *soeg* suboesophageal ganglion, *stl* stylet of ovipositor. Colors: *blue* cuticle, *green* digestive system, *yellow* central nervous system, *brown* musculature, *purple* reproductive system. Musculature see text

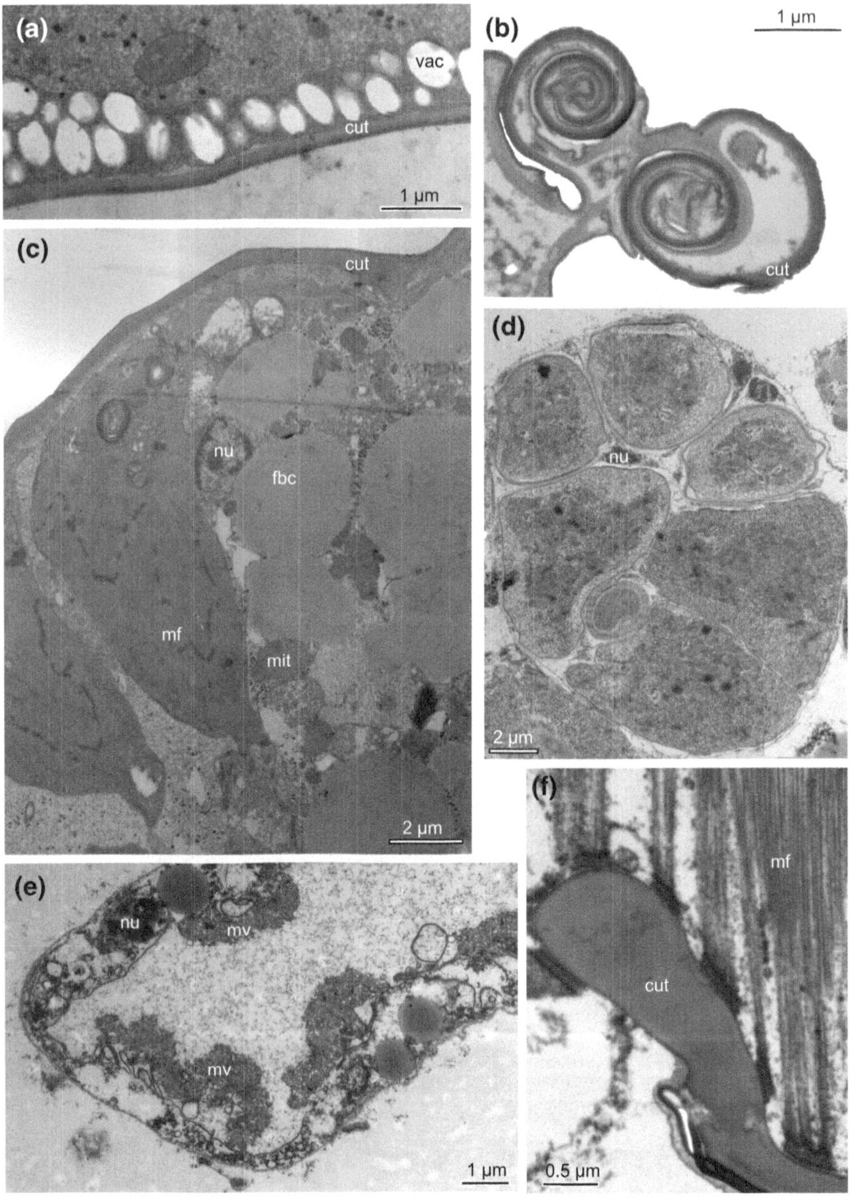

Fig. 8.10 Ultrastructure of *Megaphragma mymaripenne*, TEM, cross-sections: **a**, **b** integument, head, **c** musculature and fat body, head; **d** ovary, **e** midgut, **f** attachment site of muscles to cuticle, head; *cut* cuticle, *fbc* cell of fat body, *mf* muscle fiber, *mit* mitochondrion, *mv* microfibers, *nu* nucleus, *vac* vacuole

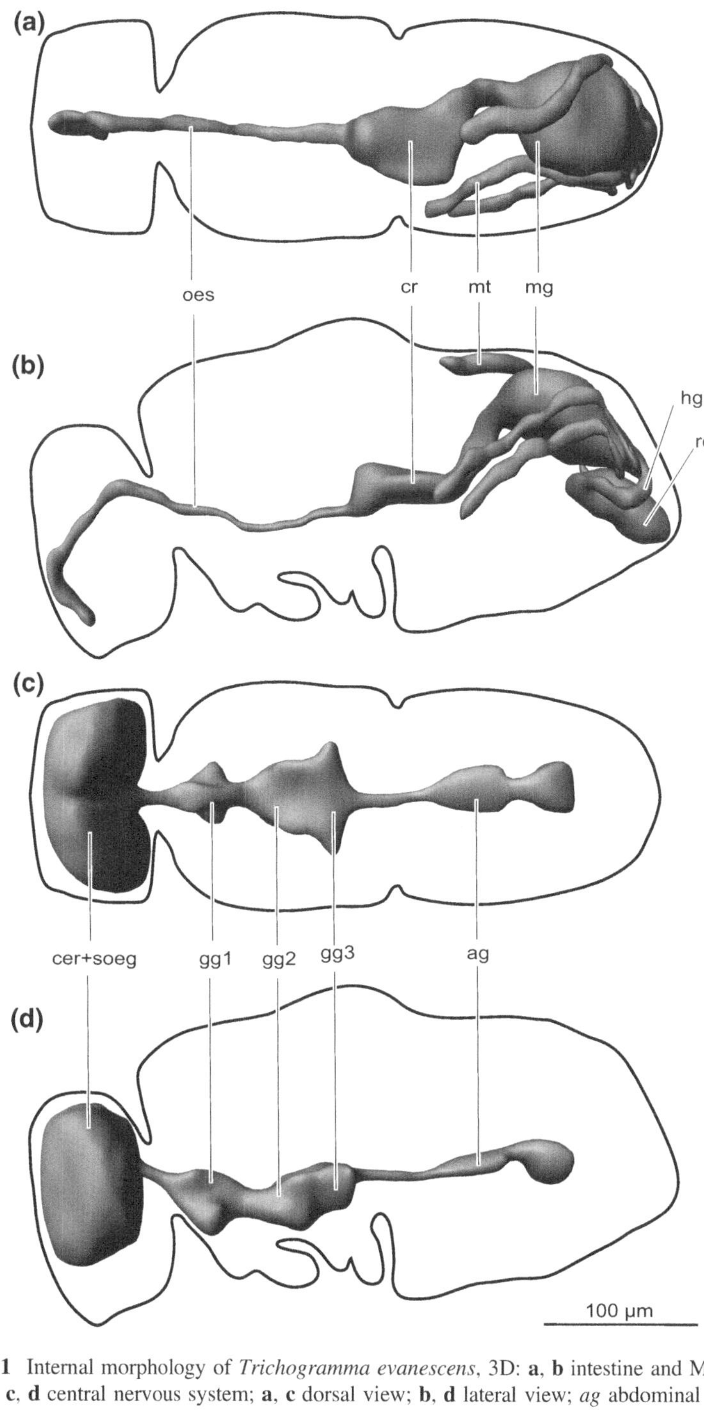

Fig. 8.11 Internal morphology of *Trichogramma evanescens*, 3D: **a**, **b** intestine and Malpighian tubules; **c**, **d** central nervous system; **a**, **c** dorsal view; **b**, **d** lateral view; *ag* abdominal ganglion, *cer* cerebrum, *cr* crop, *gg1.2.3* pro-, meso-, and metathoracic ganglia, *hg* hindgut, *mg* midgut, *mt* Malpighian tubules, *oes* oesophagus, *rc* rectum, *soeg* suboesophageal ganglion

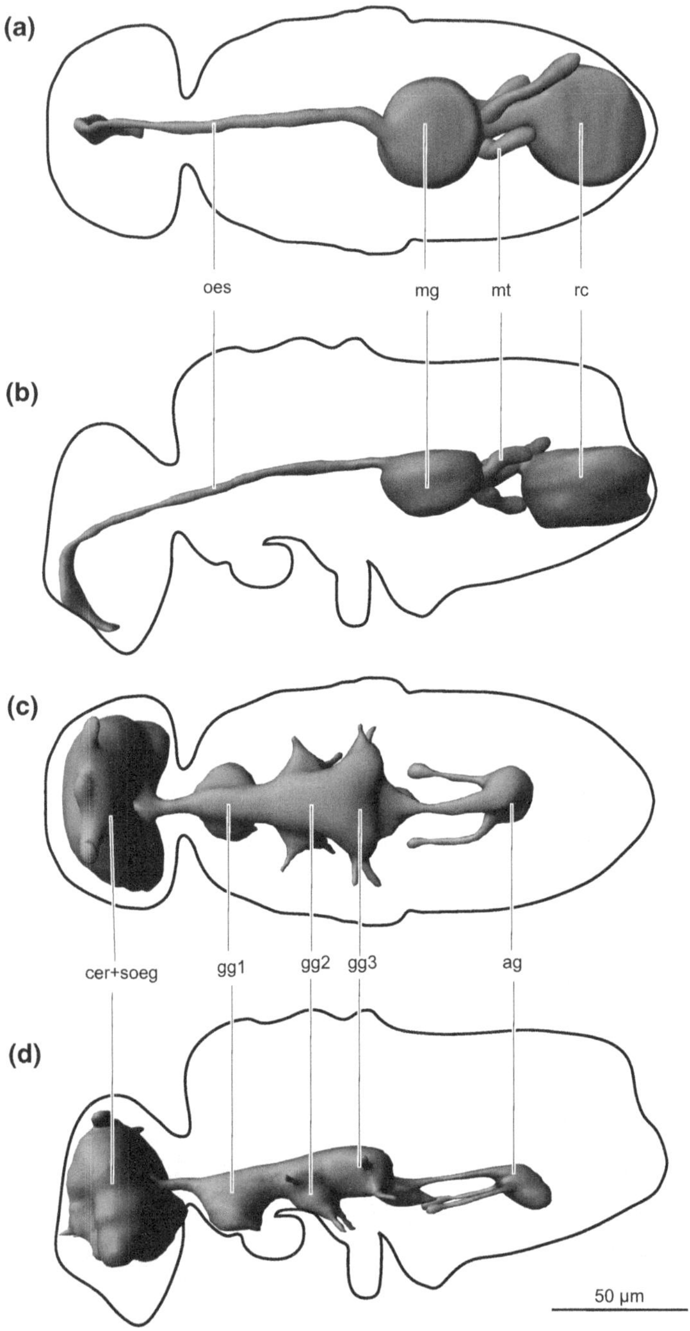

Fig. 8.12 Internal morphology of *Megaphragma mymaripenne*, 3D: **a**, **b** intestine and Malpighian tubules; **c**, **d** central nervous system; **a**, **c** dorsal view; **b**, **d** lateral view; *ag* abdominal ganglion, *cer* cerebrum, *gg1.2.3* pro-, meso-, and metathoracic ganglia, *mg* midgut, *mt* Malpighian tubules, *oes* oesophagus, *rc* rectum, *soeg* suboesophageal ganglion

hindgut with thin cuticular lining. Midgut without cuticular lining. Intestinal canal somewhat longer than body, forming loop in metasoma. Digestive glands not found.

Foregut divided into pharynx, oesophagus, and crop. Oesophagus straight, running through entire thorax. Muscles of oesophagus absent. Crop situated in metasoma.

Midgut short, wide. Walls of midgut formed by strongly flattened cells. Muscles of midgut not found, distinguishing trichogrammatids from other chalcidoids; layer of circular muscles present. Peritrophic membrane not found.

Hindgut short, divided into small intestine and rectum.

Boundary between mid- and hindgut bearing three Malpighian tubules, shaped as short tubes.

Circulatory system and fat body

Circulatory system strongly reduced. Heart and blood vessels absent, probably because of diffusion sufficing for transport of molecules in organisms of such small sizes. Fat body occupying almost all cavities between organs in metasoma and to a smaller degree in mesosoma.

Tracheal system

Tracheal system strongly simplified. Only few weakly branching tracheae present, connected to mesosomal and metasomal spiracles. Transverse stems and air sacs absent, probably because of diffusion sufficing for oxygen transport in organisms of such small sizes. Tracheae with structure typical of insects, consisting of hypoderm and intima, intima with helical thickenings—taenidia.

Nervous system

Cerebrum and suboesophageal ganglion fused into one synganglion and localized entirely in cranium. Ganglia of thoracic neural chain in *Trichogramma* separated, in *Megaphragma* mesothoracic and metathoracic ganglion fused (Figs. 8.11c, d and 8.12c, d). *Trichogramma* with pair of small abdominal ganglia, in *Megaphragma* abdominal ganglia fused into one synganglion.

Ensheathment in *Trichogramma* well developed and structured as in *Anaphes*. Neural lamella 0.05–0.3 μm thick; perineurium 1–1.5 μm thick. As in *A. flavipes*, almost entire volume of body of neuron occupied by nucleus. Heterochromatin represented by large spherical islets in center of nucleus, flattened on periphery. In spite of small volume of cytoplasm, cellular organelles of neurons clearly discernible at cellular level. Granular endoplasmic reticulum, ribosomes, mitochondria, Golgi apparatus, and lysosomes pronounced (Fig. 8.13). Neuropil containing synaptic contacts of different types, but mostly of polarized type. Vast majority of contacts represented by chemical synapses of asymmetrical type. Nonpolarized synaptic contacts occurring less often and characterized by absence of synaptic vesicles and condensation of membranes in contact areas. Axoplasm of axons and dendrites with discernible mitochondria, microtubules, neurofibrils. Neuropil with some cortical glia embedding neurons and their projections. Minimum diameter of projections in neuropil of cerebrum in *Trichogramma* 0.06 μm.

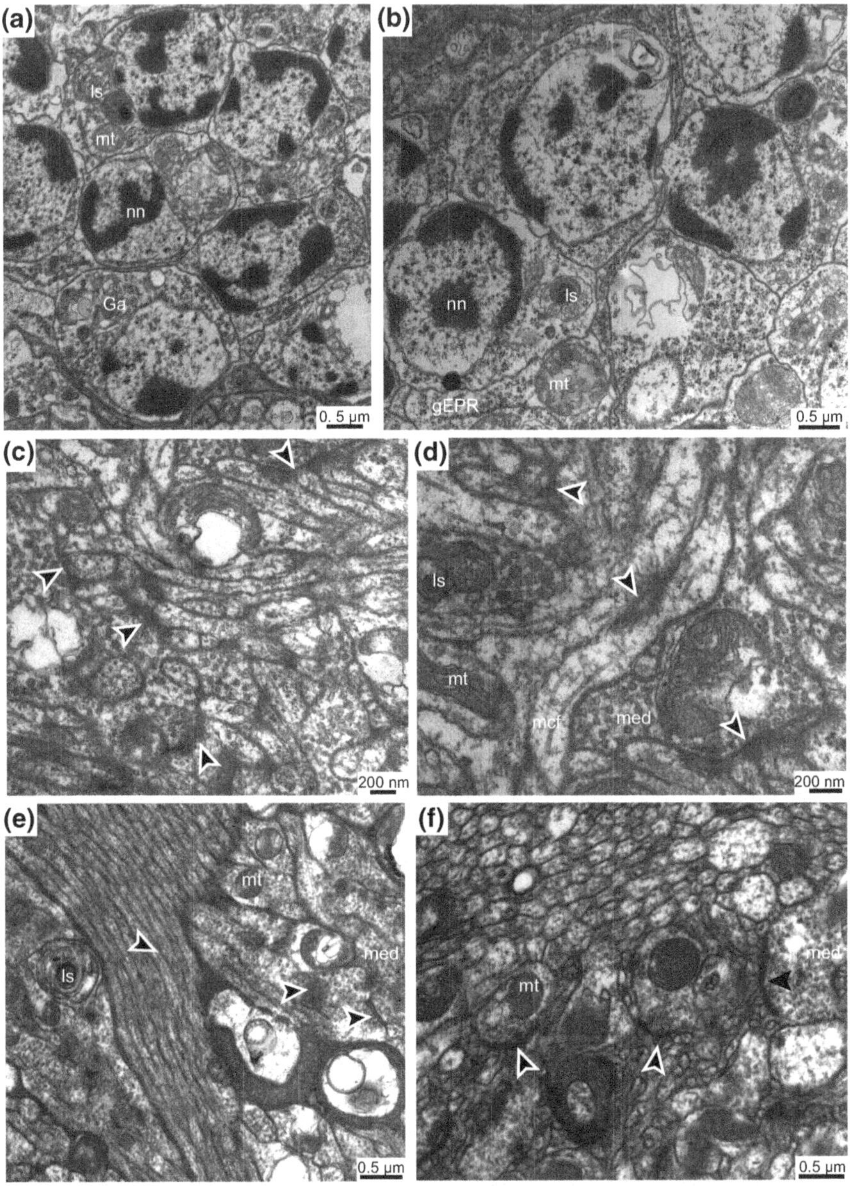

Fig. 8.13 Ultrastructure of cerebrum in *Trichogramma evanescens*, TEM: *Ga* Golgi apparatus; *gEPR* granular endoplasmic reticulum; *gl* cortical glia; *ls* lysosomes; *mcf* microfilaments; *med* granules of mediator; *mt* mitochondrion; *nn* nucleus; *syn* synaptic contact. **c–f** *arrows* indicate synaptic contacts

Cell bodies in nervous system of *T. evanescens* 1.53–3.36 μm in diameter (M = 2.4, n = 346). Cerebrum of *T. evanescens* containing only 18000 cells.

Protocerebrum larger than any other region of cerebrum in *Trichogramma*. Large central body complex occupying medial position and, as in *Anaphes*, differentiated into central body (dorsal region, consisting of dorsal arch, fan-shaped body, ventral region, and noduli) and protocerebral bridge (Fig. 8.14). Mushroom bodies well developed. Peduncles of mushroom bodies long and twisted. In dorsal distal part, near calyces, peduncles with geniculate curve. Calyces single (secondarily fused), of spherical shape and glomerular structure. Optic lobes represented by three neuropils. Lamina elongate and crescent-shaped, other optic neuropils elypsoid. Lobular complex differentiated into lobula and lobular plate. All three optic neuropils tightly adjoining each other, together forming angle, as in *Anaphes*.

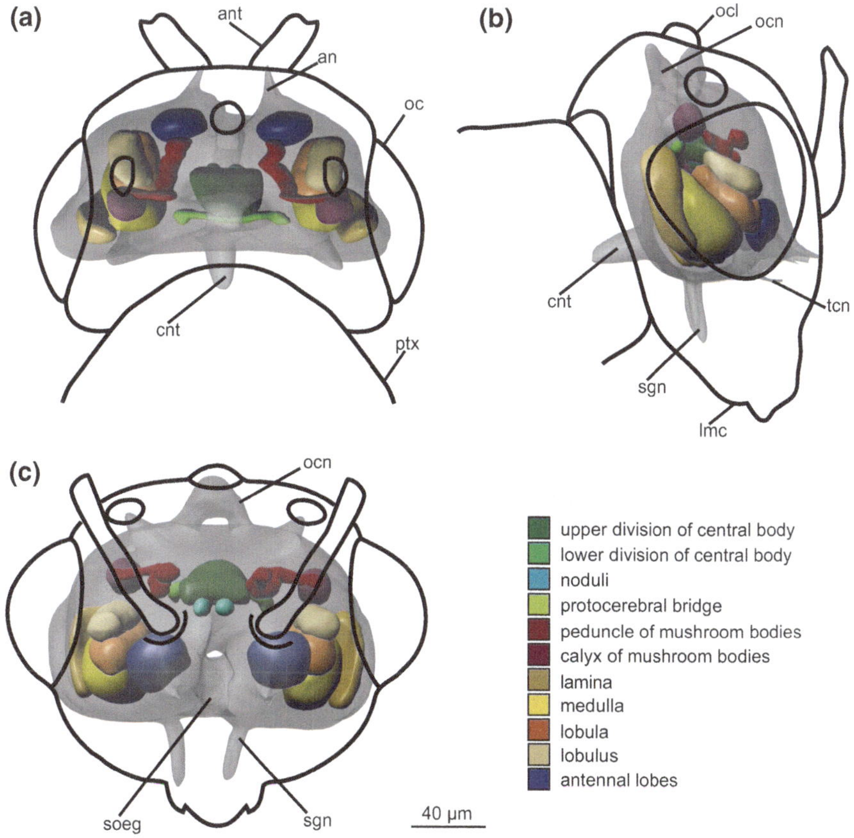

Fig. 8.14 3D reconstructions of cerebrum in *Trichogramma evanescens* (modified from Makarova and Polilov 2013): **a** dorsal view; **b** lateral view; **c** frontal view; *an* antennal nerves; *ant* antennae; *cnt* connectives linking suboesophageal ganglion with prothoracic ganglion; *oc* compound eyes; *ocl* ocelli; *ocn* ocellar nerves; *ptx* prothorax; *lmc* mouthparts; *sgn* nerves of suboesophageal ganglion, innervating mouthparts; *soeg* suboesophageal ganglion; *tcn* paired nerves of tritocerebrum

Well-developed ocellar ganglia present. Antennal lobes of spherical shape and glomerular structure. Tritocerebrum rather small, clearly discernible, appearing as dense elongate structure. Tritocerebral commissures completely sunk into common neuropilar mass of cerebrum.

Stomatogastric nervous system situated very close to head synganglion and indistinguishable.

The central nervous system of *Megaphragma* has a fundamentally different ultrastructure (Figs. 8.15 and 8.16). It was shown that in adults of *M. mymaripenne* the nervous system is almost anucleate, because over 95 % of cells in the central nervous system undergo lysis of bodies and nuclei at late stages of pupal development (Polilov 2012). All ganglia in the nervous system of adults are represented almost exclusively by neuropil, the structure of which is almost identical to that of *Trichogramma*. The neuropil of the supraesophageal ganglion contains discernible optic lobes, the central body complex, and antennal lobes. At the same time, the central nervous system contains only 339–372 (M = 360, n = 3) nuclei, 179–253 of them in cerebrum. The extraneural sheath is incomplete, consisting of separate scattered cells. The neural lamella is absent, distinguishing *Megaphragma* from other insects. The cytoplasm of neurons in *M. mymaripenne* contains organelles: mitochondria, granular endoplasmic reticulum, and many lysosomes.

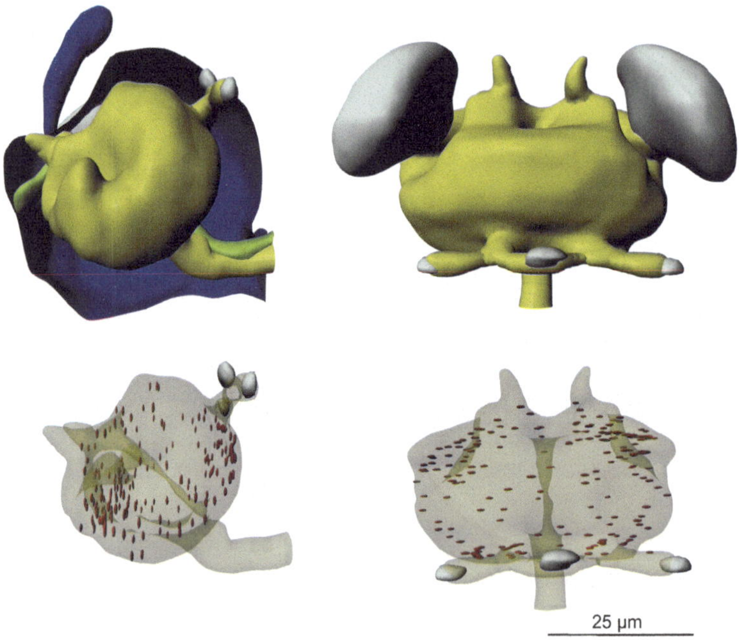

Fig. 8.15 Structure of cerebrum in *Megaphragma mymaripenne*, 3D: **a**, **c** lateral view; **b**, **d** dorsal view. Colors: *blue* cuticle, *yellow* cerebrum, *green* intestine, *white* eyes and ocelli, *red* nuclei of cells of central nervous system

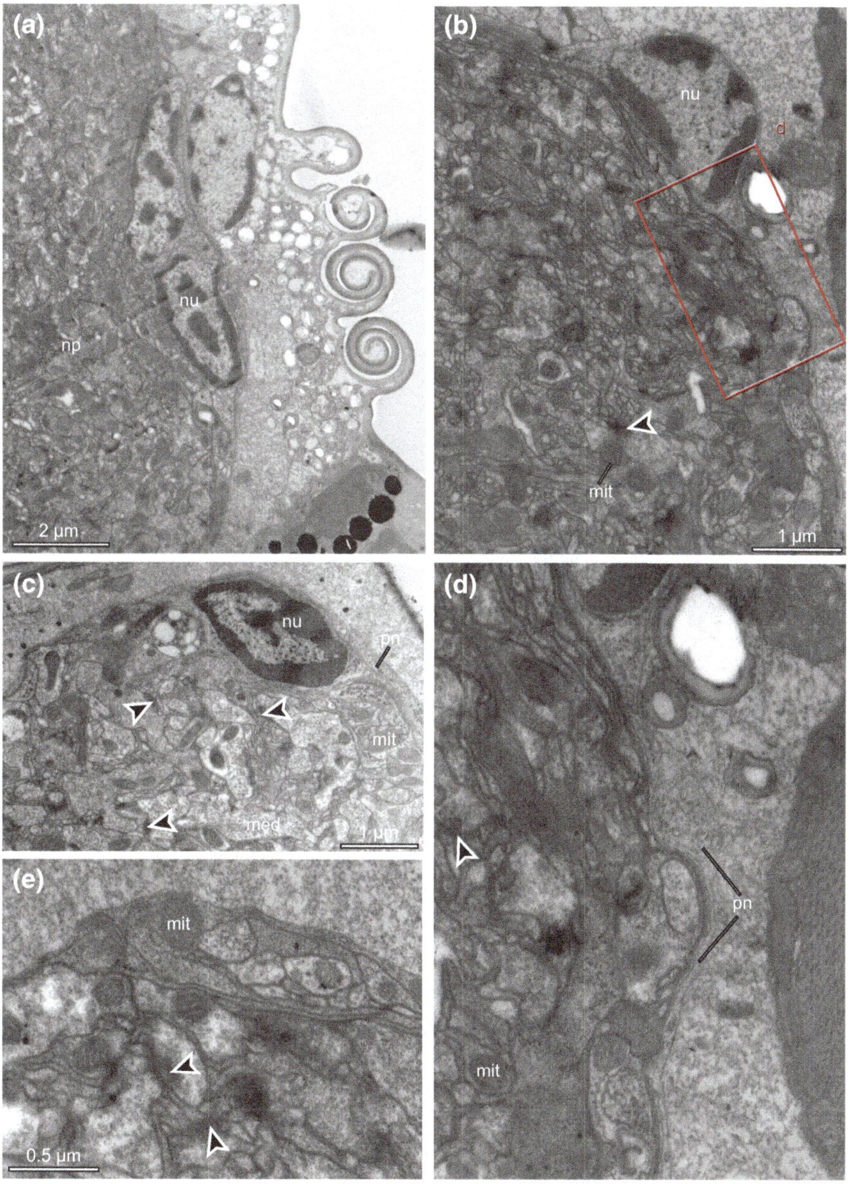

Fig. 8.16 Ultrastructure of cerebrum in *Megaphragma*, TEM: **a** *M. mymaripenne*; **b–e** *M. amalphitanum*, *med* granules of mediator, *mt* mitochondrion, *nu* nucleus, *pn* perineurium, *arrows* indicate synaptic contacts

Muscular system

Structure of muscular system in trichogrammatids, in spite of their small size, complying with general plan typical of insects. Musculature has been studied in *Trichogramma evanescens* and *Megaphragma mymaripenne* (Table 7.1). Most

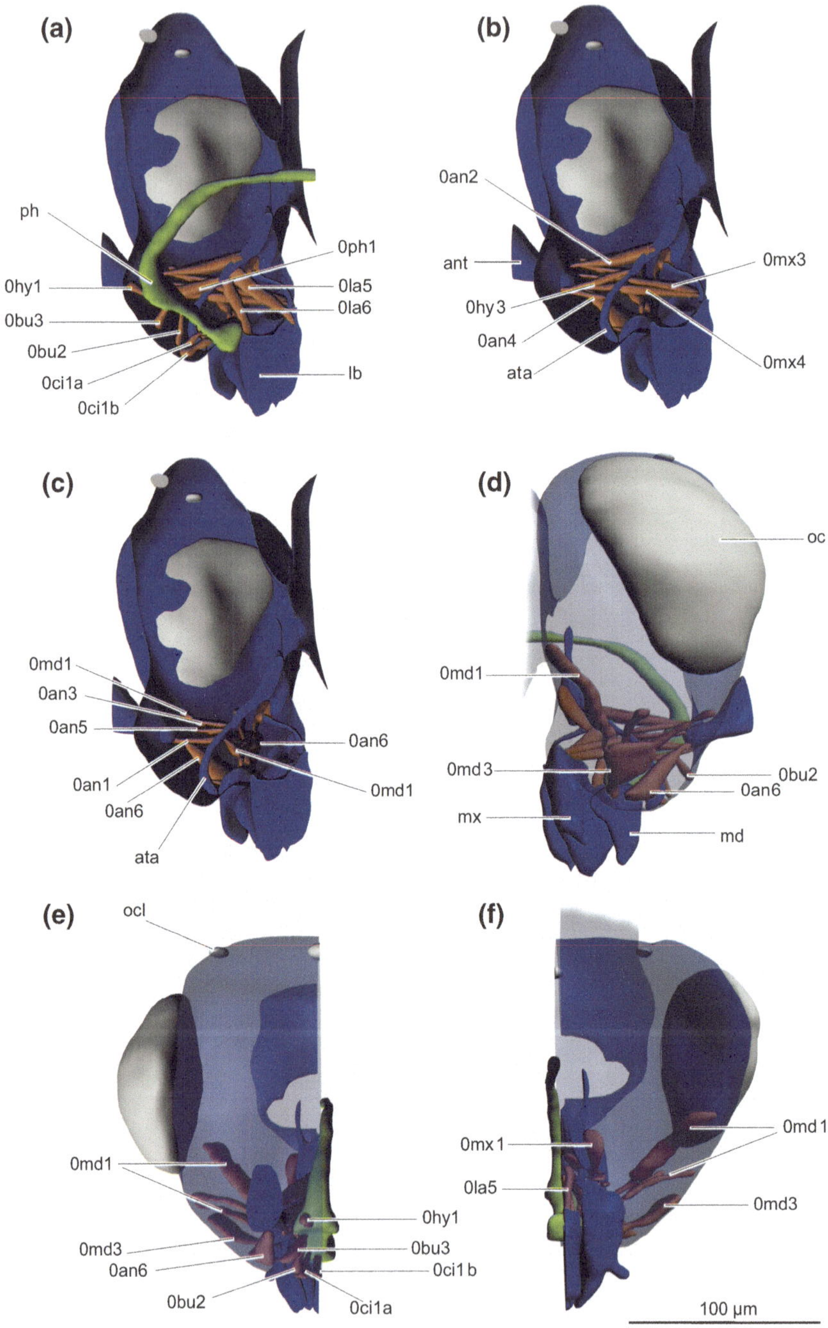

Fig. 8.17 Musculature of head in *Trichogramma evanescens*, 3D: **a–c** lateral internal view; **d** lateral external view; **e** frontal view; **f** posterior view; *ant* antenna, *ata* anterior tentorial arms, *lb* labium, *md* mandible, *mx* maxilla, *oc* eye, *ocl* ocellus, *ph* pharynx. Musculature see text

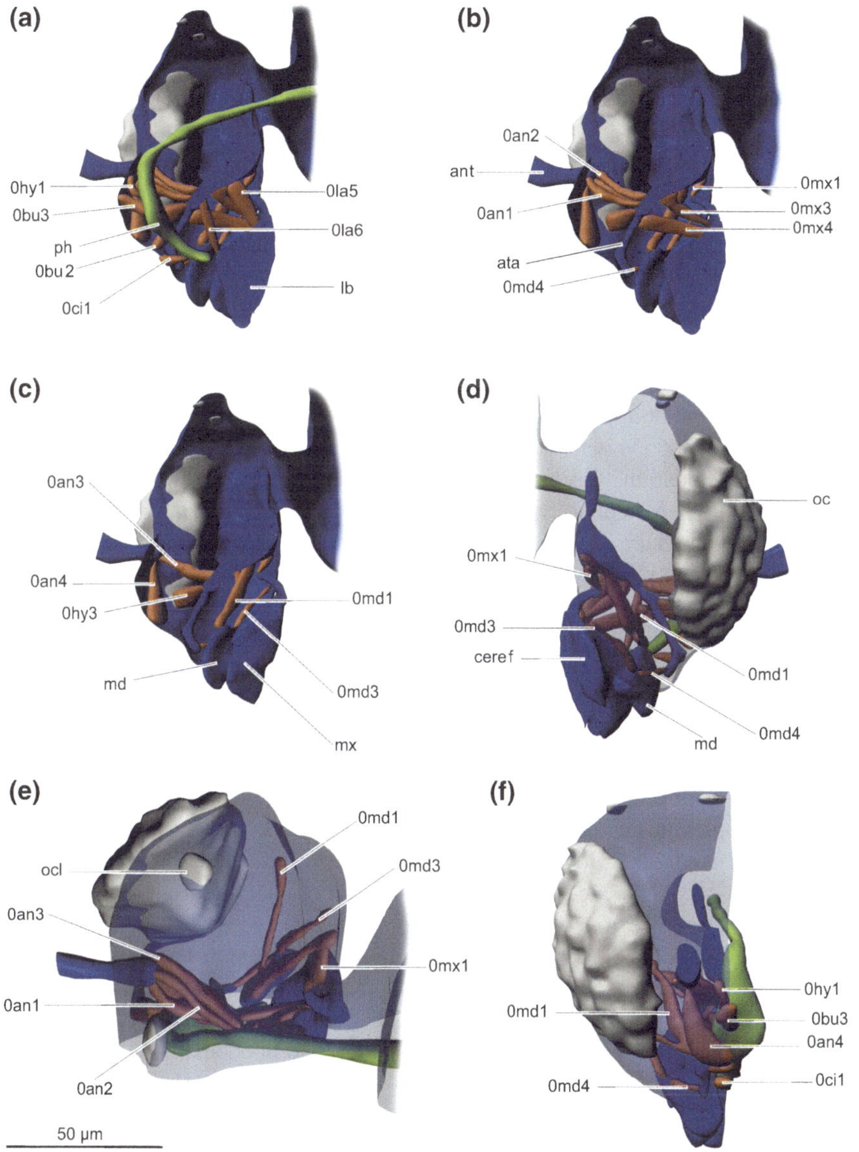

Fig. 8.18 Musculature of head in *Megaphragma mymaripenne*, 3D: **a–c** lateral internal view; **d** lateral external view; **e** dorsal view; **f** frontal view; *ant* antenna, *ata* anterior tentorial arms, *lb* labium, *md* mandible, *mx* maxilla, *oc* eye, *ocl* ocellus, *ph* pharynx. Musculature see text

muscles of *Megaphragma* are connected with the skeleton via strongly shortened tonofibril apparatus, which morphologically resembling the desmosome; a similar structure has been described in four-legged mites (Silvere and Shtein-Margolina 1976).

Musculature of head (Figs. 8.17 and 8.18). 0an1 (M. tentorioscapalis anterior): O, anterior tentorial arms; I, anterior margin of base of scape. 0an2 (M. tentorioscapalis posterior): O, anterior tentorial arms; I, posterior margin of base of scape. 0an3 (M. tentorioscapalis lateralis): O, anterior tentorial arms; I, lateral margin of base of scape. Muscle of unclear homology, possibly 0an4 (M. tentorioscapalis medialis), with atypical attachment site: O, anterolateral part of frons; I, anterior margin of base of scape. 0md1 (M. craniomandibularis internus), consisting of two subunits: O, first, lateral part of cranium; second, gular zone; I, adductor of mandible. 0md3 (M. craniomandibularis externus): O, lateral part of cranium; I, abductor of mandible. 0md4 (M. hypopharyngomandibularis): O, anterior tentorial arms; I, adductor of mandible. 0mx1 (M. craniocardinalis): O, posterior part of cranium; I, ventrolateral part of cardo. 0mx3 (M. tentoriocardinalis): O, anterior tentorial arms; I, cardo. 0mx4 (M. tentoriostipitalis anterior): O, anterior tentorial arms; I, base of stipes. 0la5 (M. tentoriopraementalis): O, anterior tentorial arms; I, posterior margin of prementum. 0la6 (M. tentorioparaglossalis): O, anterior tentorial arms; I, anterior margin of prementum. 0hy1 (M. frontooralis): O, frons; I, posterior margin of epipharynx. 0hy3 (M. craniohypopharyngealis): O, anterior tentorial arms; I, anterior margin of epipharynx. Circular muscles of hypopharynx absent. 0ci1 (M. clypeopalatalis): O, clypeus; I, dorsal part of epipharynx. 0bu2 (M. frontobuccalis anterior): O, frons; I, dorsolateral part of pharynx. 0bu3 (M. frontobuccalis posterior): O, frons; I, dorsal part of pharynx. 0st1 (M. annularis stomodaei): transverse musculature, developed only on dorsal surface of pharynx. Internal musculature of antennae and mouthparts not studied because of extremely small size.

Musculature of mesosoma (Figs. 8.19 and 8.20). **Prothorax**. Idlm5 (M. pronoto-phragmalis anterior): O, prophragma; I, medial part of pronotum. Idvm2 (M. cervico-occipitalis medialis): O, cervical region; I, occipital region. Itpm3 (M. pronoto-pleuralis anterior): O, lateral part of notum; I, propleurite, could not be determined precisely. Idvm5 (M. pronoto-cervicalis anterior) and Idvm6 (M. pronoto-cervicalis medialis) fused: O, posterior part of pronotum and prophragma; I, cervical region. Idvm7 (M. pronoto-cervicalis posterior): O, pronotum; I, cervical region. Idvm9 (M. profurca-occipitalis): O, profurca; I, occipital region. Idvm18 (M. pronoto-coxalis lateralis): O, pronotum; I, lateral margin of base of coxa. Itpm2 (M. propleuro-occipitalis): O, occipital zone; I, pleurite (*Trichogramma*), pleural apodeme (*Megaphragma*). Itpm4-5 (M. pronoto-apodemalis): O, lateral part of pronotum; I, pleurite. Ipcm3 (M. propleuro-trochantinalis): O, pleural apodeme; I, trochantin. Ipcm4 (M. propleuro-coxalis superior): O, pleural apodeme; I, anterior margin of base of coxa. Ipcm8 (M. propleuro-trochanteralis): O, pleural apodeme; I, trochanter, via fine tendon. Ivlm1 (M. profurca-cervicalis): O, profurca; I, cervical region. Ivlm3 (M. profurca-tentorialis): O, profurca; I, postoccipital region. Ivlm7 (M. profurca-mesofurcalis): O, profurca; I, mesofurca. Iscm1 (M. profurca-coxalis anterior): O, profurca; I, anterior margin of base of coxa. Iscm2 (M. profurca-coxalis posterior): O, profurca; I, posterior margin of base of coxa. Iscm3 (M. profurca-coxalis medialis): O, profurca; I, medial margin of base of coxa. Iscm5 (M. prospina-coxalis): O, fold between pro- and metathorax; I, posterior margin of base of coxa. Internal muscles of legs not studied because of extremely small size.

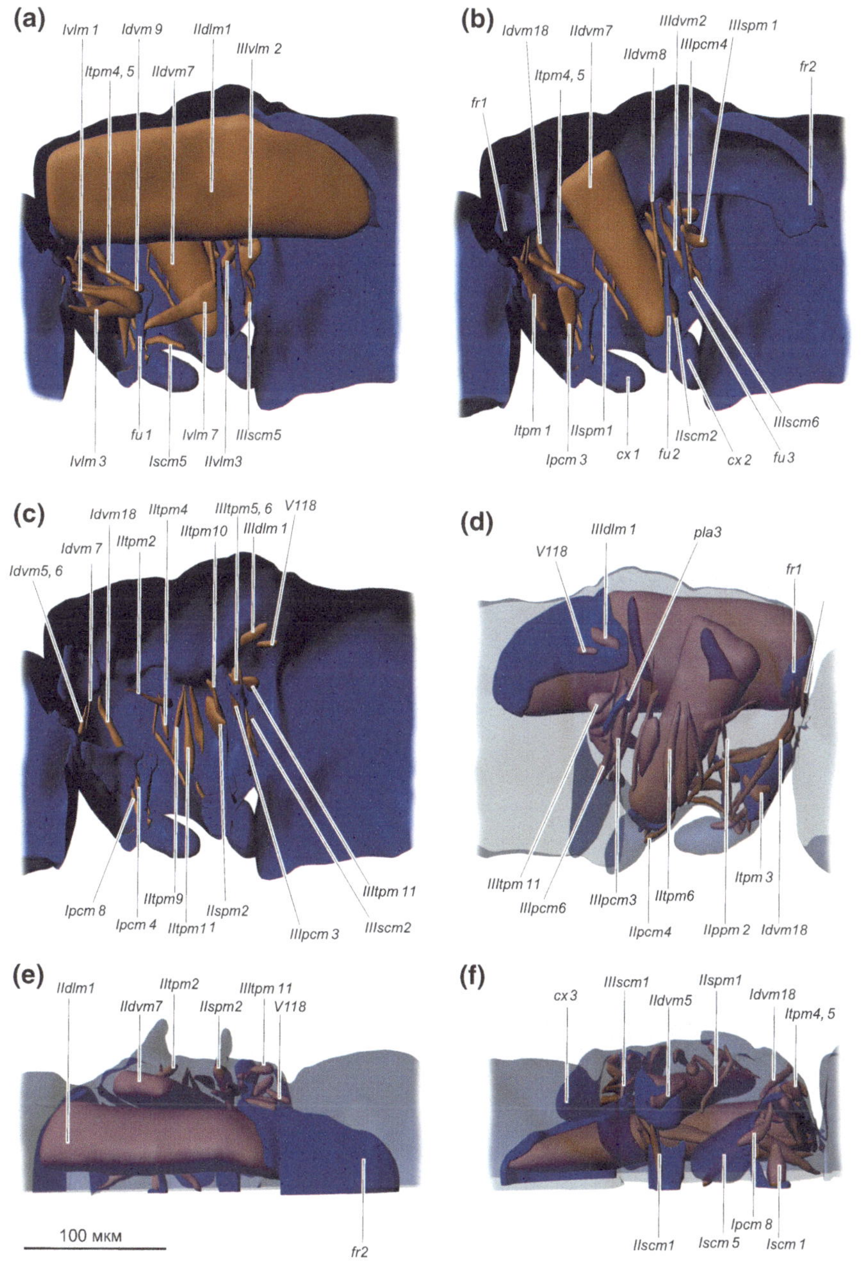

Fig. 8.19 Musculature of mesosoma in *Trichogramma evanescens*, 3D: **a–c** lateral internal view; **d** lateral external view; **e** dorsal view; **f** ventral view; *pl1* pleurite of prothorax, *pla3* pleural apodeme of metathorax, *fr1.2* pro- and mesophragma, *fu1.2.3* pro-, meso-, and metafurca, *cx1.2.3* pro-, meso-, and metacoxae. Musculature see text

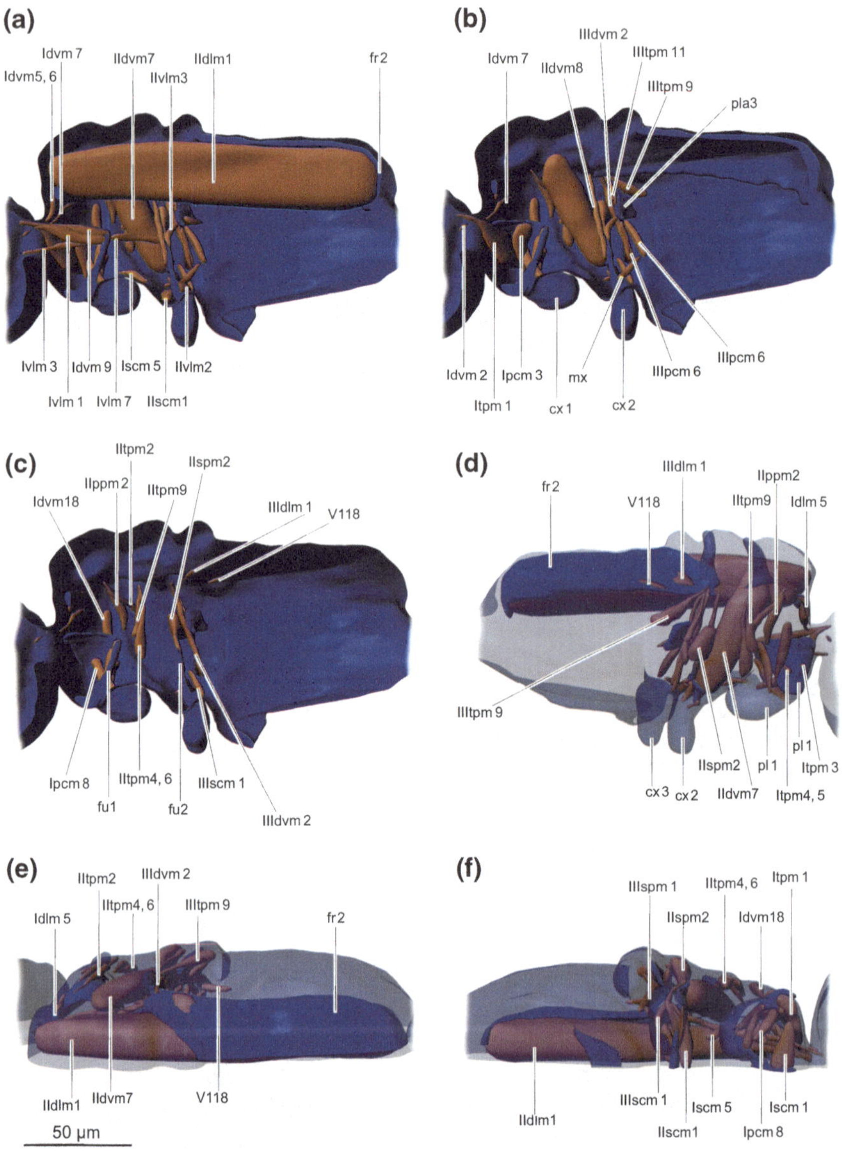

Fig. 8.20 Musculature of mesosoma in *Megaphragma mymaripenne*, 3D: **a–c** lateral internal view; **d** lateral external view; **e** dorsal view; **f** ventral view; *pl1* pleurite of prothorax, *pla3* pleural apodeme of metathorax, *fr1.2* pro- and mesophragma, *fu1.2* pro- and mesofurca, *cx1.2.3* pro-, meso-, and metacoxae. Musculature see text

Mesothorax. IIdlm1 (M. prophragma-mesophragmalis) largest muscle: O, prophragma; I, mesophragma. IIdvm7 (M. mesonoto-trochanteralis): O, mesonotum; I, apodeme of trochanter. IIdvm8 (M. mesofurca-phragmalis): O, mesofurca; I, mesophragma. IItpm2 (M. mesopleura-praealaris): O, pleurite; I, prealar zone. IItpm4 (M. mesonoto-pleuralis anterior): fused with IItpm6, O, pleurite; I, margin of mesonotum. IItpm6 (M. mesonoto-pleuralis posterior) fused with IItpm4. IItpm9 (M. mesepimero-axillaris tertius): O, pleurite; I, third axillary plate. IItpm10 (M. mesepimero-subalaris) in *Trichogramma* present: O, pleurite; I, subalare; in *Megaphragma* absent. IItpm11 (M. mesopleura-subalaris) in *Trichogramma* present: O, pleurite; I, subalare; in *Megaphragma* absent. IIppm2 (M. mesobasalare-intersegmentalis): O, intersegmental juncture; I, basalare. IIspm1 (M. mesopleura-sternalis), in *Trichogramma* present: O, anterior part of pleurite; I, sternal region; in *Megaphragma* absent. IIspm2 (M. mesofurca-pleuralis): O, apex of mesofurca, I, pleurite. IIvlm3 (M. mesofurca-metafurcalis) in *Trichogramma* present: O, mesofurca; I, metafurca. in Megaphragma with muscle of unclear homology, possibly, IIvlm3 with atypical attachment site: O, mesofurca; I, fold between meso- and metasoma. IIscm1 (M. mesofurca-coxalis anterior): O, mesofurca; I, anterior margin of base of coxa. IIscm2 (M. mesofurca-coxalis posterior): O, mesofurca; I, posterolateral margin of base of coxa. IIscm3 (M. mesofurca-coxalis medialis): O, mesofurca; I, medial margin of base of coxa. Internal muscles of legs not studied because of extremely small size.

Metathorax and propodeum. IIIdlm1 (M. mesophragma-metaphragmalis): O, mesophragma; I, metaphragma. IIIdvm2 (M. metanoto-trochantinalis anterior): O, metanotum; I, trochanter. IIItpm5 (M. metanoto-pleuralis medialis) and IIItpm6 (M. metanoto-pleuralis posterior) fused: O, pleural apodeme of metathorax; I, lateral margin of metanotum. IIItpm9 (M. metepimero-axillaris tertius): O, pleural apodeme; I, third axillary plate. IIItpm7 (M. metanepisterno-axillaris): O, pleural apodeme; I, third axillary plate. IIItpm11 (M. metapleura-subalaris): O, pleural apodeme; I, subalare. IIIspm1 (M. metapleura-sternalis): O, ventral part of propodeum; I, basalare. IIIpcm3 (M. metanepisterno-coxalis anterior): O, pleural apodeme; I, anterolateral margin of base of coxa. IIIpcm4 (M. metanepisterno-coxalis posterior): O, pleural apodeme; I, posterolateral margin of base of coxa. IIIpcm6 (M. metapleura-trochanteralis): O, pleural apodeme; I, trochanter. IIIvlm2 (M. metafurca-abdominosternalis): *in Trichogramma* present: O, metafurca; I, abdominal sternite 2; in *Megaphragma* absent. IIIscm1 (M. metafurca-coxalis anterior) in *Trichogramma*: O, metafurca; I, anterior margin of base of coxa; in *Megaphragma*: O, fold between meso- and metathorax; I, anterior margin of base of coxa. IIIscm2 (M. metafurca-coxalis posterior) in *Trichogramma* present: O, metafurca; I, posterior margin of base of coxa; in *Megaphragma* absent. IIIscm5 (M. metaspina-coxalis) in *Trichogramma* present: O, fold between meso- and metasoma; I, posterior margin of base of coxa; in *Megaphragma* absent. IIIscm6 (M. metafurca-trochanteralis) in *Trichogramma* present: O, metafurca; I, trochanter, via fine tendon; in *Megaphragma* absent. Muscle V118, described only in hymenopterans (Vilhelmsen et al. 2010: no. 118, ph3–T2. M. metaphragma-second abdominal tergal): O, metaphragma; I, tergite of abdominal segment 2. in

Megaphragma with additional muscle of unclear homology (mx): O, base of mesofurca; I, medial margin of base of metacoxae. Internal muscles of legs not studied because of extremely small size.

Musculature of metasoma. Dorsal longitudinal muscles (Mm. dorsales): O, anterior phragma; I, posterior phragma. Ventral longitudinal muscles (Mm. ventrales), several parallel fibers: O, anterior margin of segment; I, posterior

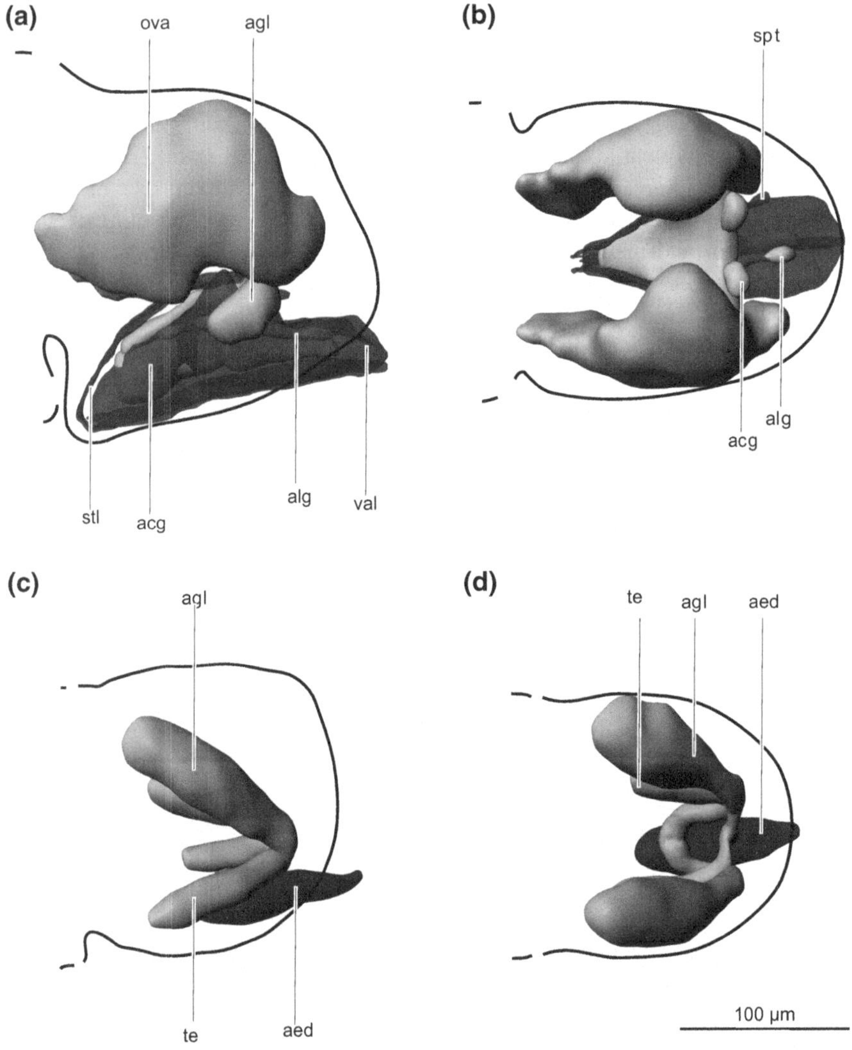

Fig. 8.21 Reproductive system of *Trichogramma evanescens*, 3D: **a**, **b** female; **c**, **d** male; **a**, **c** lateral view; **b**, **d** dorsal view; *acg* acid gland, *aed* aedeagus, *agl* accessory glands, *alg* alkaline gland, *ova* ovary, *spt* spermatheca, *stl* stylet of ovipositor, *te* testis, *val* valves of ovipositor

margin of segment. Dorsoventral muscles, several of urotergosternal muscles, homology unknown. Ovipositor with group of strong retractors (Snodgrass 1942: no. 198, 199).

Reproductive system

Male reproductive system consisting of paired testes, spermiducts, accessory glands, and copulatory apparatus (Figs. 8.21c, d and 8.22c, d).

Female reproductive system consisting of paired ovaries and paired oviducts fused into unpaired oviduct connected to vagina (Figs. 8.21a, b and 8.22a, b). Spermatheca pear-shaped (*Trichogramma*) or rounded (*Megaphragma*). Well-developed acid

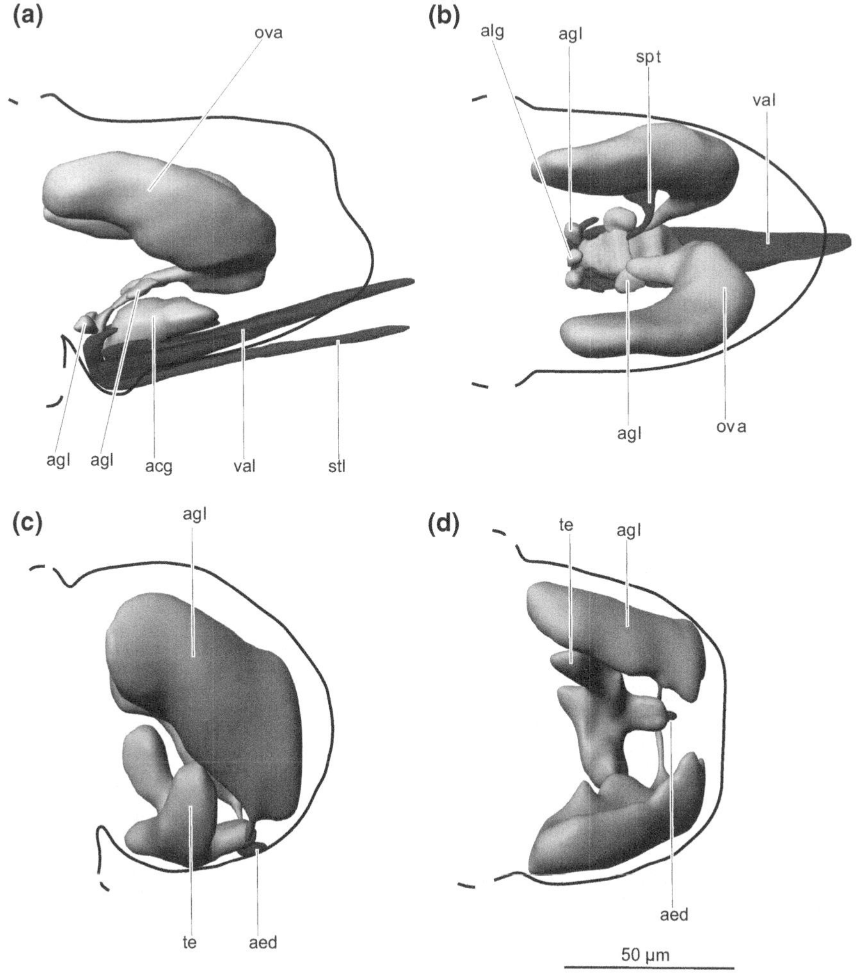

Fig. 8.22 Reproductive system of *Megaphragma mymaripenne*, 3D: **a**, **b** female; **c**, **d** male; **a**, **c** lateral view; **b**, **d** dorsal view; *acg* acid gland, *aed* aedeagus, *agl* accessory glands, *alg* alkaline gland, *ova* ovary, *spt* spermatheca, *stl* stylet of ovipositor, *te* testis, *val* valves of ovipositor

gland, alkaline gland, and paired accessory glands present. Each ovary consisting of 2 polytrophic ovarioles. Ovipositor described in section on morphology of metasoma.

8.4 Morphology of Larvae

Larvae of Trichogrammatidae lack legs, mouthparts (except for mandibles), eyes, or antennae. First instar larva represented by two morphological types (Bakkendorf 1934; Boivin 2010): mymariform, typical of *Ophioneurus* and *Poropoea* (body C-shaped, segmented; with distinguished head and caudal process; with long setae); sacciform, typical of *Trichogramma*, *Chaetostrichia*, *Oligosita*, and *Prestwichia* (body sacciform, without pronounced segmentation, head not distinguished; setae absent). Some authors have described in some species of *Trichogramma* worm-like first instar larvae (Boivin 2010) and in some cases noted weakly pronounced segmentation (Chumakova 1966). Second instar larvae sacciform, without pronounced segmentation, integument without setae. Body length in larvae of *Trichogramma* 150–300 μm at hatching; during feeding they grow to 500 μm (Boivin 2010). Larval intestine differentiated into pharynx, oesophagus, and midgut (Jarjees et al. 1998) or stomodaeum, midgut, and hindgut (midgut and hindgut not connected, Wu et al. 2000), stomodeum with four dorsal muscles and one ventral muscles (Wu et al. 2000). Structures recorded in some larvae include primordia of the reproductive system (Chumakova 1966; Ivanova-Kazas 1961), anal vesicle (Jarjees et al. 1998; Jarjees and Merritt, 2002) and primordia of the nervous system (Ivanova-Kazas 1961); at later stages of larval development primordia of the tracheal system and cutaneous musculature appear (Ivanova-Kazas 1952).

References

Amornsak, W., Cribb, B., & Gordh, G. (1998). External morphology of antennal sesilla of *Trichogramma australicum* Girault (Hym.: Trichogrammatidae). *International Journal of Insect Morphology and Embryology, 27*(2), 67–82.

Bakkendorf, O. (1934). Biological investigations on some Danish hymenopterous egg-parasites, especially in homopterous and heteropterous eggs, with taxonomic remarks and descriptions of new species. *Entomologiske Meddelelser, 19*, 1–134.

Boivin, G. (2010). Reproduction and immature development of egg parasitoids. In F. L. Consoli, J. R. P. Parra, & R. A. Zucchi (Eds.), *Egg Parasitoids in agroecosystems with emphasis on Trichogramma* (pp. 1–23). Houten: Springer.

Bradley, J. C. (1955). The wing-venation of Chalcidoidea and of some allied Hymenoptera. *Bulletin et Annales de la Societe Royale Belge d'Entomologie, 27*, 127–137.

Burks, B. D. (1938). A study of chalcidoid wing. *Annals of the Entomological Society of America, 31*, 157–160.

Chumakova, B. M. (1966). Анатомия ювенальных и дефинитивной фаз развития (The anatomy of the juvenile and definitive phases of development). *Biologicheskie metody bor'by s vreditelyami sel'skogo i lesnogo hozyajstva* (pp. 66–70). Tashkent: FAN.

Chumakova, B. M. (1968). Сравнительная анатомия и биология трихограммы обыкновенной. *Trudy VIZR, 31*, 164–182.

Cônsoli, F. L., Rossi, M. M., & Parra, J. R. P. (1999). Developmental time and characteristics of the immature stages of *Trichogramma galloi* and *T. pretiosum* (Hymenoptera: Trichogrammatidae). *A Revista Brasileira de Entomologia, 43*, 271–275.

Dahlan, A. N., & Gordh, G. (1996). Development of *Trichogramma australicum* Girault (Hymenoptera: Trichogrammatidae) on *Helicoverpa armigera* (Hübner) eggs (Lepidoptera: Noctuidae). *Australian Journal of Entomology, 35*, 337–344.

Delvare, G. (1993). Guadeloupe avec la description d'une espèce nouvelle (Hymenoptera, Trichogrammatidae). *Revue Francaise d'Entomologie, 15*, 149–152.

Fischer, S., Müller, C. H. G., & Meyer-Rochow, V. B. (2011). How small can small be: the compound eye of the parasitoid wasp *Trichogramma evanescens* (Westwood, 1833) (Hymenoptera, Hexapoda), an insect of 0.3- to 0.4-mm total body size. *Visual Neuroscience, 28*(4), 295–308.

Flanders, S. E. (1937). Notes on the life history and anatomy of the *Trichogramma*. *Annals of the Entomological Society of America, 30*, 304–308.

Hawlitzky, N., & Boulay, C. (1982). Régimes alimentaires et développement chez *Trichogramma maidis* Pintureau et Voegele (Hym. Trichogrammatidae) dans l'oeuf *d'Anagasta kuehniella* Zeller (Lep. Pyralidae). *Colloque INRA, 9*, 101–106.

Ivanova-Kazas, O. M. (1952). Постэмбриональное развитие *Prestwichia aquatica* Lubb. (Hymenoptera) (The postembryonic development of *Prestwichia aquatica* Lubb. (Hymenoptera)). *Trudy Leningradskogo obshchestva estestvoispytatelej, 71*(4), 165–213.

Ivanova-Kazas, O. M. (1954). Вопросы эволюции эмбрионального развития у перепончатокрылых (Hymenoptera) (On the evolution of the embryonic development in Hymenoptera). *Trudy Vsesoyuznogo ehntomologicheskogo obshchestva, 44*, 301–335.

Ivanova-Kazas, O. M. (1961). *Очерки по сравнительной эмбриологии перепончатокрылых (Essays on the Comparative Embryology of Hymenoptera)*. Leningrad: Leningrad Univ. Press.

Jarjees, E. A., Merritt, D. J., & Gords, G. (1998). Anatomy of the mouthparths and digestive trakt during feeding in larvae of the parasitoid wasp *Trichogramma australicum* Girault (Hymenoptera: Trichogrammatidae). *International Journal of Insect Morphology and Embryology, 22*(2), 103–110.

Jarjees, E. A., & Merritt, D. J. (2002). Development of *Trichogramma australicum* Girault (Hymenoptera: Trichogrammatidae) in Helicoverpa (Lepidoptera: Noctuidae). *Australian Journal of Entomology, 41*, 310–315.

Jarjees, E. A., & Merritt, D. J. (2004). The effect of parasitization by *Trichogramma australicum* on Helicoverpa armigera host eggs and embryos. *Journal of Invertebrate Pathology, 85*, 1–8.

Kochetova, N. I. (1969). Приспособления к паразитизму у некоторых яйцеедов рода *Trichogramma* (Hymenoptera, Trichogrammatidae) (Adaptations to parasitism in some egg parasites of the genus *Trichogramma* (Hymenoptera, Trichogrammatidae)). *Zoologicheskii Zhurnal, 48*(12), 1816–1823.

Lino-Neto, J., & Bao, S. N. (2000). Dolder H (2000) Structure and ultrastructure of the spermatozoa of *Trichogramma pretiosum* and *Trichogramma atopovirilha* (Hymenoptera: Trichogrammatidae). *Acta zoologica (Stockholm), 81*, 205–211.

Lino-Neto, J., & Dolder, H. (2001). Redescription of sperm structure and ultrastructure of *Trichogramma dendrolimi* (Hymenoptera: Chalcidoidea: Trichogrammatidae). *Acta zoologica, 82*, 159–164.

Makarova, A. A., & Polilov, A. A. (2013). Peculiarities of the brain organization and fine structure in small insects related to miniaturization. 2. The smallest Hymenoptera (Mymaridae, Trichogrammatidae). *Entomological Review, 93*(6), 714–724.

Makarova, A. A., Polilov, A. A., & Fisher, S. (2015). Comparative morphological analysis of compound eye miniaturization in minute Hymenoptera. *Arthropod Structure & Development, 44*(1), 21–32.

Nagarkatti, S., & Nagaraja, H. (1977). Biosystematics of Trichogramma and Trichogrammatoidea Species. *Annual Review of Entomology, 22*(1), 157–176.

Olson, D. M., & Andow, D. A. (1993). Antennal sensilla of female *Trichogramma nubiale* (Ertle and Davis) (Hymenoptera: Trichogrammatidae) and comparisons with other parasitic Hymenoptera. *International Journal of Insect Morphology and Embryology, 22*, 507–520.

Pinto, J. D. (2006). A review of the New World Genera of Trichogrammatidae (Hymenoptera). *Journal of Hymenoptera Research, 15*, 38–163.

Polilov, A. A. (2012). The smallest insects evolve anucleate neurons. *Arthropod Structure & Development, 41*(1), 27–32.

Polilov, A. A. (2016). Features of the structure of hymenoptera associated with Miniaturization. 2. Анатомия *Trichogramma evanescens* (Hymenoptera, Trichogrammatidae). *Entomological Review, 96*(4), 419–431.

Ruschioni, S., Romani, R., Riolo, P., & Isidoro, N. (2012). Morphology and distribution of antennal multiporous gustatory sensilla related to host recognition in some *Trichogramma* spp. *Bulletin of Insectology, 65*(2), 171–176.

Schmidt, J. M., & Smith, J. J. B. (1985). The external sensory morphology of the legs and hairplate system of female Trichogramma minutum Riley (Hymenoptera: Trichogrammatidae). *Proceedings of the Royal Society of London. Series B: Biological Sciences, 232*, 323–366.

Snodgrass, R. E. (1942). The skeleto-muscular mechanisms of the honey bee. Washington: The *Smithsonian Institution.*

Sorokina, A. P. (1993). Определитель видов рода *Trichogramma* Westw. (Hymenoptera, Trichogramma) мировой фауны (Guide to Species of the Genus *Trichogramma* Westw. (Hymenoptera, Trichogramma) of the World). Moscow: Kolos.

Sylvere, A. P., & Stein-Margolina, V. V. (1976). Tetrapodili — четырехногие клещи: Электронномикроскопическая анатомия, проблемы эволюции и взаимоотношения с возбудителями болезней растений (Tetrapodili: Fore-legs Mites; Electron Microscopic Anatomy, Evolution Problems and Mutual Relations with Plant Pathogenic Organisms). Tallinn: Valgus.

Vilhelmsen, L., Miko, I., & Krogmann, L. (2010). Beyond the Wasp-Waist: Structural Diversity and Phylogenetic Significance of the Mesosoma in Apocritan Wasps (Insecta: Hymenoptera). *Zoological Journal of the Linnean Society, 159*(1), 22–194.

Voegelt, J., Cals-Usciati, J., Pihan, J. P., & J. Daumal (1975). Structure de l'antennae female des Trichogrammes. *Entomophaga, 20*(16), 1–169.

Volkoff, A. N., Daumal, J., Barry, P., François, M. C., Hawlitzky, N., & Rossi, M. M. (1995). Development of *Trichogramma cacoeciae* Marchal (Hymenoptera: Trichogrammatidae): Time table and evidence for a single larval instar. *International Journal of Insect Morphology and Embryology, 24*, 459–466.

Wu, Z. X., Cohen, A. C., & Nordlund, D. A. (2000). The feeding behavior of *Trichogramma brassicae*: New evidence for selective ingestion of solid food. *Entomologia Experimentalis et Applicata, 96*, 1–8.

Zhang, S., Zhang, Z., Kong, X., Wang, H., Zhou, G., & Yu, J. (2012). External morphology of *Trichogramma dendrolimi* Matsumura (Hymenoptera: Trichogrammatidae) organ and ultrastructure of the sensilla. *Microscopy Research and Technique, 75*(11), 1513–1521.

Chapter 9
Peculiar Miniature-Related Structural Features of Different Organ Systems of Insects

9.1 Introduction

First special studies on the effects of size on structure in insects were published almost 70 years ago (Rensch 1948; Goossen 1949). Those studies described some peculiar features of structure found in relatively small insects (about 2 mm long) and were followed later by studies on the structure of larvae in miniature insects (Beutel and Haas 1998; Grebennikov and Beutel 2002), but large-scale work on investigating the effects of miniaturization on structure in insects started 10 years ago (Beutel et al. 2005; Polilov 2005), and later grew into a special field of research represented by many publications annually. To date, the external and internal morphology of adults or larvae/nymphs has been studied in detail in microinsects of five orders (Psocoptera, Thysanoptera, Coleoptera, Strepsiptera, and Hymenoptera). Not only many publications of results of research, but also several reviews have been published in recent years (Polilov 2015a, b).

9.2 Cuticular Structures (Integument, Skeleton, and Appendages)

The plan of structure of the integument in microinsects is rather similar to that of larger insects, but decreasing body size is accompanied by thinning of the cuticle to small fractions of its original thickness. For instance, the average cuticle thickness of adults is 45.5 μm in *Staphylinus*, 25.8 in *Aleochara*, 14.3 in *Ochthebius*, 6.23 in *Acrotrichis*, and 1.9 μm in *Mikado*. Furthermore, in larvae the cuticle is even thinner: on average 1.0 μm thick in last instar larvae of *Mikado*, and 0.68 μm thick in first instar larvae of *Mikado*. At the same time, in spite of decreased thickness, the cuticle of many microinsects is divided into the epi-, exo-, and endocuticle (in some microinsects only division into epi- and procuticle is traceable). It also has to be

A.A. Polilov, *At the Size Limit—Effects of Miniaturization in Insects*,
DOI 10.1007/978-3-319-39499-2_9

noted that the syncytial structure of the hypoderm described in the smallest four-legged mites (Silvere and Shtein-Margolina 1976) has not been found in insects. The hypoderm of all studied microinsects is represented by an uneven layer, and in many areas it is so thin that it is barely discernible.

As I have shown in the previous chapters, generally body size diminution in the studied insects is accompanied by reduction of the number of fully formed elements of the exoskeleton, simplification of the endoskeleton, and reduction of the number of elements in some articulate structures. The same trends are found in other small insects (Beutel et al. 1999; Pohl 2000; Grebennikov and Beutel 2002). Because skeletal structures are very diverse and the homology of particular elements in different orders is sometimes difficult to trace, analysis of their changes is very difficult. Therefore, I will focus below on a number of skeletal elements rather than on the evolution of particular parts of the skeleton.

The degree of fusion of cranial elements is quite different in microinsects of different orders. Adults of Ptiliidae and Corylophidae have fused cranium without sutures, distinguishing them from larger representatives of related groups of Coleoptera, in which at least the frontoclypeal suture and gular sutures are present. The cranium of Mymaridae and Trichogrammatidae also lacks distinguished parts, except for the frontoclypeal zone, separated by trabecula in Mymaridae. The only exception among Mymaridae is the male of *Dicopomorpha*, in which the cranium lacks any sutures. Adults of other studied microinsects have at least three distinguished elements of the cranium. Larvae display a similar picture: cranium has no sutures in larvae of Ptiliidae and Corylophidae and in first instar larvae of Trichogrammatidae and Mymaridae, and in late instar the larvae of Trichogrammatidae and Mymaridae the cranium is not clearly distinguished (Bakkendorf 1934; Boivin 2010). In nymphs of psocopterans and thrips the cranium is divided similar to that of adults, but it is not so strongly pronounced.

The tentorium of the studied microinsects demonstrates different degrees of simplification. Adults of Liposcelididae have the same structure of tentorium as in larger psocopterans. All thrips display strong reduction and asymmetry of tentorium, associated with the reorganization of mouthparts. Adults of Mymaridae and Trichogrammatidae have tentorium similar to that of other chalcidoids, but differ from most other hymenopterans in the absence of dorsal arms. The tentorium of adult ptiliids differs from that of other staphylinoids in the absence of the laminitentorium, and in Nanosellini it also differs in the absence of dorsal arms. The most strongly reduced tentorium is found in adult Corylophidae, in which it is represented only by the posterior arms and tentorial bridge. The larvae of *Liposcelis* differ from adults in the absence of dorsal hands; ptiliid larvae also have no dorsal hands. The larvae of Trichogrammatidae and Mymaridae have no tentorium (Ivanova-Kazas 1961).

In spite of their small body size, in most studied insects the number of antennomeres is not strongly reduced. Only in adults of some genera of Ptiliidae and Corylophidae the number of antennomeres is reduced from 11 to 8; and males of *Dicopomorpha* have 1-segmented antennae.

The structure of mouthparts in most microinsects displays no considerable simplifications. The mouthparts of adults and nymphs of *Liposcelis* are not different from those of larger psocopterans. The mouthparts of thrips are peculiarly structured, but they cannot be considered simplified, and their reorganization is related rather to the very peculiar type of feeding found in thrips (Heming 1978). The mouthparts of adult Ptiliidae, in spite of their extremely small body size, not only lack simplifications, but also have several novel formations, not described in other staphylinoids. Mouthparts of ptiliid larvae are identical in structure to those of other staphylinoids. Corylophids display no considerable differences in the structure of mouthparts from larger representatives of related groups of beetles. Some of the representatives of Scydmaeninae (Coleoptera: Staphylinidae) have adapted to feeding oribatid mites (Fig. 9.1) (Jałoszyński and Olszanowski 2013, 2015, 2016). Only Mymaridae and Trichogrammatidae display simplifications in some mouthparts: labrum is weakly discernible and represented by membranous fold, and maxillary and labial palps are strongly reduced, 1-segmented. In males of *Dicopomorpha* the mouthparts are absent. In the larvae of Trichogrammatidae and Mymaridae only rather small mandibles or their primordia are present (Bakkendorf 1934; Jackson 1961; Boivin 2010).

The structure of prothorax strongly varies in the studied microinsects. Among adults one extreme is represented by thrips, which have eight differentiated sclerites (basisternite, furcosternite, spinasternite, right and left episterna and epimera, and pronotum), and the other extreme is represented by Ptiliidae and Corylophidae, in which the entire exoskeleton of prothorax is represented by a single formation (pronotum with hypomeron). In others 3–5 fully formed elements are found. Larvae and nymphs of most microinsects have weakly pronounced sclerites typical of adults (nymphs of thrips and psocopterans), without fully formed sclerites, except for the weakly sclerotized notal sclerite (larvae of Ptiliidae and Corylophidae), or lack sclerites altogether (late instar larvae of Mymaridae and Trichogrammatidae, Bakkendorf 1934; Jackson 1961; Boivin 2010).

It is convenient to discuss the meso- and metathorax together, because in many microinsects they are strongly fused. The minimum number of fully formed sclerites among adults—10 (excluding wing plates, because they are indiscernible in most microinsects)— is found in Liposcelididae (in large psocopterans their number is at least 16, but in this case the differences reflect not only decreased body size, but also the fact that *Liposcelis* is wingless) and Ptiliidae (in larger representatives of related groups 14, because of distinguished mesepisterna and mesepimera). In *Heliothrips* the number is 11, but in many thrips of the same size it is at least 14 (Matsuda 1970). In Corylophidae it is 12, and they differ from large Cucujoidea only in the fusion of the mesepisternum and mesepimeron. In Mymaridae it is 15, and in Trichogrammatidae 16, as in most chalcidoids. A fundamentally different picture of the structure of meso- and metathorax is found in males of *Dicopomorpha*, in which these segments are weakly sclerotized without additional differentiation. It can be suggested that most microinsects share the trend to fusion of pleural elements of the thorax. The meso- and metathorax of nymphs display the same peculiar features as prothorax, described above.

Fig. 9.1 *Cephennium ruthenum* (Coleoptera: Staphylinidae) during feeding on a non-ptyctimous (**a**) and a ptyctimous (**b**) mite, SEM (modified after Jałoszyński and Olszanowski 2016)

The endoskeleton of the thorax in adults of most microinsects is strongly simplified, the furci are unbranched; Ptiliidae and Corylophidae differ from their large relatives in the absence of common stem of the metendosternite (metafurca), in *Megaphragma* the metafurca is absent. The furci of the larvae and nymphs of microinsects are even less developed than those of adults. One peculiar feature of Trichogrammatidae is the enlarged mesophragma sunk into the body and reaching the middle of the abdomen in *Trichogramma* or almost reaching the apex of the abdomen in *Megaphragma*.

The legs of all microinsects are divided into parts typical of insects. In some (*Liposcelis*) the trochanters are fused with the femora. The number of tarsomeres in most microinsects is smaller than in larger representatives of related groups. In ptiliids the tarsi are 2–3-segmented (in larger staphylinoids they are 4–5-segmented); in Corylophidae they are 3-segmented (in larger cucujoids 3–5-segmented), in thrips 1–2-segmented, in males of *Dicopomorpha* tarsomeres are even absent. The larvae of Trichogrammatidae and Mymaridae are legless.

The wing apparatus of most microinsects is characterized by ptiloptery (Fig. 9.2) and reduced number of veins. The reduction of venation related to decreased body size is found in many hymenopterans (Rasnitsyn 1980; Antonov 2009). The wing blade of all flying microinsects is narrow and contains at most three veins, and even

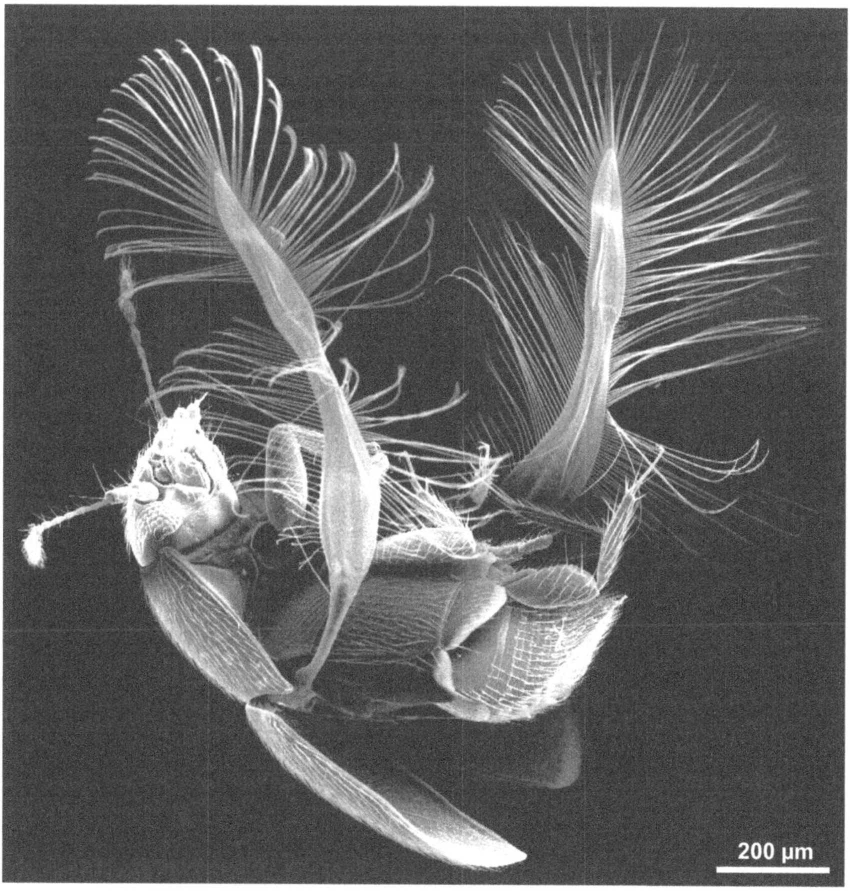

Fig. 9.2 Ptiliid *Acrotrichis grandicollis* with unfolded wings, SEM

they are often developed very poorly or their number is reduced to one (hindwings of Thripidae, Mymaridae, and Trichogrammatidae). The main plane of the wing is formed by long setae on the perimeter of the narrow wing blade. Interestingly, in Ptiliidae these setae bear numerous processes, probably for increasing the total area of the wing. One of the most important challenges faced by insects of decreasing size is the decreased efficiency of flapping flight. This efficiency is determined by the Reynolds number, which is directly proportional to the size of the wing and velocity of flight and inversely proportional to the kinematic viscosity of air. This parameter is much lower in microinsects than in larger insects (Horridge 1956). As wing size decreases, laminar separations emerge, also decreasing the efficiency of flapping flight (Svidersky 1980b). It is unclear how microinsects solve these aerodynamic problems; it is possible that they do so using a special trajectory of wing movement, in which each phase of lowering the wing begins from a position of wings joined together above the body, and lift is created as a result of rarefied air formed between them as they are separated (Weiss-Fough 1973). Some authors also believe that at such small body sizes the relative viscosity of air is so high that the flight of microinsects is rather similar to swimming (Walker 2002).

The skeleton of the abdomen in adult microinsects demonstrates two opposite trends: on one hand, in some of the smallest hymenopterans the number of distinguished elements of the abdominal skeleton is reduced (in *Megaphragma* sternites are almost indiscernible; in males of *Dicopomorpha* only four abdominal segments are present); on the other hand, in Ptiliidae and Thripidae ten well developed tergites have been found, a state rather rare in larger insects. The larvae or nymphs of all studied microinsects lack fully formed elements of abdominal skeleton.

The external genitalia of Liposcelididae and Thripidae are rather similar in complexity of organization to those of large psocopterans. The copulatory apparatuses of male Ptiliidae and Corylophidae are different from those of larger representatives of related groups in the absence of parameres. The ovipositors and male copulatory apparatuses of the smallest hymenopterans display no considerable changes, except for the reduction of parameres in *Megaphragma*.

9.3 Digestive and Excretory Systems

The digestive canals of all microinsects are similar in structural organization and different from larger representatives of related groups in somewhat weaker differentiation of regions. In most studied microinsects both the crop and the muscular stomach are absent. In Ptiliidae and Corylophidae the muscle 0st2 is reduced, and all studied microinsects lack musculature of the midgut. In ptiliids numerous diverticula of the midgut have been found, in the anterior region they are especially large and reaching into the cranium. In first instar larvae of *Liposcelis* the midgut is shifted anteriad and partly lies in the head. The number of Malpighian tubules of all studied microinsects is as small as the smallest, described in larger representatives of related groups, and in Mymaridae and Trichogrammatidae the number of Malpighian tubules is reduced compared to larger chalcidoids. In many microinsects the salivary glands are reduced. Therefore, it can be seen that body size diminution does not lead to strong simplification of the Digestive and excretory systems. At the same time, the shape of cells of the midgut changes: in larger insects they are mostly columnar, whereas in the studied microinsects they are strongly flattened. The same picture has been described also in some other small insects (Rensch 1948).

9.4 Circulatory System and Fat Body

The circulatory system of the smallest insects is strongly simplified; none of the studied microinsects have any blood vessels or pulsating organs, except for a simple heart and short aorta; in Ptiliidae and Trichogrammatidae heart and blood vessels are absent. This simplification is explained by the fact that with decreasing body size capillary forces hinder efficient circulation of hemolymph over the body. In Ptiliidae hemolymph is replaced by parenchyma-like fat body (Fig. 9.3).

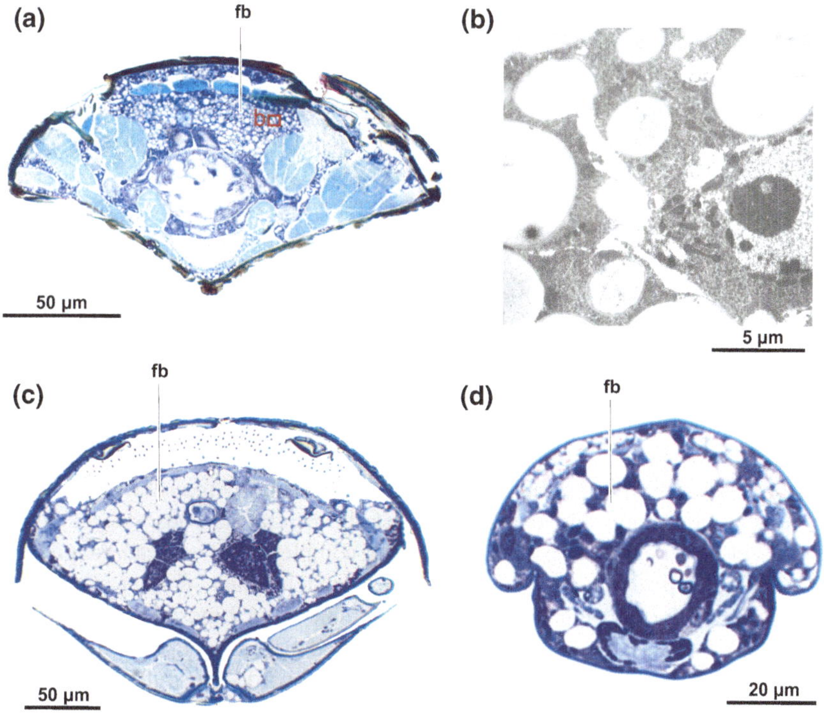

Fig. 9.3 Fat body of Ptiliidae: **a**, **b** *Acrotrichis montandoni*, adult, mesothorax, cross-section; **a** azure and eosin; **b** TEM; **c**, **d** *Mikado* sp., cross-section, toluidine blue, and pyronin; **c** adult, base of abdomen, **d** last instar larva, abdominal segment 3; *fb* fat body

The absence of a transport system is compensated by the high efficiency of diffusion at such small sizes. A similar picture is found in many Acari, especially small ones (Silvere and Shtein-Margolina 1976). Simplification of vascular system with decreasing body size, up to complete loss in the smallest representatives, has also been described in Archiannelida (Rensch 1948). Lack of heart is also known in some smaller mollusks (Brenzinger et al. 2013) and crustaceans (Boxshall 1982).

9.5 Tracheal System

The structure of the tracheal system displays considerable simplification with decreasing body size. In most microinsects it is represented only by longitudinal stems and weakly branching tracheae. The number of spiracles is smaller than in larger insects, and in some cases only one pair of spiracles is present (larvae of Ptiliidae) or spiracles are absent (larvae of Trichogrammatidae and Mymaridae). The larvae of Trichogrammatidae and Mymaridae develop in eggs of their hosts, have no pronounced tracheal system, and breathe through the integument. Ptiliid larvae are free-living, but with only one pair of spiracles (mesothoracic), without tracheal stems or tracheae in the abdomen, and probably employ mixed, tracheal-cutaneous type of respiration. Respiratory system is absent in many collembolans (Ruppel 1953) and four-legged mites (Silvere and Shtein-Margolina 1976). Simplification and reduction of the respiratory system with decreasing body size has also been recorded in many other invertebrates (Gorodkov 1984).

9.6 Central Nervous System

All microinsects display strong oligomerization and condensation of ganglia. These phenomena, known also in other insects (Niven et al. 2008; Chaika 2010), in many cases are associated with body size (Heath and Evans 1990). Miniaturization considerably increases their degree, and in extreme cases oligomerization and condensation of ganglia in the central nervous system transforms it into a single formation with weakly discernible ganglia. Adult beetles demonstrate the highest degrees of compaction and condensation of the central nervous system among microinsects; beetles are also characterized by complete shift of head ganglia into the thorax in adults (*Sericoderus*), which is unknown in adults of other insects. The abdominal ganglia in adult Ptiliidae, Corylophidae, and Liposcelididae are also shifted into the thorax and fused with the metathoracic ganglion The smallest degree of compaction among microinsects has been recorded in adult Thripidae and Trichogrammatidae. Compared to adults, the larvae and nymphs of microinsects display two opposite phenomena. The central nervous system of beetle larvae demonstrates markedly smaller degrees of convergence and fusion of ganglia, but their cerebrum is always situated outside of the cranium (in first instar larvae of *Mikado* the cerebrum reaches abdominal segment 2). By contrast, nymphs of thrips and psocopterans have considerably higher degrees of oligomerization and condensation of ganglia compared to adults, but their cerebrum invariably remains in head. The larvae of Trichogrammatidae and Mymaridae have only primordial of the central nervous system (Ivanova-Kazas 1961).

Another important peculiar feature of the central nervous system of microinsects is its asymmetry, found in many of the smallest insects, but pronounced to different degrees. Adults of the smallest ptiliids (Nanosellini) have asymmetrical distal processes of the protocerebrum (the left one larger than the right one). Adult females of Mymaridae have the abdominal synganglion shifted into the right half of the abdomen. The cerebrum of first instar larvae of the smallest beetles is strongly asymmetrical (the right half is markedly larger than the left half). The cerebrum of first instar nymphs of *Liposcelis* has asymmetrical distal processes (left larger than right). The suboesophageal ganglion and thoracic ganglia of first instar larvae of *Heliothrips* have asymmetrical processes running into the coxal zone of legs. The unusual shape of the brain is also described for Scydmaeninae (Coleoptera: Staphylinidae) (Jałoszyński et al. 2012). The shift of part of the suboesophageal ganglion into the coxae has also been described in nymphs of smaller spiders (Quesada et al. 2011).

In spite of their small body sizes, the cerebrum of all studied microinsects demonstrates complete differentiation into all cerebral centers typical of larger insects. The only exceptions are the loss and strong reduction of calyces in the mushroom bodies in *Heliothrips* and *Nanosella*, respectively. The cerebrum of the smallest insects displays strong compaction and convergence of cerebral centers, as well as changes of their spatial organization, such the convergence and changes of the position of optic lobes found in most of the smaller coleopterans and hymenopterans. In addition to the complete set of cerebral centers, the cerebrum almost preserves the proportions of particular parts. The neuropilar zones of particular cerebral centers change strictly isometrically (e.g., the central body complex in beetles) or displays minor allometry. One common trend of almost all groups of microinsects is the decrease, although rather small, in the relative volume of zones responsible for co-ordination and processing of sensory signals (optic and antennal lobes). Allometric changes of particular structures of the cerebrum have been described in vertebrates (Stephan et al. 1970, 1981, 1988; Platel 1976; Boire and Baron 1994); for insects only scattered data are available on volumes of particular cerebral centers in some model organisms: *Drosophila* (Rein et al. 2002), *Apis* (Rohlfi et al. 2004; Brandt et al. 2005), *Schistocerca* (Kurylas et al. 2008), *Manduca* (el Jundi et al. 2009), and *Tribolium* (Dreyer et al. 2010), but it is impossible to use these data in analysis, because in the above-listed studies the relative volume was calculated not relative to the volume of the entire cerebrum, but relative to the total volume of al neuropilar zones used in the analysis.

One important difference between hemimetabolous and holometabolous insects is the timing of the formation of cerebral centers. It has been shown that in most holometabolous insects the formation of the structure of the cerebrum mainly takes place at the pupal stage (Panov 1957, 1959, 1960a, b, 1961; Hinke 1961; Wegerhoff and Breidbach, 1992). In hemimetabolous insects the differentiation of structures in the nervous system is mainly completed already in embryogeny (Panov 1957, 1959, 1960a, b, 1961; Bate 1976; Doe and Goodman 1985). The same picture is found also in the smallest insects: the cerebrum of beetle larvae is not differentiated into cerebral zones or only the earliest stages of the formation of cerebral centers can be observed. The structure of cerebrum in first instar nymphs of psocopterans and thrips is already quite similar to that of adults.

Body size diminution is accompanied in all studied insects by considerable diminution of cell bodies in the nervous system (Figs. 9.4 and 9.5a). For instance, the average diameter of cell bodies of the cerebrum is 6.2 μm in *Aleochara*, 3.9 μm in *Acrotrichis grandicollis*, and 1.7 μm in *Nanosella*. This tendency can be traced in all studied orders; it has also been revealed in Strepsiptera (Beutel et al. 2005). The size of ganglionic cells in four-legged mites is comparable to that of the neurons of the smallest insects (Silvere and Shtein-Margolina 1976). Decreasing neuron size with decreasing body size has also been recorded in other animals (Schmalhausen 1984). The diminution of cells is accompanied by averaging: in larger insects neurons strongly vary in size, from 5 to 50 μm (Plotnikova 1979; Svidersky 1980a), whereas in the smallest insects all cells of the central nervous system are subequal in size (e.g., in *Nanosella* the size range is only from 1.19 to 1.98 μm). The decrease of the cell body size of neurons is accompanied by considerable changes in the nuclear-cytoplasmic ratio: the volume of the cytoplasm decreases and in the smallest insects the nucleus occupies up to 90 % of the volume of cell bodies. Considerable increase in the degree of chromatin compaction has been recorded in the smallest of Ptiliidae, Mymaridae, and Trichogrammatidae. In spite of the decreasing size of cells and decreasing volume of cytoplasm of cell bodies and projections of neurons, the neurons of the studied microinsects retain the complete set of organelles typical of nerve cells. The composition and structure of the ensheathment of microinsects are also identical to those of larger insects, except for the incomplete extraneural sheath (Ptiliidae, *Megaphragma*) and absence of neural lamella in *Megaphragma*. Absence of neural lamella has also been found in the smallest mites (Silvere and Shtein-Margolina 1976). The ganglia of the studied microinsects contain all main types of glia found in larger insects. Therefore, it can be concluded that the ultrastructure of the central nervous system in the smallest insects remains unchanged as body size decreases (except for *Megaphragma*).

Since the process of reducing the size of cell bodies of neurons (which involves reducing the size of the nucleus and main organelles) is obviously complicated, it would be logical to assume that the relative volume of the cortical layer of the cerebrum increased with decreasing body size. Nevertheless, judging by the results of studies, the relative volume of neuropil in all studied insects is almost constant (the average value for all studied insects is 59.4 %), and changes isometrically relative to the cerebrum with decreasing body size. According to mathematical models, the diameter of projections cannot be smaller than 80 nm (Faisal et al. 2005). However, the minimum diameter of fibers in *Megaphragma* is 30 nm; axons

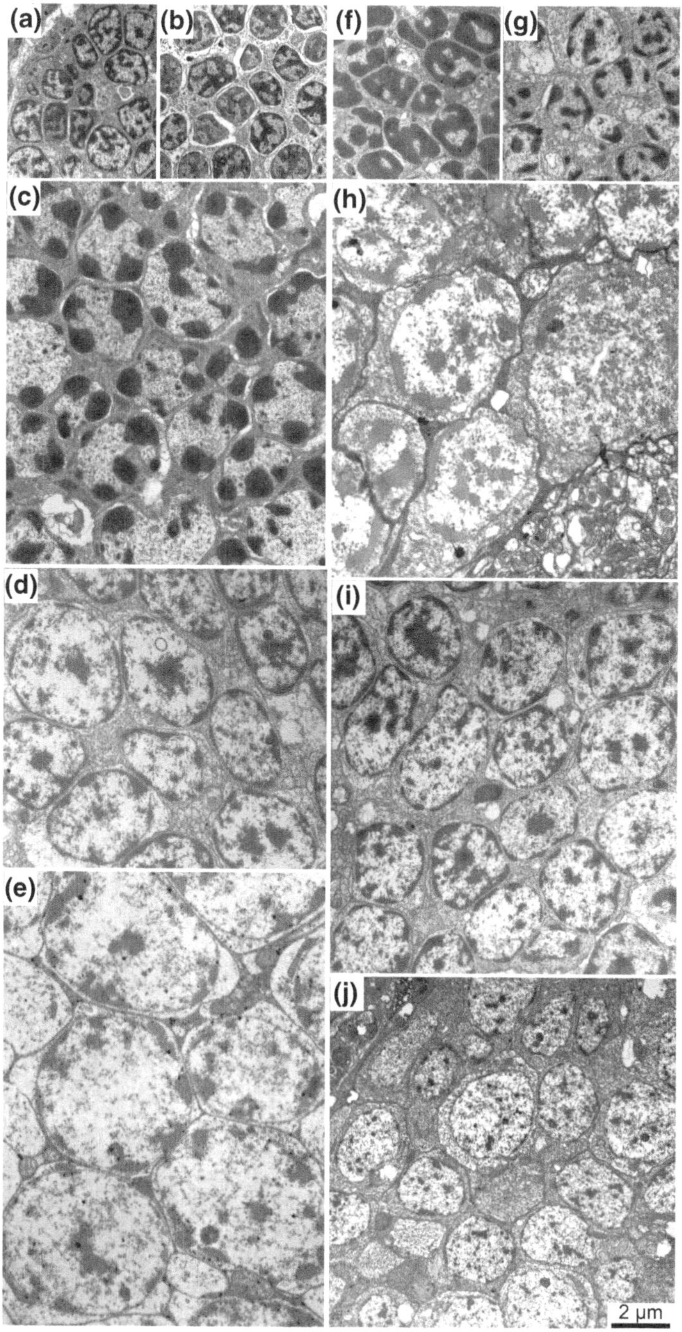

Fig. 9.4 Comparative size of neurons, TEM: **a** *Mikado* sp.; **b** *Nanosella* sp.; **c** *Acrotrichis grandicollis*; **d** *Sericoderus lateralis*; **e** *Aleochara* sp.; **f** *Anaphes flavipes*; **g** *Trichogramma evanescens*; **h** *Hemiptarsenus* sp.; **i** *Heliothrips haemorrhoidalis*; **j** *Liposcelis* sp

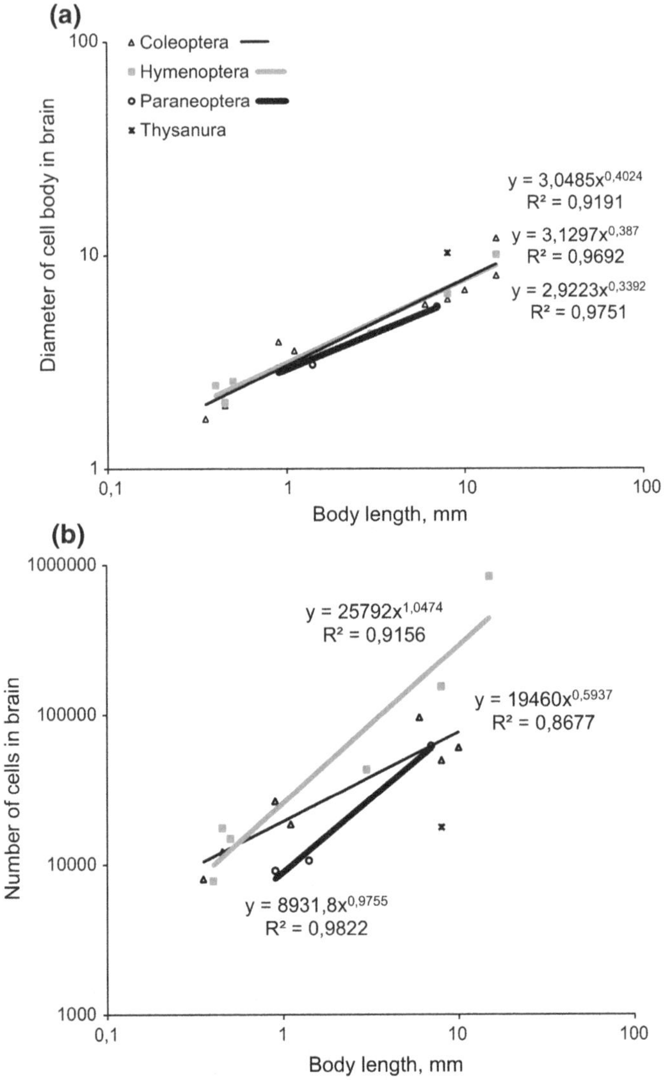

Fig. 9.5 Changes in diameter of cell bodies (**a**) and number of cells (**b**) in cerebrum of adult insects. For list of studied species see Polilov 2014

45 nm in diameter have been found in *Encarsia* (Hymenoptera, Aphelinidae; Hustert 2012). Although the actual minimum diameter of nerve cell projections in microinsects is smaller than the theoretical minimum, the limits implied by the models proposed by Faisal do not allow the relative volume of neuropil to decrease considerably.

The number of cells in the nervous system of animals varies from 10^2 in rotifers and nematodes to 10^{14} in some mammals (Meinertzhagen 2010). Among invertebrates, insects have some of the highest numbers of neurons: the number of neurons in the cerebrum can reach 850,000 in workers of *Apis* and 1 200,000 in males of *Apis* or cockroaches of the genus *Periplaneta* (Strausfeld 1976). Similar numbers of neurons have been found only in larger crustaceans and cephalopods (Meinertzhagen 2010). Microinsects display considerably decreased numbers of neurons compared to larger representatives of related groups (Fig. 9.5b). For instance, in adult *Nanosella* the cerebrum contains only about 8000 cells. It is also interesting that the number of cells in the cerebrum is higher in larger hymenopterans than in other insects of the same size class, but the difference levels out in the smallest hymenopterans.

Unique structure of the nervous system is found in some of the smallest representatives of the family Trichogrammatidae, *Megaphragma mymaripenne* and *M. amalphitanum*. The central nervous system of *M. mymaripenne* contains only 339–372 nuclei (M = 360, n = 3), 179–253 of them (M = 215, n = 3) in the supraoesophageal ganglion. In spite of the fact that the central nervous system of adult *M. mymaripenne* almost lacks nuclei and cell bodies of neurons, it occupies 6 % of the body volume (the supraoesophageal ganglion occupies 2.9 %), much more than in larger hymenopterans—e.g., supraoesophageal ganglion occupies from 0.35 % to 1.02 % of body volume in *Apis* (Strausfeld 1976; Wigglesworth 1953) and 0.57 % in *Formica*—but similar to the parameters found in average-sized representatives of related groups of Chalcidoidea.

The study of pupae has shown that at the pupal stage the central nervous system of *M. mymaripenne* contains 7199–7593 nuclei (M = 7396, n = 2) and has a structure typical of insects, i.e., each ganglion consists of neuropil surrounded by the cortical layer of cell bodies of neurons. The pupal central nervous system occupies 19 % of body volume. The supraoesophageal ganglion of pupa contains 4600 cells and occupies 11 % of body volume. At final stages of pupal development, large-scale lysis of cells takes place in the central nervous system, so that in adults only few nuclei of neurons and some cell bodies of neurons at different stages of lysis can be seen. Because the volume and structure of neuropil are almost identical in adults and pupae of *M. mymaripenne*, and the number of neurons remains almost unchanged from pupa to adult of the closely related species *T. evanescens*, it can be assumed that in adults of *Megaphragma* the number of neurons is also the same as in pupae.

Therefore, nuclei and cell bodies of more than 95 % neurons undergo lysis prior to the emergence of the adult (Fig. 9.6). As a result of this lysis, the absolute and relative size of the nervous system is markedly reduced, and especially strongly so in the cerebrum, which has a volume of 93,600 μm^3 in pupae and 52,200 μm^3 in adults. Considerable changes of volume of the cerebrum are accompanied by a peculiar transformation of the cranium at late stages of pupal development. The cranium is considerably larger in pupae than in adults, because of the strongly convex occipital area, and at last stages of pupal development and in adults the occipital region is contracted by many cuticular folds (Fig. 9.7). Each of these folds

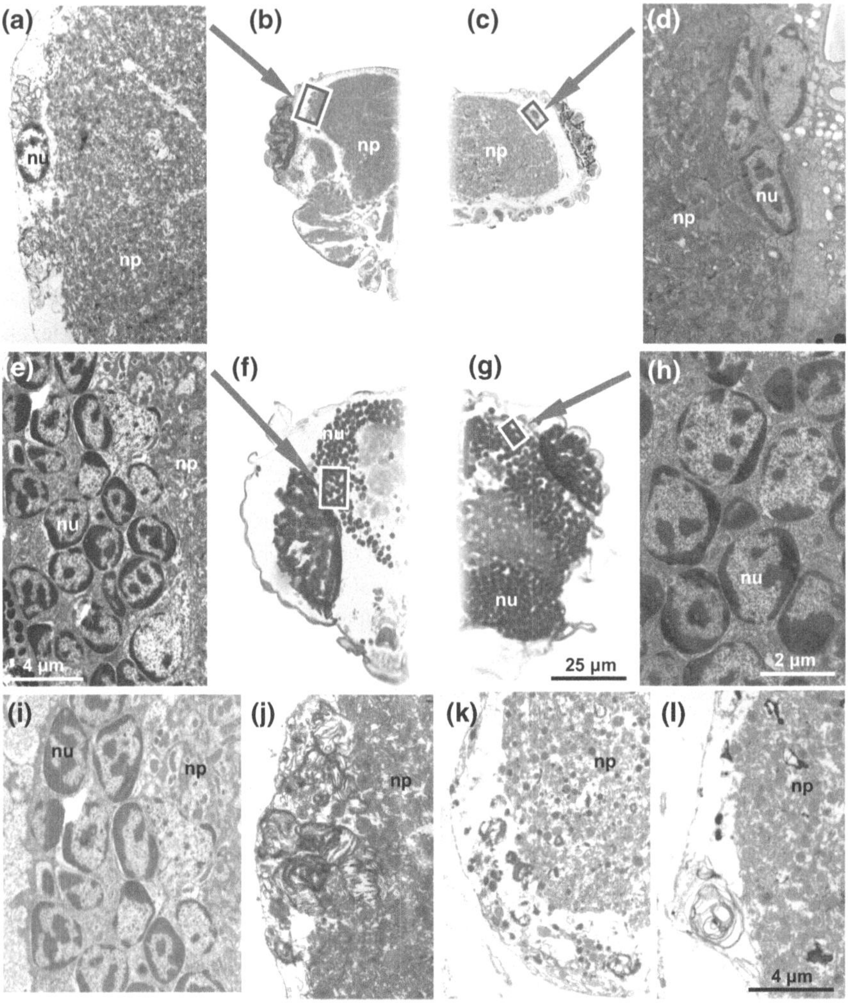

Fig. 9.6 Lysis of cell bodies and nuclei of neurons in *Megaphragma mymaripenne*: **a–d**, **l** adult; **e–k** pupa; **a**, **b**, **e**, **f**, **i–l** cross-section; **c**, **d**, **g**, **h** longitudinal section; **a–e**, **h–l** TEM; **f**, **g** histology, toluidine blue, pyronin; **i–l** process of lysis of cell bodies and nuclei of neurons; *np* neuropil, *nu* nuclei

is a helical area of the cuticle. This transformation of the cranium is probably a rather widespread phenomenon among Trichogrammatidae: many representatives of this family have folds on the parietal and occipital parts, but the functions of these folds have remained unclear.

The nervous system *M. mymaripenne* is unique not only in the lysis of 95 % nuclei, but also in containing the smallest number of neurons known in all insects and all flying animals. The number of cells in central nervous system is estimated as

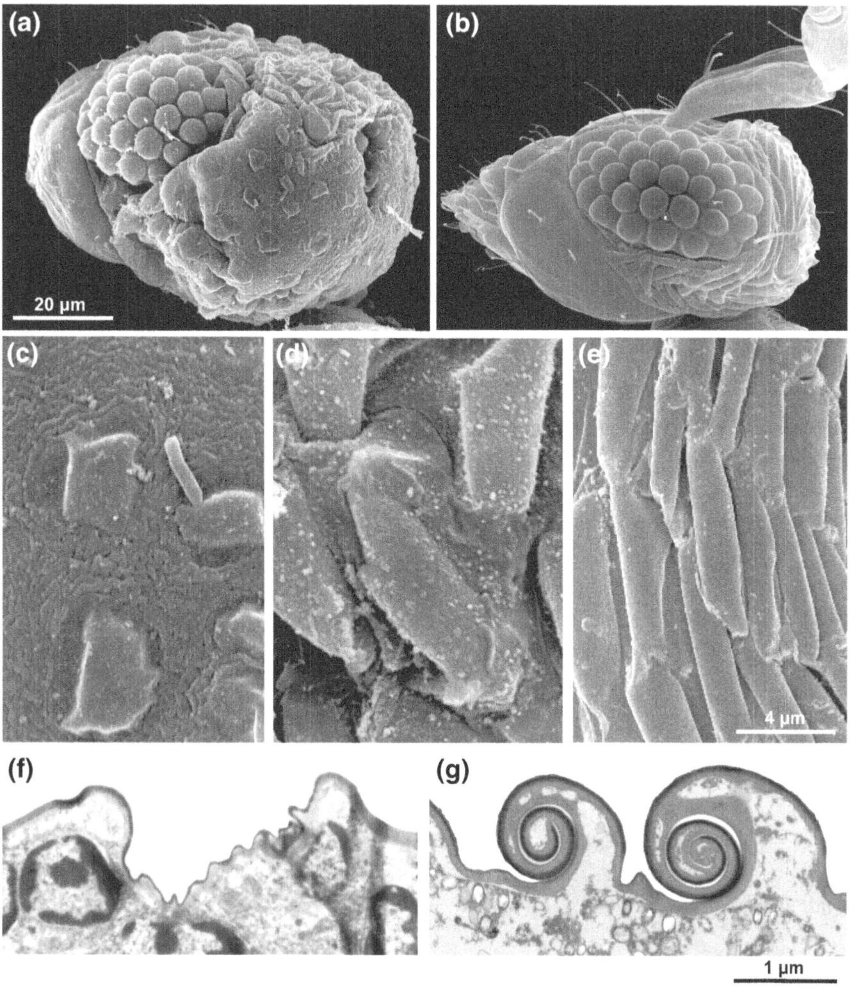

Fig. 9.7 Transformation of cranium after formation of cerebrum in *Megaphragma mymaripenne*: **a–e** SEM; **f**, **g** TEM; **a**, **c**, **f** pupa; **d** late pupa; **b**, **e**, **g** adult; **a**, **b** head, lateral view; **c–e** area cuticle of occipital area. **f**, **g** section through cuticular folds of occipital area

about 7400, and 4600 are situated in the supraoesophageal ganglion. Of course, all smaller insects display considerably reductions in the number of neurons, but even the closely related *T. evanescens* has 18000 neurons in the supraoesophageal ganglion. In spite of the extremely small number of neurons, adults of *M. mymaripenne* retain all vital behaviors, including flight, feeding, and search for host to lay eggs. The possibility of functioning anucleate neurons is a key point in the study of the regeneration of neurons. It has been shown during experiments in vitro that axons separated from the nucleus can function (Bittner, 1988, 1991; Sotnikov et al. 2010), but intact anucleate neurons in vivo have never been described before. The average lifespan of adults in *M. mymaripenne* is five days at 25 °C and 8.8 days at 15 °C (Bernardo and Viggiani 2002), which is comparable to many larger chalcidoids. Therefore, the almost anucleate nervous system of adults can retain all functions during the entire rather long adult life of these insects.

9.7 Sense Organs

The number of sensilla becomes considerably reduced with decreasing body size. This reduction is especially clear on the antennae. Each antenna bears only 38 sensilla in *Megaphragma mymaripenne*, about 260 in *Trichogramma* (Amornsak et al. 1998), 560 in *Anaphes* (van Baaren et al. 1999), 850 in *Metaphycus* (Encyrtidae; Zhou et al. 2013), 1000 in *Pteromalus* (Pteromalidae; Onagbola and Fadamiro 2008; Dweck 2009), 6900 in *Microplitis*—and 6600 *Cotesia* in (Braconidae; Das et al. 2011). Moreover, it has been shown that the relative area of antennae increases with decreasing body size (Symonds and Elgar 2013). Reduced number of sensilla is found also on other structures; e.g., the number of sensilla on each cercal plate is four in *Hemiptarsenus*, three in *Anaphes* and *Trichogramma*, and one in *Megaphragma*.

The size of particular sensilla also decreases with decreasing body size, but the scale of changes is much smaller than in the case of the number of sensilla (Table 9.1).

Apparently, the size of sensilla, especially their length, is determined not only by body size, because, e.g., in *Metaphycus* (Encyrtidae, body size about 1.4 mm; Zhou et al. 2013), the average length of sensilla of some types is closer to the size of sensilla of *Megaphragma*. However, the relative diameter of the base of in sensilla is much greater in *Megaphragma* considerably than in most Chalcidoidea. The smallest sensilla have a diameter of at least 0.6 μm, similar to the minimum

Table 9.1 Average length (L) and diameter of base (D) of antennal sensilla in Chalcidoidea (μm)

		Megaphragma mymaripenne	*Trichogramma australicum*	*Trichogramma dendrolimi*	*Pteromalus puparum*	*Pteromalus cerealellae*
			Amornsak et al. (1998)	Zhang et al. (2012)	Dweck (2009)	Onagbola and Fadamiro (2008)
Chaetoid	L	11.72 (5.33)	23.16 (4.63)	19.69 (3.28)	21.4 (0.82)	48 (2.40)
	D	0.98 (0.45)	–	1.54 (0.26)	2.1 (0.08)	3.5 (0.18)
Trichoid (type 1)	L	13.41 (6.10)	8.52 (1.70)	11.93 (1.99)	36 (1.38)	88.9 (4.45)
	D	0.95 (0.43)	–	0.87 (0.15)	2.8 (0.11)	4.6 (0.23)
Placoid	L	29.64 (13.47)	38.83 (7.77)	–	39.4 (1.52)	112.8 (5.64)
	D	2.55 (1.16)	2.17 (0.43)	–	5.4 (0.21)	11.6 (0.58)

Numbers in brackets indicate size relative to body length, %

diameter of sensilla in *Trichogramma* and other chalcidoids. Interestingly, both mechanoreceptor (chaetoid) and olfactory (trichoid) sensilla of *Megaphragma* have similar minimum diameter of the base.

The smallest beetles also display strongly reduced numbers of sensilla compared to larger representatives of related groups. Adults of *Primorskiella* have about 160 sensilla on each antenna, whereas adults of *Aleochara* have at least 2300 (Staphylinidae; Skilbeck and Anderson 1996), and adults of *Oxelytrum* have about 40 000 (Silphidae; Oliva 2012). The smallest diameter of the base of sensillum recorded in Ptiliidae is 0.65 μm.

Insect sensillum is a complex receptor consisting of the cuticular part and several cells (Ivanov 2000; Sinitsyna and Chaika 2006). This structure is probably sensitive to scaling and limits further diminution of sensilla (to less than 0.6 μm in diameter at the base).

The number of ommatidia in the eyes of insects is reduced with decreasing body size to small fractions of its initial values (Fig. 9.8): the smallest insects have only about 30 ommatidia in each eye (36 ± 2 in *Nanosella*; 29 ± 1 in *Megaphragma*). The numbers of ommatidia in the eye differ by a factor of almost 60 between larger chalcidoids and *Megaphragma* and more than 130 between larger ichneumonids and *Megaphragma* (Table 9.2). At the same time, the diameter of one ommatidium in the studied chalcidoids can differ only by a factor of 2 (while body size differs by a factor of over 25), and large ichneumonids differ from trichogrammatids in facet size by a factor of at most 4 (while body size differs by a factor of over 70).

Similar difference in sizes and similar minimum size of facets are found also in beetles; e.g., facet diameter is 7.4 ± 0.2 μm in *Nanosella* sp., 13.7 ± 0.2 in *Hydraena* sp., and 15.5 ± 0.3 in *Aleochara* sp. Reduction of the number of ommatidia to small fractions of its initial values and quite small reduction of the size of facets with decreasing body size has also been shown in other insects (Fischer et al. 2012, 2014).

Thus, the limit to miniaturization of facet diameter is about 6 μm. This limit is determined by several factors. One factor is structural. Each ommatidium, even in some of the smallest insects (e.g., *Megaphragma)* consists of 21 cells: nine retinal cells (photoreceptor cells), four crystalline cone cells (Semper cells), and two primary and six secondary pigment cells (Makarova et al. 2015). All these cells have nuclei. Although almost all chromatin in these cells is compacted, the diameter of most nuclei is more than 1 μm (1.2 ± 0.3 in retinal cells; 1.9 ± 0.3 in crystalline cone cells; 1.2 ± 0.3 in primary pigment cells; and 1.5 ± 0.3 in secondary pigment cells). Most nuclei of retinal cells are shifted to the distal part of the ommatidium. The nuclei in adjacent ommatidia are arranged in staggered order and occupy much of cell diameter. Apparently, their size makes it impossible to decrease ommatidium size further. Another factor limiting the size of ommatidia is optic. The light-gathering power of ommatidium is determined by the size of the lens (facet) and size (area) of the rhabdom (Warrant and McIntyre 1993; Warrant and Nilsson 1998; Land and Nilsson 2012), and therefore these parameters cannot be infinitely decreased. Rhabdom length is considerably reduced with decreasing body size; e.g., rhabdom length is 22.7 ± 0.9 μm in *Megaphragma*; 34.9 ± 3.8 in *Trichogramma*

Fig. 9.8 Structure of eyes in Chalcidoidea, SEM: **a**, **b** *Megaphragma mymaripenne*; **c**, **d** *Trichogramma evanescence*; **e**, **f** *Anaphes flavipes*; **g**, **h** *Hemiptarsenus* sp.

(Fischer et al. 2011); and 67.2 ± 7.7 in *Hemiptarsenus*; for comparison, in *Formica polyctena* it is about 140 (Menzel 1972); and in *Apis* it is about 400 (Perrelet 1970). Rhabdom diameter and facet diameter are greater in *Megaphragma* than in *Trichogramma*, in spite of the opposite difference in body size. This can be

Table 9.2 Number of ommatidia and size of ommatidial facets in some Chalcidoidea and Ichneumonoidea

Species	Family	Body size (mm)	Number of ommatidia	Facet diameter (μm)
*Labena grallator**	Ichneumonidae	16	Over 4000	25 ± 0.8
*Dusona egregia**	Ichneumonidae	8	2100	17.8 ± 0.5
*Lymeon orbum**	Ichneumonidae	6	2200	17.7 ± 0.9
*Biosteres carbonarius**	Braconidae	4	670	14.4 ± 0.3
*Eurytoma gigantea**	Eurytomidae	6	1700	12.9 ± 0.3
Hemiptarsenus sp.	Eulophidae	3	462 ± 22	9.2 ± 0.3
*Cirrospilus coachellae**	Eulophidae	1.2	320	7.8 ± 0.4
Anaphes flavipes	Mymaridae	0.45	121 ± 14	8.1 ± 0.6
Trichogramma evanescens	Trichogrammatidae	0.4	148 ± 8	6.4 ± 0.2
Megaphragma mymaripenne	Trichogrammatidae	0.22	29 ± 1	8.1 ± 0.3

*Calculated from SEM images from Morphbank (2014)

explained by the need to preserve the optical power of the ommatidium. It has also been shown that small insects approach the diffraction limit to vision (Fischer et al. 2011).

Therefore, the effects of miniaturization on sense organs include the reduction of the numbers of structural units (sensilla and ommatidia) to small fractions of their initial values and only minor reduction of the size of sensilla and ommatidia. The constructive and functional organization of sensilla and ommatidia limits their miniaturization and diminution of sense organs in general. In addition to constructive limits, there is probably also the energy limit to decreasing size of receptors, because it has been shown that relative energy expenses increase with decreasing size (Niven et al. 2007; Niven and Laughlin 2008).

9.8 Musculature

The musculature of head and of thorax has now been studied in detail in many microinsects and compared on a large scale to that of larger representatives of related groups of insects (Chaps. 3–8). The musculature of the abdomen has been studied only at the level of the main groups of muscles preserved in all microinsects. Particular muscles of abdomen have not been studied because of technical difficulties (the abdominal musculature of microinsects is represented by very fine muscles, and after fixation sclerites of abdomen and their musculature have different

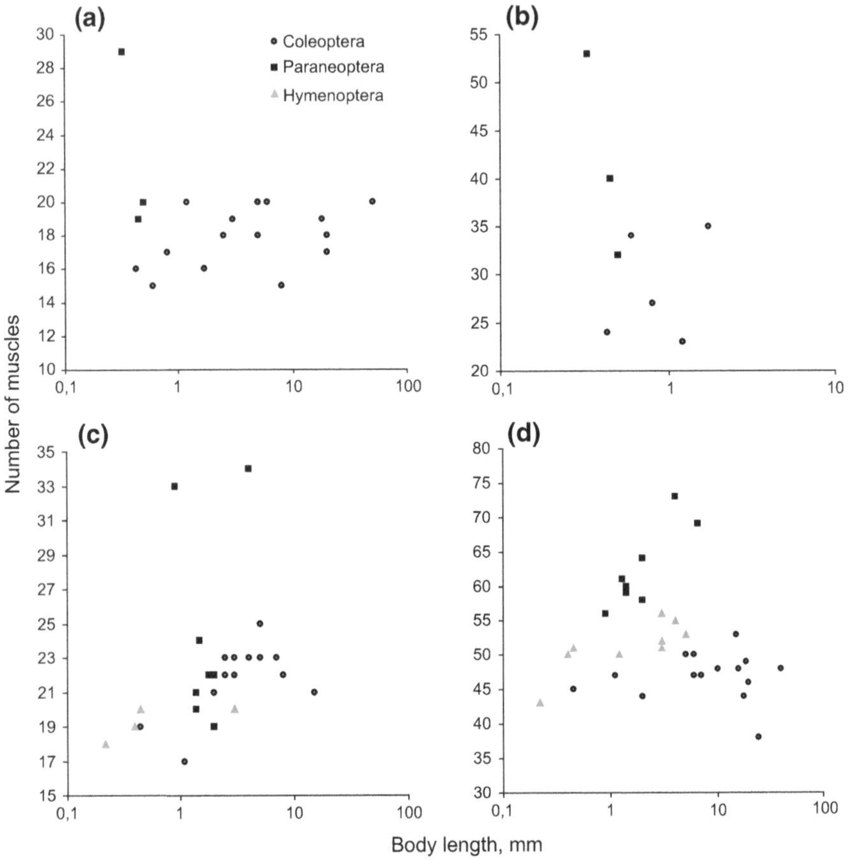

Fig. 9.9 Dependence of number (of pairs) of muscles on body size in insects: **a** head of larva or nymph; **b** thorax of larva or nymph; **c** head of adult; **d** thorax of adult. For list of studied species see Polilov 2014

positions in different samples of the same species, strongly hindering analysis) and because of the absence of a single homology of musculature of abdomen in insects, which makes interpretation of the data impossible.

No clear common dependence of the number of muscles on body size can be recognized in all analyzed groups of insects (Fig. 9.9). The number of muscles depends on many factors: type of feeding, presence of flight, general locomotor activity, etc. The number of muscles of the head in larvae of the studied microbeetles corresponds to the minimum found in larger representatives of related groups. The number of muscles of the head in adult beetles and hymenopterans is reduced approximately by 10–20 % if body size decreases by a factor of 10 or more. The number of thoracic muscles in adults decreases with decreasing body size. The smallest coleopterans retain the minimum number of muscles typical of their larger relatives. A similar picture is found also in hymenopterans, except for

Megaphragma, in which the number of muscles is reduced almost by 20 % even compared to *Trichogramma*. Considerable decrease in the number of muscles of thoracic segments is typical of Paraneoptera, but in this case we compare representatives of different orders and winged and wingless forms.

Comparison of the set of musculature in adult microinsects and representatives of related groups reveals only sporadic reductions not found in large relatives. For instance, in Ptiliidae only three muscles are absent (0hy9, 0st2. IIIscm4) and IIIdlm1 and IIIdlm2, present in all larger representatives of related groups, are fused. In Corylophidae the head lacks 0an3, 0hy9, 0bu5, 0ph9, and 0st2, but only the musculature of a single related species is known for comparison. The thorax of Corylophidae lacks IItpm10. IIpcm6, IIvlm3, present in all flying relatives of this family. Ipcm2 is also present in all large chalcidoids. Trichogrammatidae differ from all studied chalcidoids in the absence of 0lb2 and IIdvm1. Marked reduction of musculature has been recorded only in *Megaphragma*, which lacks 0ph1, IItpm10, IItpm11, IIspm1, IIIvlm2, IIIscm2, IIIscm5, and IIIscm6, in addition to the above-listed reductions. Liposcelididae also have a smaller set of musculature, compared to other psocopterans, but this is above all the result of their winglessness. As for muscles unrelated to flight, it should be noted that *Liposcelis* lacks 0hy5, Iscm5, IIdvm4, IIIpcm4, and IIIscm1, which are present in all studied psocopterans.

Nymphs of thrips have markedly fewer muscles than adults, but above all this is determined by the undeveloped flight muscles. The nymphs of *Liposcelis* differ from adults in the markedly smaller difference in the number of muscles than nymphs of thrips: the absence 0of mx1, 0hy7, 0hy8, Itpm3, and IIpcm1 and fusion of 0bu5 and 0bu6 are recorded. Unfortunately, no data are available on the musculature of the larvae of large thrips. The head musculature of ptiliids larvae differ from that of all larger relatives only in the absence of M4 in first instar larvae of *Mikado*. No data are available on the thoracic musculature in larvae of Staphylinoidea. Musculature of the head in Corylophidae differ from those of all studied cucujiform beetles in the fusion of M17 and M18 and absence of M48. No data are available on the thoracic musculature of larvae in large staphylinoids and cucujoids absent.

By analyzing peculiar features of musculature found in different groups of microinsects, it is impossible to find a single reduction shared by all groups of microinsects, and only three modifications shared by several groups can be found: the absence of 0hy9 and 0st2 in adult Ptiliidae and Corylophidae and absence of IItpm10 in adult Corylophidae and *Megaphragma*. 0hy9 and 0st2 participate in movements of the hypopharynx and pharynx, and their main function is moving of food; if they are reduced, this function is performed by the remaining musculature of hypopharynx and pharynx. IItpm10 participates in opening the forewings (elytra); and IIdvm5 has the same function. Considering also the above-discussed absence of clear dependence of the number of muscles on body size and single unique reductions found in particular groups, one can conclude that miniaturization weakly influences musculature, reflecting the fact that the efficiency of muscles is determined by the area of their cross-section, and the relative force of muscles increases with decreasing body size (Gorodkov 1984); as a result, musculature is not very sensitive to miniaturization.

Considerable reduction of the number of muscles in thoracic segments in *Mengenilla chobauti* (Strepsiptera) compared to *Dytiscus marginalis,* described by Osswald et al. (2010), can be explained by the fact that they belong to both taxonomic and ecological groups very remote from each other, and by the fact the nomenclature of larval muscles is poorly developed and in early studies large subunits of muscles are treated as separate muscles. For instance, 92 pairs of muscles were described in larvae of *Dytiscus* in a study (Korschelt 1923) referenced by Osswald et al. (2009), whereas in adults, according to modern data, only 52 muscles have been found (Beutel and Hass 2000).

Structure of the muscular system is much more conserved than the skeleton (Matsuda 1965, 1970, 1976), making it especially important for the study of functional morphology and taxonomy. The high degree of stability of the set of musculature, in spite of considerable changes in body size, make this set of characters irreplaceable for taxonomic purposes, including the purposes of macrotaxonomy, because it allows to compare large taxa strongly different in size.

The ultrastructure of musculature of most microinsects is distinguished by the localization of nuclei and many organelles on the periphery of muscles. The transition to morphologically smooth musculature, described in four-legged mites (Silvere and Shtein-Margolina 1976), is atypical of the smallest insects, although some very small muscles have only weakly pronounced pattern of cross-striation. The muscles of *Megaphragma* are connected with the skeleton via the strongly shortened tonofibril apparatus, which is morphologically similar to a desmosome; and this is precisely how a similar structure in four-legged mites was interpreted (Silvere and Shtein-Margolina 1976). However, it is unlikely that this structure can actually be a desmosome; it is much more likely that this is a case of considerably shortened tonofibril apparatus because of thinned hypoderm and absence of fully formed endocuticle. All musculatures of Psocoptera and Thysanoptera belong to the densely packed type. The flight musculatures in the studied beetles and hymenopterans are of the fibrillare type; the other musculature is densely packed.

One interesting peculiar feature of musculature was found in adults of *Sericoderus lateralis* (Coleoptera, Corylophidae). All studied populations include individuals with reduced flight musculature (Polilov 2011). In all samples the individuals lacking considerable parts of wing musculature were females with two almost formed or completely formed eggs or males with completely developed reproductive system (Fig. 9.10). No degeneration of musculature was found in juvenile males and females. Other changes, except for musculature degeneration, have not been found in the structure of the wing apparatus. The musculature of the pro- and mesothorax is identical in all individuals. In the metathorax of some individuals IIIdlm1, IIIdlm2, IIIdvm1, IIIspm1, IIIdvm2, IIIdvm4, IIIdvm5, and IIIdvm6 are absent. Therefore, such individuals lack all large muscles moving the wing up and down, and all such individuals are incapable of flight, while almost all muscles necessary for folding and holding of the wing remain. The part of volume of the metathorax freed from musculature is occupied by the reproductive system.

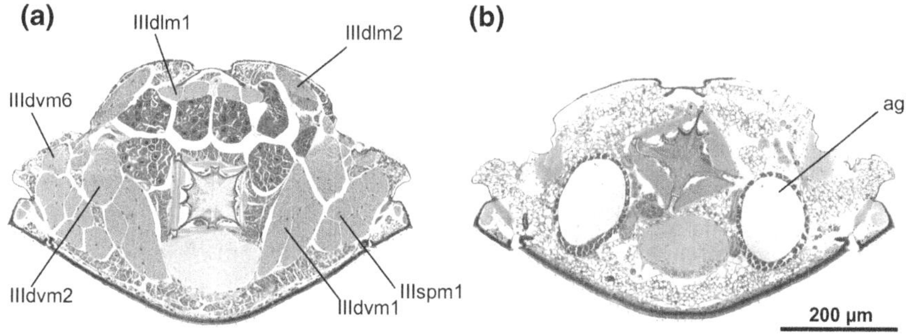

Fig. 9.10 Cross-section of metathorax in beetle *Sericoderus lateralis* with normally developed musculature (**a**), and reduced flight muscles (**b**): *agl* accessory glands. Musculature see text

The absence of muscle degeneration in juvenile (weakly pigmented) individuals, males and females with developing reproductive system, gives evidence that adults of *S. lateralis* emerge having a complete set of thoracic musculature and suggests that individuals of this species undergo degeneration of musculature during maturation caused by the development of the reproductive system.

This phenomenon is termed the oogenesis-flight syndrome; it has been shown in some bark beetles (Chapman 1956; Reid 1958; Atkins and Farris 1962; Borden and Slater 1968, 1969; Bhakthan et al. 1970), weevils (Muda et al. 1981; Rankin et al. 1994; Linders et al. 1995), ground beetles (Desender 2000), leaf beetles (Stegwee et al. 1963) and scarab beetles (Tada et al. 1991). For *S. lateralis* degeneration of musculature doubtlessly has high adaptive value: young individuals actively disperse between substrates and, having found sufficient substrate, start breeding, which is accompanied by lysis of much of flight muscles and by occupation of the freed space by the reproductive system. This is especially topical in cases of miniaturization of the body, because the reproductive system of *S. lateralis* occupies a much greater relative volume than the reproductive system of larger beetles. Interestingly, in males of *S. lateralis* development of the reproductive system causes a strong degeneration of musculature as in females. The rather small proportion of individuals with degenerated flight muscles (13.3 % of all samples together) may give evidence of the ability of *S. lateralis* to regenerate lost musculature, as in *Leptinotarsa decimlineata* (Stegwee et al. 1963), *Ips confusus* (Bhakthan et al. 1971). and *Lissorhoptrus oryzophilus* (Muda et al. 1981), or else it may give evidence that breeding adults live less than juvenile or immature adults, and therefore occur less frequently. This question requires further studies.

9.9 Reproductive System

The reproductive system of most microinsects is rather similar to that of larger insects: it consists of paired gonads, all glands typical for the related group of insects (including the complete set of glands found in the smallest hymenopterans), copulatory apparatus, and ovipositor. However, in some of the smallest insects the reproductive system becomes asymmetrical, with one gonad markedly larger than the other, or, as in males and females of Ptiliidae, with unpaired gonads. Of course, asymmetry of the reproductive system is typical of many insects and sometimes unrelated to miniaturization, but in most cases it is the need to save space that leads to the reduction or loss of one of the gonads (Will et al. 2005), and therefore in the case of microinsects it can be assumed that the asymmetry of their reproductive systems is directly linked to miniaturization. In some of the smallest arachnids, the four-legged mites, the gonads are also unpaired (Silvere and Shtein-Margolina 1976). Unpaired gonads have also been described in tardigrades and rotifers (Westheide and Rieger 2004). The reproductive system becomes asymmetrical as a result of miniaturization also in some fishes (Parenti 1986a, b). The number of ovarioles in microinsects is decreased, and the smallest insects have only two ovarioles in each ovary (*Megaphragma*) or only two ovarioles in a single ovary (Nanosellini). Peculiar features of the structure of the ovipositor and copulatory apparatus are discussed in one of the previous sections. With decreasing body size strong disproportion can be seen in the size of spermatozoa; in some species of Ptiliidae spermatozoa are sometimes longer than body (Dybas and Dybas 1981, 1987; De Marzo 1992).

References

Amornsak, W., Cribb, B., & Gordh, G. (1998). External morphology of antennal sesilla of *Trichogramma australicum* Girault (Hym.: Trichogrammatidae). *International Journal of Insect Morphology and Embryology, 27*(2), 67–82.

Antonov, A. V. (2009). Миниатюризация размеров тела и направления эволюции жилкования крыльев у жалящих перепончатокрылых насекомых (Insecta: Hymenoptera: Aculeata: Apoidea) (опыт функциональной интерпретации (Miniaturization of body side and trends of the evolution of wing venation in Apoidea (Insecta: Hymenoptera: Aculeata)). *Sbornik Trudov Zoologicheskogo Muzeya MGU, 50*, 247–274.

Atkins, M. D., & Farris, S. H. (1962). A contribution to the knowledge of flight muscle changes in the Scolytidae (Coleoptera). *Canadian Entomologist, 94*(1), 25–32.

Bakkendorf, O. (1934). Biological investigations on some Danish hymenopterous egg-parasites, especially in homopterous and heteropterous eggs, with taxonomic remarks and descriptions of new species. *Entomologiske Meddelelser, 19*, 1–134.

Bate, C. M. (1976). Embryogenesis of an insect nervous system I. A map of the thoracic and abdominal neuroblasts in *Locusta migratoria*. *Journal of Embryology and Experimental Morphology, 35*, 107–123.

Bernardo, U., & Viggiani, G. (2002). Biological data on *Megaphragma amalphitanum* Viggiani and *Megaphragma mymaripenne* Timberlake (Hymenoptera: Trichogrammatidae), egg-parasitoid of *H. haemorrhoidalis* (Bouché) (Thysanoptera: Thripidae) in southern Italy. *Bollettino del Laboratorio di Entomologia Agraria Filippo Silvestri, 58*, 77–85.

Beutel, R. G., Maddison, D. R., & Haas, A. (1999). Phylogenetic analysis of Myxophaga (Coleoptera) using larval characters. *Systematic Entomology, 24*, 1–23.

Beutel, R. G., Pohl, H., & Hunefeld, F. (2005). Strepsipteran brain and effect of miniaturization (Insecta). *Arthropod Structure & Development, 34*(3), 301–313.

Beutel, R. G., & Haas, A. (1998). Larval head morphology of *Hydroscapha natans* LeConte 1874 (Coleoptera, Myxophaga, Hydroscaphidae) with special reference to miniaturization. *Zoomorphology, 118*(2), 103–116.

Beutel, R. G., & Haas, A. (2000). Phylogenetic releationships of the suborders of Coleoptera (Insecta). *Cladistics, 16*, 103–141.

Bhakthan, N. M. G., Borden, J. H., & Nair, K. K. (1970). Fine structure of degenerating and regenerating flight muscles in a bark beetle, *Ips confusus*. I. Degeneration. *Journal of Cell Science, 6*, 807–819.

Bhakthan, N. M. G., Borden, J. H., & Nair, K. K. (1971). Fine structure of degenerating and regenerating flight muscles in a bark beetle, *Ips confusus*. II. Regeneration. *Canadian Journal of Zoology, 49*(1), 85–89.

Bittner, G. D. (1988). Long term survival of severed distal axonal stumps in vertebrates and invertebrates. *American Zoologist, 28*, 1165–1179.

Bittner, G. D. (1991). Long-term survival of anucleate axons and its implications for nerve regeneration. *Trends in Neurosciences, 14*, 188–193.

Boire, D., & Baron, G. (1994). Allometric comparison of brain and main brain subdivisions in birds. *Brain Research, 35*, 49–66.

Boivin, G. (2010). Reproduction and immature development of egg parasitoids. In F. L. Consoli, J. R. P. Parra, & R. A. Zucchi (Eds.), *Egg Parasitoids in Agroecosystems with Emphasis on Trichogramma* (pp. 1–23). Houten: Springer.

Borden, J. H., & Slater, C. E. (1968). Induction of flight muscle degeneration by synthetic juvenile hormone in *Ips confusus* (Coleoptera: Scolytidae). *Zeitschrift für vergleichende Physiologie, 61*, 366–368.

Borden, J. H., & Slater, C. E. (1969). Flight muscle volume change in *Ips confusus* (Coleoptera: Scolytidae). *Canadian Journal of Zoology, 47*, 29–32.

Boxshall, G. A. (1982). On the anatomy of the misophirioid copepods, with special reference to *Benthomisophria palliata* Sars. *Philosophical Transactions of the Royal Society of London B, 297*(1086), 125–181.

Brandt, R., Rohlfi, T., Rybak, J., Krofczik, S., Maye, A., Westerhoff, M., et al. (2005). Threedimensional average-shape atlas of the honeybee brain and its applications. *Journal of Comparative Neurology, 492*, 1–19.

Brenzinger, B., Haszprunar, G., & Schrodl, M. (2013). At the limits of a successful body plan - 3D microanatomy, histology and evolution of Helminthope (Mollusca: Heterobranchia: Rhodopemorpha), the most worm-like gastropod. *Frontiers in Zoology, 10*(1), 37.

Chaika, S. Y. (2010). *Нейроморфология насекомых (Insect neuromorphology)*. Moscow: Tipografiya Rossel'hozakademii.

Chapman, J. A. (1956). Flight-muscle changes during adult life in a scolytid beetle. *Nature, 177*, 1183–1184.

Das, P., Chen, L., Sharma, K. R., & Fadamiro, H. Y. (2011). Abundance of antennal chemosensilla in two parasitoid wasps with different degree of host specificity may explain sexual and species differences in their response to host-related volatiles. *Microscopy Research and Technique, 74*(10), 900–909.

De Marzo, L. (1992). Osservationi anatomiche sui genitali interni maschili in alcuni ptilidi (Coleoptera). *Entomologica Bari, 27*, 107–115.

Desender, K. (2000). Flight muscle development and dispersal in life cycle of carabid beetles: Patterns and processes. *Bulletin de l'Institut Royal des Sciences Naturelles de Belgique Entomologie, 70*, 13–31.

Doe, C. Q., & Goodman, C. S. (1985). Early events in insect neurogenesis. I. Development and segmental differences in the pattern of neuronal precursor cells. *Development Biology, 111*, 193–205.

Dreyer, D., Vitt, H., Dippel, S., Goetz, B., el Jundi, B., Kollmann, M., et al. (2010). 3D standard brain of the red flour beetle Tribolium castaneum: a tool to study metamorphic development and adult plasticity. *Frontiers in Systems Neuroscience, 4*(3), 1–13.

Dweck, K. M. (2009). Antennal sensory receptors of *Pteromalus puparum* female (Hymenoptera: Pteromalidae), a gregarious pupal endoparasitoid of Pieris rapae. *Micron, 40*, 769–774.
Dybas, L. K., & Dybas, H. S. (1981). Coadaptation and taxanomic differentiation of sperm and spermathecae in featherwing beetles. *Evolution, 35*(1), 168–174.
Dybas, L. K., & Dybas, H. S. (1987). Ultrastructure of mature spermatozoa of a minute featherwing beetle from Sri Lanka (Coleoptera, Ptiliidae: Bambara). *Journal of Morphology, 191*, 63–76.
el Jundi, B., Huetteroth, W., Kurylas, A. E., & Schachtner, J. (2009). Anisometric brain dimorphism revisited: implementation of a volumetric 3D standard brain in *Manduca sexta*. *Journal of Comparative Neurology, 517*, 210–225.
Faisal, A. A., White, J. A., & Laughlin, S. B. (2005). Channel noise places limits on the miniaturization of the brain's wiring. *Current Biology, 12*(12), 1143–1149.
Fischer, S., Müller, C. H. G., & Meyer-Rochow, V. B. (2011). How small can small be: the compound eye of the parasitoid wasp *Trichogramma evanescens* (Westwood, 1833) (Hymenoptera, Hexapoda), an insect of 0.3- to 0.4-mm total body size. *Visual Neuroscience, 28*(4), 295–308.
Fischer, S., Meyer-Rochow, V. B., & Müller, C. H. G. (2012). Challenging limits: ultrastructure and size-related functional constraints of the compound eye of *Stigmella microtheriella* (Lepidoptera: Nepticulidae). *Journal of Morphology, 273*(9), 1064–1078.
Fischer, S., Meyer-Rochow, V. B., & Müller, C. H. G. (2014). Compound Eye Miniaturization in Lepidoptera: a comparative morphological analysis. *Acta Zoologica, 95*(4), 438–464.
Goossen, H. (1949). Untersuchungen an gehirnen verschieden grosser, jeweils verwandter Coleopteren- und Hymenopteren. *Arten Zoologische Jahrbücher Abteilung für Allgemeine Zoologie, 62*, 1–64.
Gorodkov, K. B. (1984). Олигомеризация и эволюция систем морфологических структур: 2. Олигомеризация и уменьшение размеров тела (Oligomerization and evolution of the morphological structure systems. 2. Oligomerization and body size decrease. *Zoologicheskii Zhurnal, 63*, 1765–1778.
Grebennikov, V. V., & Beutel, R. G. (2002). Morphology of the minute larva of *Ptinella tenella*, with special reference to effects of miniaturisation and the systematic position of Ptiliidae (Coleoptera: Staphylinoidea). *Arthropod Structure & Development, 31*(2), 157–172.
Heath, R. V., & Evans, M. E. G. (1990). The relationship between the ventral nerve cord, body size and phylogeny in ground beetles (Coleoptera: Carabidae). *Zoological Journal of the Linnean Society, 98*(3), 259–293.
Heming, B. S. (1978). Structure and function of the mouthparts in larvae of *Haplothrips verbasci* (Osborn) (Thysanoptera, Tubulifera, Phlaeothripidae). *Journal of Morphology, 156*(1), 1–37.
Hinke, W. (1961). Das relative postembryonale wachstum der hirnteile von Culex pipiens, Drosophila melanogaster und Drosophila-mutanten. *Zeitschrift fuer Morphologie und Oekologie der Tiere, 50*(1), 81–118.
Horridge, G. A. (1956). The flight of very small insects. *Nature, 178*, 1334–1335.
Hustert, R. (2012). Giant and dwarf axons in a miniature insect, *Encarsia formosa* (Hymenoptera, Calcididae). *Arthropod Structure & Development, 41*(6), 535–543.
Ivanov, V. P. (2000). *Органы чувств насекомых и других членистоногих*. Moskva: Nauka.
Ivanova-Kazas, O. M. (1961). *Очерки по сравнительной эмбриологии перепончатокрылых (Essays on the comparative embryology of Hymenoptera)*. Leningrad: Leningrad Univ. Press.
Jackson, D. J. (1961). Observations on the biology of *Caraphractus cinctus* Walker (Hymenoptera, Mymaridae), a parasitoid of the eggs of Dytiscidae. II. Immature stages and seasonal history with a review of mymarid larvae. *Parasitology, 51*, 269–294.
Jałoszyński, P., Hünefeld, F., & Beutel, R. G. (2012). The evolution of "deformed" brains in ant-like stone beetles (Scydmaeninae, Staphylinidae). *Arthropod Structure & Development, 41* (1), 17–28.
Jałoszyński, P., & Olszanowski, Z. (2013). Specialized feeding of *Euconnus pubicollis* (Coleoptera: Staphylinidae, Scydmaeninae) on oribatid mites: Prey preferences and hunting behaviour. *European Journal of Entomology, 110*(2), 339–353.

Jałoszyński, P., & Olszanowski, Z. (2015). Feeding of *Scydmaenus rufus* (Coleoptera: Staphylinidae, Scydmaeninae) on oribatid and uropodine mites: Prey preferences and hunting behaviour. *European Journal of Entomology, 112*(1), 151–164.

Jałoszyński, P., & Olszanowski, Z. (2016). Feeding of two species of Scydmaeninae 'hole scrapers', *Cephennium majus* and *C. ruthenum* (Coleoptera: Staphylinidae), on oribatid mites. *European Journal of Entomology, 113*, 372–386.

Korschelt, E. (1923). Der Gelbrandkäfer *Dytiscus marginalis*. *Bearbeitung einheimischer Tiere, 2*, 1–863.

Kurylas, A. E., Rohlfi, T., Krofczik, S., Jenett, A., & Homberg, U. (2008). Standardized atlas of the brain of the desert locust, Schistocerca gregaria. *Cell and Tissue Research, 333*, 125–145.

Land, M. F., & Nilsson, D.-E. (2012). *Animal eyes* (2nd ed.). Oxford: Oxford University Press.

Linders, E. G. A., Turin, H., & Bruggemann, C. G. (1995). Biology of the weevil *Trichosirocalus troglodytes* and impact on its host *Plantago lanceolata*. *Acta Oecologica, 16*(6), 703–718.

Makarova, A. A., Polilov, A. A., & Fisher, S. (2015). Comparative morphological analysis of compound eye miniaturization in minute Hymenoptera. *Arthropod Structure & Development, 44*(1), 21–32.

Matsuda, R. (1965). Morphology and evolution of the insect head. *Memoirs of the American Entomological Institute, 4*, 1–334.

Matsuda, R. (1970). Morphology and evolution of the insect thorax. *Memoirs of the Entomological Society of Canada, Suppl 76*, 431.

Matsuda, R. (1976). *Morphology and evolution of the insect abdomen*. New York: Pergamon.

Meinertzhagen, I. A. (2010). The organisation of invertebrate brains: cells, synapses and circuits. *Acta Zoologica Stockholm, 91*(1), 64–71.

Menzel, R. (1972). The fine structure of the compound eye of Formica polyctena — Functional morphology of a hymenopteran eye. In R. Wehner (Ed.), *Information Processing in Visual Systems of Arthropods* (pp. 37–47). Berlin: Springer.

Morphbank. (2014). Biological imaging. Florida State University, Department of Scientific Computing, Tallahassee, http://www.morphbank.net. Accessed 2014, November 20.

Muda, A. R. B., Tugwell, N. P., & Haizlip, M. B. (1981). Seasonal history and indirect flight muscle degeneration and regeneration in the rice water weevil. *Environmental Entomology, 10* (5), 685–690.

Niven, J. E., Anderson, J. C., & Laughlin, S. B. (2007). Fly photoreceptors demonstrate energy-information trade-offs in neural coding. *PLoS Biology, 5*(4), e116.

Niven, J. E., & Laughlin, S. B. (2008). Energy limitation as a selective pressure on the evolution of sensory systems. *Journal of Experimental Biology, 211*, 1792–1804.

Niven, J. E., Graham, C. M., & Burrows, M. (2008). Diversity and evolution of the insect ventral nerve cord. *Annual Review of Entomology, 53*, 253–271.

Oliva, A. (2012). The Antennal Sensilla of *Oxelytrum erythrurum* (Blanchard) and *Oxelytrum apicale* (Brullé) (Coleoptera: Silphidae). *Neotropical Entomology, 41*(5), 395–403.

Onagbola, E. O., & Fadamiro, H. Y. (2008). Scanning electron microscopy studies of antennal sensilla of *Pteromalus cerealellae* (Hymenoptera: Pteromalidae). *Micron, 39*, 526–535.

Osswald, J., Pohl, H., & Beutel, R. G. (2010). Extremely miniaturised and highly complex: the thoracic morphology of the first instar larva of *Mengenilla chobauti* (Insecta, Strepsiptera). *Arthropod Structure & Development, 39*(4), 287–304.

Panov, A. A. (1957). Строение головного мозга насекомых на последовательных этапах постэмбрионального развития (Structure of the insect cerebrum at consecutive stages of postembrynic development). *Entomologiceskoe Obozrenie, 36*(2), 269–284.

Panov, A. A. (1959). О формировании гломерулярной структуры нейропиля обонятельной доли мозга насекомых (On the formation of glomerular structure of the neuropil of the olfactory lobe of the insect cerebrum). *Zoologicheskii Zhurnal, 38*(5), 775–777.

Panov, A. A. (1960a). Строение головного мозга насекомых на последовательных этапах постэмбрионального развития: 2. Центральное тело (Structure of the cerebrum at consecutive stages of postembryonic development: 2. Central body). *Entomologiceskoe Obozrenie, 38*(2), 86–105.

Panov, A. A. (1960b). Строение головного мозга насекомых на последовательных этапах постэмбрионального развития: 3. Зрительные доли (Structure of the cerebrum at consecutive stages of postembryonic development: 3. Optic lobes). *Entomologiceskoe Obozrenie, 39* (1), 86–105.

Panov, A. A. (1961). Строение головного мозга насекомых на последовательных этапах постэмбрионального развития: 4. Обонятельный центр (Structure of the cerebrum in insects at consecutive stages of postembryonic development: 4. Olfactory center). *Entomologiceskoe Obozrenie, 40*(2), 259–271.

Parenti, L. R. (1986a). Bilateral asymmetry in phallostethid fishes (Atherinomorpha) with description of a new species from Sarawak. *Proceedings of the California Academy of Sciences, 44*, 225–236.

Parenti, L. R. (1986b). *Homology of pelvic fin structures in female phallostethid fishes (Atherinomorpha, Phallostethidae)* (pp. 305–310). Copeia.

Perrelet, A. (1970). The fine structure of the retina of the honey bee drone. An electron microscopial study. *Z Zellforsch, 108*, 530–562.

Platel, R. (1976). Analyse volumetrique compaire des principales subdividions enciphaliques chez les reptiles sauriens. *Journal fur Hirnforschung, 17*, 513–537.

Plotnikova, S. I. (1979). *Структурная организация центральной нервной системы насекомых (Structural organization of the central nervous system of insects)*. Leningrad: Nauka.

Pohl, H. (2000). Die Primärlarven der Fächerflügler – evolutionäre Trends (Insecta, Strepsiptera). *Kaupia, 10*, 1–144.

Polilov, A. A. (2005). Anatomy of the feather-winged beetles *Acrotrichis montandoni* and *Ptilium myrmecophilum* (Coleoptera, Ptiliidae). *Entomological Review, 85*(5), 467–475.

Polilov, A. A. (2011). Thoracic musculature of *Sericoderus lateralis* (Coleoptera, Corylophidae): Miniaturization effects and flight muscle degeneration related to development of reproductive system. *Entomological Review, 91*(6), 735–742.

Polilov, A. A. (2014). Морфологические особенности мельчайших насекомых (Morphological features of the smallest insects). Doctor Dissertation, Lomonosov Moscow State University.

Polilov, A. A. (2015a). Small is beautiful: features of the smallest insects and limits to miniaturization. *Annual Review of Entomology, 60*, 103–121.

Polilov, A. A. (2015b). Consequences of Miniaturization in Insect Morphology. *Moscow University Biological Sciences Bulletin, 70*(3), 136–142.

Quesada, R., Triana, E., Vargas, G., Douglass, J. K., Seid, M. A., Niven, J. E., et al. (2011). The allometry of CNS size and consequences of miniaturization in orb-weaving and cleptoparasitic spiders. *Arthropod Structure & Development, 40*(6), 521–529.

Rankin, M. A., Hampton, E. N., & Summy, R. R. (1994). Investigations of the oogenesis-flight syndrome in *Anthonomus grandis* (Coleoptera: Curculionidae) using tethered flight tests. *Journal of Insect Behavior, 7*(6), 798–810.

Rasnitsyn, A. P. (1980). Происхождение и эволюция перепончатокрылых насекомых (Origin and evolution of hymenoptera). *Trudy Paleontologicheskogo instituta Akademii nauk SSSR, 174*, 1–192.

Reid, R. W. (1958). Internal changes in the female mountain pine beetle *Dendroctonus monticolae* Hopk. associated with egg laying and flight. *Canadian Entomologist, 90*(8), 464–468.

Rein, K., Zöckler, M., Mader, M. T., Grübel, C., & Heisenberg, M. (2002). The Drosophila standard. *Current Biology, 12*, 227–231.

Rensch, B. (1948). Histological changes correlated with evolutionary changes in body size. *Evolution, 2*, 218–230.

Rohlfi, T., Brandt, R., Menzel, R., & Maurer, C. R., Jr. (2004). Evaluation of atlas selection strategies for atlas-based image segmentation with application to confocal microscopy images of bee brains. *Neuroimage, 21*(4), 1428–1442.

Ruppel, H. (1953). Physiologische untersuchungen über die bedeutung des ventraltubus und die atmung der collembolen. *Zoologische Jahrbucher, 64*, 429–469.

Schmalhausen, I. I. (1984). *Рост и дифференцировка (Growth and Differentiation)*. Kiev: Naukova dumka.

Silvere AP, Shtein-Margolina VV (1976) Tetrapodili — четырехногие клещи: Электронномикроскопическая анатомия, проблемы эволюции и взаимоотношения с возбудителями болезней растений (Tetrapodili: Fore-legs mites; Electron microscopic anatomy, evolution problems and mutual relations with plant pathogenic organisms) Tallinn: Valgus.

Sinitsyna, E. E., & Chaika, S Yu. (2006). *Атлас электроно-микроскопической морфологии хеморецепторных органов насекомых (Atlas of electron-microscopic morphology of chemoreceptor organs of insects)*. Moscow: Tipografiya Rossel'hozakademii.

Skilbeck, C. A., & Anderson, M. (1996). The ultrastructure, morphology and distribution of sensilla on the antennae of the adult parasitoids *Aleochara bilineata* gyll. and Aleochara bipustulata L. (Coleoptera: staphylinidae). *International Journal of Insect Morphology and Embryology, 25*(3), 261–280.

Sotnikov, O. S., Laktionova, A. A., Solovieva, I. A., & Krasnova, T. V. (2010). Neuron division or enucleation. *Neuroscience and Behavioral Physiology, 40*, 841–847.

Stegwee, D., Kimmel, E. C., De Boer, J. A., & Henstra, S. (1963). Hormonal control of reversible degeneration of flight muscle in the colorado potato beetle, *Leptinotarsa decemlineata* Say (Coleoptera). *Journal of Cell Biology, 19*, 519–527.

Stephan, H., Bauchot, R., & Andy, O. J. (1970). Data on the size of the brain and of various parts in insectivores and primates. In C. R. Noback & W. Montagna (Eds.), *The primate brain* (pp. 289–297). NY: Appleton-Century-Crofts.

Stephan, H., Frahm, H., & Baron, G. (1981). New and revised data on volumes of brain structure in insectivores and primates. *Folia Primatologica, 35*, 1–29.

Stephan, H., Baron, G., Frahm, H. (1988). Comparative size of brain and brain components. In J. Erwin & H. D. Steklis (Eds.), Comparative primate biology (Vol 4, pp. 1–38). Alan R. Liss, NY.

Strausfeld, N. J. (1976). *Atlas of an insect brain*. NY: Springer.

Svidersky, V. L. (1980a). *Основы нейрофизиологии насекомых (Fundamentals of insect neurophysiology)*. Leningrad: Nauka.

Svidersky, V. L. (1980b). *Полет насекомого (Insect flight)*. Moscow: Nauka.

Symonds, M. R. E., & Elgar, M. A. (2013). The evolution of body size, antennal size and host use in parasitoid wasps (Hymenoptera: Chalcidoidea): A phylogenetic comparative analysis. *PLoS ONE, 8*(10), e78297.

Tada, S., Yamamoto, A., & Nishigaki, J. (1991). Flight-muscle dimorphism of female-adults in the yellowish elongate chafer, *Heptophylla picea* Motschulsky (Coleoptera, Scarabaeidae). *Applied Entomology and Zoology, 26*, 515–521.

van Baaren, J., Boivin, G., & Le, Lannic J. (1999). Nénon J-P (1999), Comparison of antennal sensilla of Anaphes victus and A. listronoti (Hymenoptera, Mymaridae), egg parasitoids of Curculionidae. *Zoomorphology, 119*(1), 1–8.

Walker, J. A. (2002). Functional morphology and virtual models: Physical constraints on the design of oscillating wings, fins, legs, and feet at intermediate Reynolds numbers. *Integrative and Comparative Biology, 42*, 232–242.

Warrant, E. J., & McIntyre, P. D. (1993). Arthropod eye design and the physical limits to spatial resolving power. *Progress in Neurobiology, 40*, 413–461.

Warrant, E. J., & Nilsson, D. E. (1998). Absorption of white light in photoreceptors. *Vision Research, 38*, 195–207.

Wegerhoff, R., & Breidbach, O. (1992). Die entwicklung des zentral komplexes beim käfer. studie zur ontogenie eines zentralen hirnbereiches. *Verhandlungen der Deutschen Zoologischen Gesellschaft, 85*, 1–92.

Weis-Fogh, T. (1973). Quick estimates of flight fitness in hovering animals, including novel mechanisms for lift production. *Journal of Experimental Biology, 59*(1), 169–230.

Westheide, W., & Rieger, (2004). *Spezielle Zoologie*. Berlin: Spektrum Akademischer Verlag Heidelberg.

Wigglesworth, V. B. (1953). *The principles of insect physiology*. London: E. P. Dutton.

Will, K. W., Liebherr, J. K., Maddison, D. R., & Galin, J. (2005). Absence asymmetry: The evolution of monorchid beetles (Insecta: Coleoptera: Carabidae). *Journal of Morphology, 264* (1), 75–93.

Zhang, S., Zhang, Z., Kong, X., Wang, H., Zhou, G., & Yu, J. (2012). External morphology of *Trichogramma dendrolimi* Matsumura (Hymenoptera: Trichogrammatidae) organ and ultrastructure of the sensilla. *Microscopy Research and Technique, 75*(11), 1513–1521.

Zhou, H., Wu, W. J., Zhang, F. P., & Fu, Y. G. (2013). Scanning Electron Microscopy Studies of the Antennal Sensilla of *Metaphycus parasaissetiae* Zhang & Huang (Hymenoptera: Encyrtidae). *Neotropical Entomology, 42*(3), 278–287.

Chapter 10
Changes in Relative Size of Organs that Accompany Decrease in Body Size

10.1 Introduction

Body size is a parameter of utmost importance; it largely determines the morphology, physiology, and biology of living organisms. Allometry of organs that accompanies changes in body size has been described in many vertebrates (Schmidt-Nielsen 1984), but only fragmentary data on different organs of particular species were (Polilov and Beutel 2009, 2010) and on brain size (Eberhard and Wcislo 2011) were available for insects until recently. At present, volumes of organs have been studied in 22 insect species of 11 families and five orders (Thysanura, Psocoptera, Thysanoptera, Coleoptera, and Hymenoptera) including the smallest insects *Megaphragma* (Hymenoptera: Trichogrammatidae) and *Nanosella* (Coleoptera: Ptiliidae) and large representatives of related taxa (the body length and body volume differed by a factor of up to 300 and over 150 000, respectively).

10.2 Changes in Relative Size of Organs that Accompany Decrease in Body Size

The volume of the skeleton in adult insects generally changes proportionally to the body volume, i.e., isometrically (slope of SMA regression is not significantly different from 1, Fig. 10.1, Table 10.1). Analysis of changes in this parameter in different groups shows (Fig. 10.2a) that the relative volume of the skeleton slightly increases in Coleoptera and Paraneoptera and slightly decreases in Hymenoptera as body size decreases. In spiders the weight of the skeleton markedly decreases with decreasing body size (Anderson et al. 1979). Smaller birds and mammals also have smaller relative weight of the skeleton (Prange et al. 1979).

The volume of the musculature in insects generally changes isometrically (Fig. 10.1, Table 10.1). Decreasing relative volume with decreasing body size is

A.A. Polilov, *At the Size Limit—Effects of Miniaturization in Insects*,
DOI 10.1007/978-3-319-39499-2_10

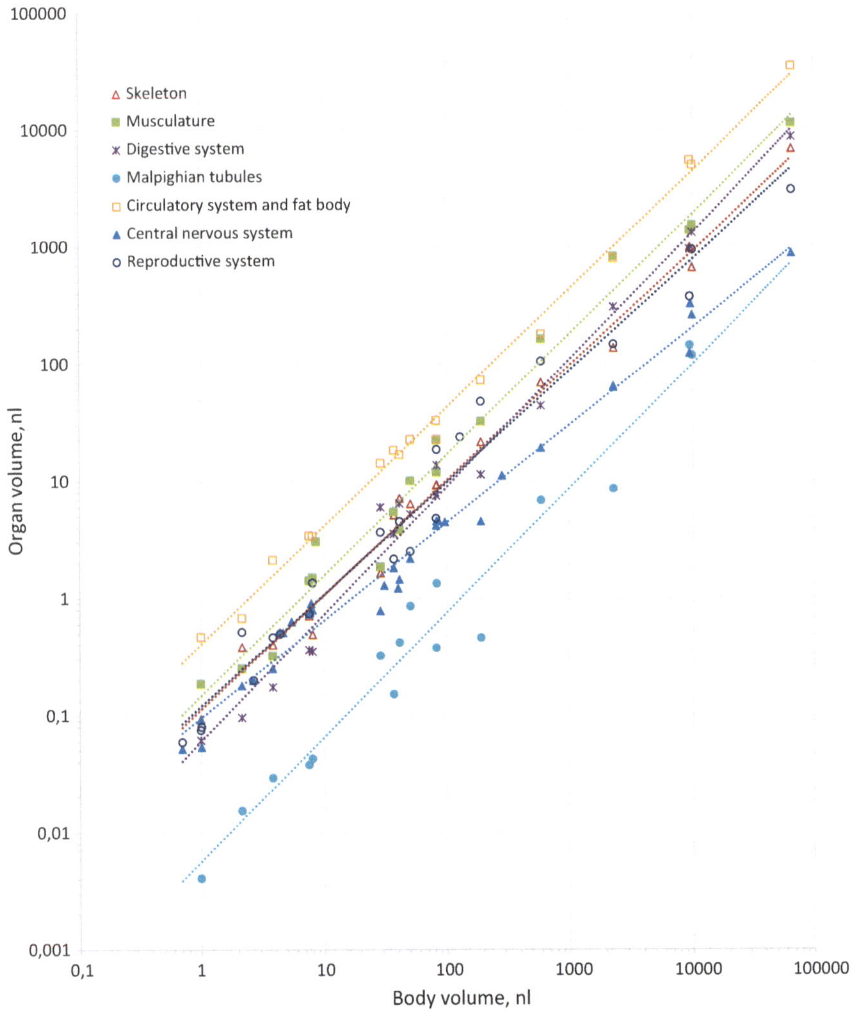

Fig. 10.1 Scaling of organ size in insects. For list of studied species see Polilov (2014)

found in Coleoptera and Paraneoptera separately (Fig. 10.2b). Since relative muscle strength is determined by cross-section area, which changes with linear body size at a lower rate than volume, relative muscle strength increases as body size decreases (Hiestend 1928), making it possible to increase the relative volume of the musculature. Hymenoptera are exceptional: the volume of their musculature slightly increases as body size decreases (Fig. 10.2b). Increased relative area of flight musculature in cross sections has been shown for small dipterans (Rensch 1948), but since no data on volumes are available for these insects, it is difficult to make conclusions about allometry.

Table 10.1 Scaling of organ volume at body volume in insects analyzed with SMA regression method in the SMATR 3 package for R (Warton et al. 2012)

Taxon	Elevation	Slope	CI 95 % for slope	R^2
Skeleton				
Insects, all	−0.957	0.983	0.926–1.042	0.988
Coleoptera	−0.848	0.968*	0.931–1.007	0.998
Hymenoptera	−1.110	1.055	0.788–1.412	0.990
Paraneoptera	−0.909	0.923	0.138 –6.144	0.934
Musculature				
Insects, all	−0.844	1.037	0.963–1.116	0.980
Coleoptera	−0.891	1.049	0.960–1.145	0.990
Hymenoptera	−0.654	0.980**	0.733–1.312	0.974
Paraneoptera	−1.672	1.367*	0.888–2.106	0.998
Digestive system				
Insects, all	−1.230	1.091**	1.020–1.168	0.984
Coleoptera	−1.229	1.081*	0.996–1.173	0.991
Hymenoptera	−1.286	1.010	0.815–1.251	0.994
Paraneoptera	−0.618	0.923	0.473–1.802	0.996
Malpighian tubules				
Insects, all	−2.276	1.079	0.967–1.205	0.963
Coleoptera	−2.233	1.108	0.935–1.314	0.970
Hymenoptera	−2.274	0.887	0.599–1.312	0.982
Paraneoptera	−1.581	0.749***	0.741–0.758	0.999
Circulatory system and fat body				
Insects, all	−0.390	1.01	0.974–1.049	0.995
Coleoptera	−0.435	1.028	0.960–1.101	0.994
Hymenoptera	−0.316	0.955**	0.915–0.997	0.999
Paraneoptera	−0.239	0.934	0.602–1.448	0.998
Central nervous system				
Insects, all	−0.986	0.822***	0.779–0.868	0.980
Coleoptera	−1.014	0.818***	0.785–0.853	0.996
Hymenoptera	−0.702	0.665***	0.577–0.768	0.974
Paraneoptera	−1.416	0.960	0.872–1.057	0.995
Brain				
Insects, all	−1.090	0.698***	0.648–0.752	0.959
Coleoptera	−1.167	0.707***	0.658–0.760	0.986
Hymenoptera	−0.889	0.662***	0.538–0.816	0.934
Paraneoptera	−1.713	0.929	0.829–1.042	0.993
Reproductive system				
Insects, all	−0.941	0.967*	0.894–1.016	0.971
Coleoptera	−0.872	0.924	0.793–1.077	0.958
Hymenoptera	−1.078	1.212***	1.120–1.312	0.995
Paraneoptera	−0.697	0.852**	0.641–1.132	0.999

For list of studied species see Polilov (2014)

*p-value for slope different from 1, $0.05 \le p < 0.1$; ** $0.01 \le p < 0.05$; *** $p < 0.001$

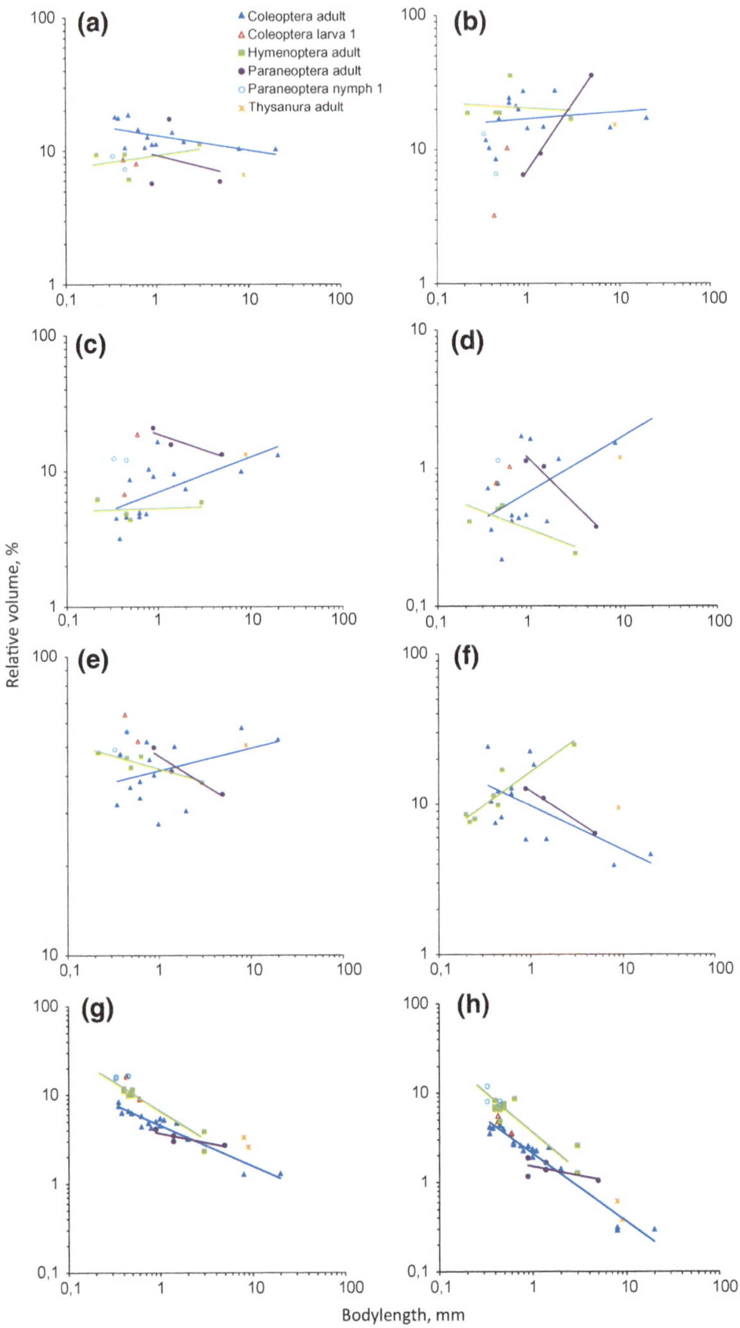

Fig. 10.2 Changes in relative volume of organs in adults and larvae of insects of different taxa: **a** skeleton, **b** musculature, **c** digestive system, **d** Malpighian tubules, **e** circulatory system and fat body, **f** reproductive system, **g** central nervous system, **h** brain. For list of studied species see Polilov (2014)

The volume of the digestive system in insects generally changes allometrically (slope of SMA regression is significantly different from one, Fig. 10.1, Table 10.1). The relative volume decreases in Coleoptera, almost unchanged in Hymenoptera, and increases in Paraneoptera as body size decreases (Fig. 10.2c). Isometry and decrease can be explained by the gut efficiency, determined by area, which changes with body size at a lower rate than volume, so that efficiency increases as body size decreases (Huxley 1932). The relative gut weight similarly decreases as body size decreases in many vertebrates (Franz et al. 2009). Malpighian tubules generally show similar trends, except in Hymenoptera, in which their relative volume increases as body size decreases (Figs. 10.1 and 10.2d).

The volume of the circulatory system and fat body in insects generally changes isometrically (Fig. 10.1, Table 10.1). Analysis of separate taxa shows that the relative volume of this system somewhat increases in Hymenoptera and Paraneoptera and decreases in Coleoptera as body size decreases (Fig. 10.2e). The decrease found in Coleoptera probably reflects the almost absent circulatory system of Ptiliidae (Polilov 2015). Since the volume of the circulatory system in insects depends on many factors (amount of reserve substances, degree of reproductive products development, age, etc.; Carstens and Storch 1980), data on the types of changes within orders can change as new, larger material becomes available.

Because of the extremely small diameter of the tracheae, **the volume of the respiratory system** could not be calculated, but considering the strong reduction of this system in microinsects (Polilov 2015), it can be assumed that its relative volume decreases as body size decreases, which is compensated by the higher passive respiration efficiency in smaller insects.

The volume of the central nervous system strongly changes allometrically as body size decreases (Fig. 10.1, Table 10.1); in insects generally and all studied taxa, the relative volume of this system strongly increases as body size decreases (Fig. 10.2g). The greatest relative volume of this system among adult insects is found in Hymenoptera (about 12 %). First-instar larvae of Coleoptera and first-instar nymphs of Paraneoptera have considerably greater relative volumes of this system compared to adults. The relative volume of this system in the first-instar nymph of the thrips *Heliothrips haemorrhoidalis* is about 17 %, which is almost five times as great as in the adult. Increasing relative size of this system has also been shown in other small arthropods (Rensch 1948; Goossen 1949; Beutel and Haas 1998; Beutel et al. 2005; Eberhard and Wcislo 2011; Quesada et al. 2011).

The relative **volume of the brain** (cerebral index), widely used in discussions of the evolution of neural activity in animals (Schmidt-Nielsen 1984), should be considered separately. The relative weight of the human brain is 2.5 %; it was long believed that cerebral index among animals was the highest in hummingbirds, 8.33, but it reaches 8.36 % in the miniature hymenopteran *Trichogramma*, 11.95 % in first-instar nymphs of the psocopteran *Liposcelis,* and 15 % in *Brachymyrmex,* some of the smallest ants (Seid et al. 2011). The rule of brain size changing allometrically with body size, known as Haller's rule or brain–body allometry, has been confirmed for many vertebrate (von Bonin 1937; Gould 1975; Platel 1976; Martin 1981; Bennett and Harvey 1985; Striedter 2005), insects (Rensch 1948; Mares et al. 2005; Wehner et al. 2007; Riveros and Gronenberg 2010; Seid et al. 2011; Eberhard and Wcislo 2011), spiders (Quesada et al. 2011), other invertebrate (Eberhard and Wcislo 2011); it is fully realized in the smallest insects; moreover, our results considerably broaden the scope of this rule. The only exceptions to Haller's rule are several laboratory lines of *Trichogramma* (Woude et al. 2013). *Megaphragma* also has markedly smaller relative brain volume than larger representatives of related taxa, because of the anucleate neurons, a unique feature of this genus (Polilov 2012). Data on the link between the complexity of behavior and brain size in vertebrates are extremely interesting (Bernstein and Bernstein 1969), but among insects such data are available only for social species (Cole 1986; Eberhard and Wcislo 2011). The behavior of microinsects never became subject of special studies, but all principal behaviors typical of large representatives of related groups are found also in microinsects. It has also been shown that in microscopic spiders body size diminution does not result in simplification of behavior (Eberhard 2007, 2011).

The reproductive system in insects generally changes allometrically and increases in relative volume as body size decreases (Fig. 10.1, Table 10.1). The same trend is found in Coleoptera and Paraneoptera separately (Fig. 10.2f). Increasing relative volume of the reproductive system in free-living insects is associated with the relative egg size strongly increasing as body size decreases (Polilov 2015). The relative volume of the reproductive system varies considerably among coleopterans, apparently because of the dependence of this parameter on the reproductive stage. Negative allometry has also been shown in linear measurements of the genital apparatus of many insects and spiders (Eberhard 2009). Among hymenopterans, this parameter markedly decreases as body size decreases, because in egg parasitoids the relative size of the egg shows no considerably increase: their larvae, which develop in the host egg, are strongly de-embryonized and consume little or no yolk (Ivanova-Kazas 1961).

Thus, most insect organs show huge scaling potentials, retaining their organization and even relative volume as body size decreases to small fractions of its initial values. The skeleton, musculature, and circulatory systems change isometrically over the entire studied series of insects. The relative volume of those organs the efficiency of which is determined by area (the digestive system and Malpighian tubules) or diffusion rate (the tracheal system), parameters that increase as body size decreases, is smaller in smaller insects, but the relative volumes of the reproductive and nervous systems strongly increases as body size decreases. These systems can geometrically restrain miniaturization in insects and determine the limits to the smallest possible body size. The greatest rate of relative volume increase at decreasing body size is found in the nervous system; in the smallest insects this parameter reaches nearly one-fifth, making the cerebral index (relative brain weight) considerably greater than in any animals for which it is known, including humans.

References

Anderson, J. F., Rahl, H., & Prange, H. D. (1979). Scaling of Supportive Tissue Mass. *Quarterly Review Biology, 54*(2), 139–148.

Bennett, P. M., & Harvey, P. H. (1985). Relative brain size and ecology in birds. *Journal of Zoology, 207*, 151–169.

Bernstein, S., & Bernstein, R. A. (1969). Relationships between foraging efficiency and the size of the head and component brain and sensory structures in the red wood ant. *Brain Research, 16*(1), 85–104.

Beutel, R. G., & Haas, A. (1998). Larval head morphology of *Hydroscapha natans* LeConte 1874 (Coleoptera, Myxophaga, Hydroscaphidae) with special reference to miniaturization. *Zoomorphology, 118*(2), 103–116.

Beutel, R. G., Pohl, H., & Hunefeld, F. (2005). Strepsipteran brain and effect of miniaturization (Insecta). *Arthropod Structure & Development, 34*(3), 301–313.

Carstens, Von S., & Storch, V. (1980). Beeinflussung der Ultrastruktur von Fettkorper und Mitteldarm des Staphyliniden *Atheta fungi* (Grav.) durch Umwelteinflusse. *Zoologische Jahrbuecher Abteilung fuer Anatomie und Ontogenie der Tiere, 103*, 73–84.

Cole, B. J. (1986). Size and behavior in ants: Constraints on complexity. *Proceedings of the National Academy of Sciences of the United States of America, 82*, 8548–8551.

Eberhard, W. G. (2007). Miniaturized orb-weaving spiders: behavioural precision is not limited by small size. *Proceedings of the Royal Society B, 274*, 2203–2209.

Eberhard, W. G. (2009). Static allometry and animal genitalia. *Evolution, 63*(1), 48–66.

Eberhard, W. G. (2011). Are smaller animals behaviourally limited? Lack of clear constraints in miniature spiders. *Animal Behaviour, 81*, 813–823.

Eberhard, W. G., & Wcislo, W. T. (2011). Grade changes in brain–body allometry: morphological and behavioural correlates of brain size in miniature spiders, insects and other invertebrates. *Advances in Insect Physiology, 40*, 155–214.

Franz, R., Hummel, J., Kienzle, E., Kölle, P., Gunga, H. C., & Clauss, M. (2009). Allometry of visceral organs in living amniotes and its implications for sauropod dinosaurs. *Proceedings of the Royal Society B, 276*, 1731–1736.

Goossen, H. (1949). Untersuchungen an gehirnen verschieden grosser, jeweils verwandter Coleopteren- und Hymenopteren. *Arten Zoologische Jahrbücher Abteilung für Allgemeine Zoologie, 62*, 1–64.

Gould, S. J. (1975). Allometry in primates, with emphasis on scaling and evolution of brai. *Contributions to primatology, 5*, 244–292.

Hiestend, W. A. (1928). Strength and weight of insects. *Annals of the Entomological Society of America, 21*, 601–606.

Huxley, J. (1932). *Problems of relative growth*. London: Methuen.

Ivanova-Kazas, O. M. (1961). *Очерки по сравнительной эмбриологии перепончатокрылых (Essays on the comparative embryology of Hymenoptera)*. Leningrad: Leningrad Univ. Press.

Mares, S., Ash, L., & Gronenberg, W. (2005). Brain allometry in bumblebee and honeybee workers. *Brain, Behavior and Evolution, 66*, 50–61.

Martin, R. D. (1981). Relative brain size and basal metabolic rate in terrestrial vertebrates. *Nature, 29*, 57–60.

Platel, R. (1976). Analyse volumetrique compaire des principales subdividions enciphaliques chez les reptiles sauriens. *Journal fur Hirnforschung, 17*, 513–537.

Polilov, A. A. (2012). The smallest insects evolve anucleate neurons. *Arthropod Structure & Development, 41*(1), 27–32.

Polilov, A. A. (2014). Морфологические особенности мельчайших насекомых (Morphological features of the smallest insects). Doctor Dissertation, Lomonosov Moscow State University.

Polilov, A. A. (2015). Small is beautiful: features of the smallest insects and limits to miniaturization. *Annual Review of Entomology, 60*, 103–121.

Polilov, A. A., & Beutel, R. G. (2009). Miniaturization effects in larvae and adults of *Mikado* sp. (Coleoptera: Ptiliidae), one of the smallest free-living insects. *Arthropod Structure & Development, 38*(3), 247–270.

Polilov, A. A., & Beutel, R. G. (2010). Developmental stages of the hooded beetle *Sericoderus lateralis* (Coleoptera: Corylophidae) with comments on the phylogenetic position and effects of miniaturization. *Arthropod Structure & Development, 39*(1), 52–69.

Prange, H. D., Anderson, J. F., & Rahn, H. (1979). Scaling of skeletal mass to body mass in birds and mammals. *American Naturalist, 113*, 103–122.

Quesada, R., Triana, E., Vargas, G., Douglass, J. K., Seid, M. A., Niven, J. E., et al. (2011). The allometry of CNS size and consequences of miniaturization in orb-weaving and cleptoparasitic spiders. *Arthropod Structure & Development, 40*(6), 521–529.

Rensch, B. (1948). Histological changes correlated with evolutionary changes in body size. *Evolution, 2*, 218–230.

Riveros, A. J., & Gronenberg, W. (2010). Brain allometry and neural plasticity in the bumblebee *Bombus terrestris*. *Brain, Behavior and Evolution, 75*, 138–148.

Schmidt-Nielsen, K. (1984). *Scaling: Why is animal size so important?*. Cambridge: Cambridge Univ. Press.

Seid, M. A., Castillo, A., & Wcislo, W. T. (2011). The allometry of brain miniaturization in ants. *Brain, Behavior and Evolution, 77*(1), 5–13.

Striedter, G. F. (2005). *The principles of brain evolution*. Inc, MA: Sinaur Assoc.

von Bonin, G. (1937). Brain-weight and bodyweight of mammals. *The Journal of General Psychology, 16*, 379–389.

Warton, D. I., Duursma, R. A., Falster, D. S., & Taskinen, S. (2012). smatr 3 - an R package for estimation and inference about allometric lines. *Methods in Ecology and Evolution, 3*(2), 257–259.

Wehner, R. T., Fukushi, T., & Isler, K. (2007). On being small: brain allometry in ants. *Brain Behav Evo., 69*, 220–228.

Woude, E., Smid, H. M., Chittka, L., & Huigens, M. E. (2013). Breaking Haller's rule: brain-body size isometry in a minute parasitic wasp. *Brain, Behavior and Evolution, 81*(2), 86–92.

Chapter 11
Effects of Miniaturization at Different Stages of the Life Cycle

11.1 Introduction

The embryonic development of microinsects and structure of their eggs remain almost unstudied, except for the interesting data on the relative sizes of eggs in insects of different sizes and different groups, discussed in detail in Chap. 13 of this book, and several studies on the development of particular representatives of parasitoid hymenopterans (Ivanova-Kazas 1952, 1954, 1961; Boivin 2010). In these studies, strong desembryonization is described in Mymaridae and Trichogrammatidae; this desembryonization is associated with egg parasitism and decreased body size. A unique feature of the ontogeny of these insects is their alecithal egg and complete cleavage, which distinguish them from all other insects.

Different groups of insects considerably differ in miniaturization-related features that manifest themselves at different stages of postnatal development. The nature of these differences depends above all on the type of metamorphosis and mode of life. It is especially revealing to compare free-living and parasitic insects of the same size class or holometabolous and hemimetabolous insects of similar size. The morphology of all postnatal stages of development has been described in some minute beetles (Polilov and Beutel 2009, 2010), psocopterans (Polilov 2016), and thrips (Moritz 1988a, b, 1989a, b, c, d; Polilov and Shmakov 2016). It is especially interesting to compare the triungulins (first instar larvae, Fig. 11.1) of Strepsiptera, many of which are extremely small (about 200 μm long), an insect order characterized by hypermetamorphosis and parasitism (Pohl 2000; Beutel et al. 2005; Osswald et al. 2010).

A.A. Polilov, *At the Size Limit—Effects of Miniaturization in Insects*,
DOI 10.1007/978-3-319-39499-2_11

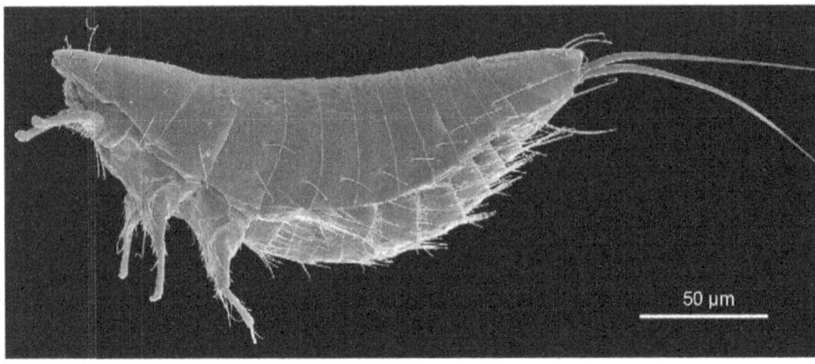

Fig. 11.1 First instar larva of *Mengenilla chobauti* (Strepsiptera), lateral view, SEM (modified after Osswald et al. (2010))

11.2 Peculiar Morphological Differences Between Microinsects at Different Stages of Postnatal Development

The larvae or nymphs of microinsects of different orders are distinguished by considerable reduction in the number of fully developed skeletal elements both from the larvae or nymphs of related groups of insects and from the adults of their own species. The larvae of Ptiliidae and Corylophidae beetles have crania without sutures, as in adults. The nymphs of psocopterans and thrips have the same division of crania as their adults, although in the adults it is more pronounced. The nymphs of psocopterans of the genus *Liposcelis* differ from adults in the absence of the dorsal tentorial arms; the larvae of strepsipterans also have no dorsal arms. The larvae of Trichogrammatidae and Mymaridae lack the tentorium (Ivanova-Kazas 1961). The larvae or nymphs of most microinsects have poorly pronounced thoracic sclerites typical of adults (thrips and psocopterans) or have no fully developed sclerites, except for the poorly sclerotized total sclerite (Ptiliidae and Corylophidae). The furcae of larvae or nymphs in most microinsects are even more poorly developed than in adults. In the first instar larvae of strepsipterans, thoracic endoskeleton is absent (Osswald et al. 2010). The larvae or nymphs of all studied microinsects lack fully developed skeletal elements in the abdomen.

Adults and larvae or nymphs of microinsects are almost indistinguishable by the morphology of their intestines. Only in the first instar nymphs of psocopterans of the genus *Liposcelis* and some first instar strepsiperan larvae the midgut is shifted anteriad and partly lies in the head (Osswald et al. 2010). The intestine of first instar strepsipterans is fundamentally distinguished by the isolated mid- and hindgut (Fig. 11.2) and absent Malpighian tubules (Pohl 2000).

The tracheal system of larvae or nymphs of microinsects is simplified compared to that of adults; spiracles are reduced in number, sometimes to a single pair (in larvae of Ptiliidae), or absent (larvae of Trichogrammatidae and Mymaridae and first instar larvae of Strepsiptera). The larvae of Strepsiptrera also differ from adults

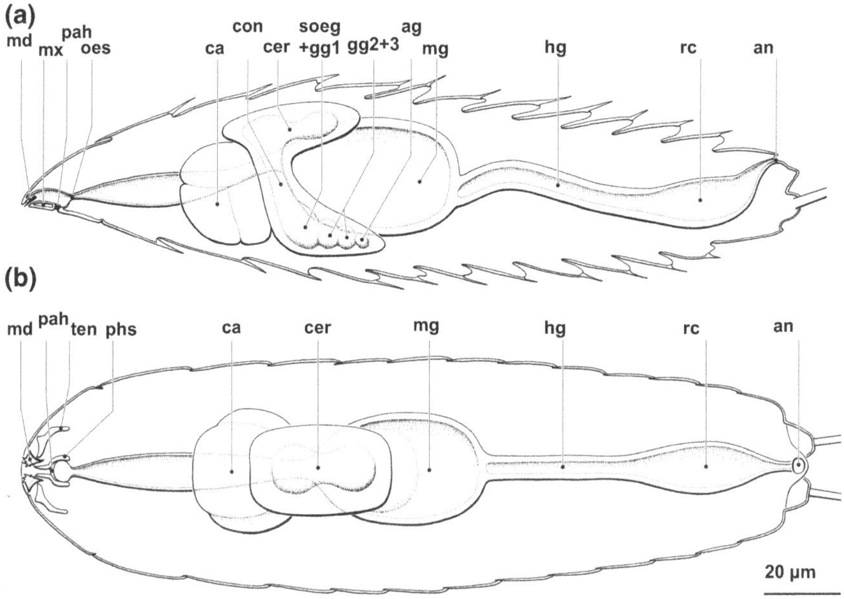

Fig. 11.2 Anatomy of first instar larva of *Stylops melittae* (Strepsiptera) (modified after Pohl (2000)). *ag* abdominal ganglia, *an* anus, *ca* corpus adiposum, *cer* cerebrum, *con* connective, *gg1,2,3* pro- meso and metathoracic ganglion, *hg* hindgut, *md* mandible, *mg* midgut, *mx* maxilla, *pah* preoral hole, *phs* pharynxspange, *rc* rectum, *soeg* suboesophageal ganglion, *ten* tentorium

in the absence of tracheal stems or tracheae in the abdomen and probably employ mixed tracheal-cutaneous respiration.

Fist instar larvae of Sterpsiptera differ from adults in the absence of heart (Pohl 2000).

Compared to adults, larvae and nymphs of microinsects display two opposite phenomena. The central nervous system of beetles shows considerably smaller degrees of condensation and fusion of ganglia, but the cerebrum is always situated outside the cranium (in first instar larvae of *Mikado* the brain reaches the second abdominal segment). By contrast, the nymphs of thrips and psocopterans demonstrate considerably greater degrees of oligomerization and condensation of ganglia, compared to adults, but the cerebrum always remains in the head. The central nervous systems of first instar larvae of most microinsects are distinguished from those adults by strong asymmetry. The first instar larvae of Strepsiptera have extremely compacted central nervous systems (Pohl 2000). The differences in the number of cells in the central nervous system between adults and first instar larvae or nymphs are very different in miniature holometabolous and miniature hemimetabolous insects. For instance, in the beetle *Sericoderus* the cerebrum contains 18,000 cells in adults and 7000 cells in first instar larvae, while in the psocopteran *Liposcelis* the cerebrum contains 9000 cells in adults and 8000 cells in first instar nymphs. The brain of the larvae of miniature beetles is not differentiated into cerebral zones, or the earliest stages of the formation of cerebral centers are observed; the main differentiation takes place at the pupal stage. By contrast, in first instar nymphs of psocopterans and thrips the structure of the cerebrum already shows little difference from that of adults. Thus, differences in the timing of cerebral zone differentiations coincide with the processes of development described in larger holometabolous and hemimetabolous insects (Panov 1957, 1959, 1960a,, b, 1961; Hinke 1961; Bate 1976; Doe and Goodman 1985; Wegerhoff and Breidbach 1992).

The nymphs of thrips have markedly fewer muscles than adults, but this difference is determined above all by the undeveloped flight muscles of nymphs. The nymphs of *Liposcelis* are distinguished from adults by a considerably smaller difference in the number of muscles than the nymphs of thrips: the recorded differences include the absence of 0mx1, 0hy7, 0hy8, and Itpm3, IIpcm1, and fusion of 0bu5 and 0bu6. The musculature of the head in the larvae of Ptiliidae differs from those of all larger related beetles only in the absence of M4 in the first instar larva of *Mikado*. The musculature of the head in Corylophidae differs from those of all studied representatives of Cucujiformia in the fusion of M17 and M18 and absence of M48. On the whole, fist instar nymphs of hemimetabolous insects have a considerably greater number of muscles than first instar larvae of holometabolous insects. For instance, the head contains 29 pairs of muscles in first instar nymphs of psocopterans of the genus *Liposcelis* and only 16 pairs of muscles in first instar larvae of beetles of the genus *Mikado*. The number of muscle pairs in thoracic segments is 53 in *Liposcelis* and 24 in *Mikado*. The thorax of first instar larva of the strepsipteran species *Mengenilla chobauti* (Fig. 11.3) contain 94 muscles (Osswald et al. 2010).

In the smallest insects, as in other animals, the relative volumes of organs change during ontogeny, and different groups of microinsects strongly differ in patterns of this allometry (Fig. 11.4).

The relative volume of the skeleton and musculature increases during postnatal development in all studied insects except *Liposcelis*. The increasing volume of the musculoskeletal system is explained by flight capacity of adults and, possibly, by the increasing general locomotor activity. The adults of *Liposcelis* are wingless.

In the studied holometabolous insects, the greatest relative volume of the digestive system is found in first instar larvae, probably because of the need to accumulate large reserves for metamorphosis. In *Liposcelis* and *Heliothrips* the relative volume of the digestive system increases during ontogeny, which is explained by the decreasing relative efficiency of the intestine with increasing body size, because this efficiency is determined by surface area.

The relative volume of Malpighian tubules changes little, and the patterns of its changes are different in all studied microinsects.

The relative volume of the central nervous system in all studied in all studied insects is the greatest in first instar larvae or nymphs. The relative volume of the central nervous system in first instar nymphs of *Heliothrips* is greater than in adults by a factor of almost 5.

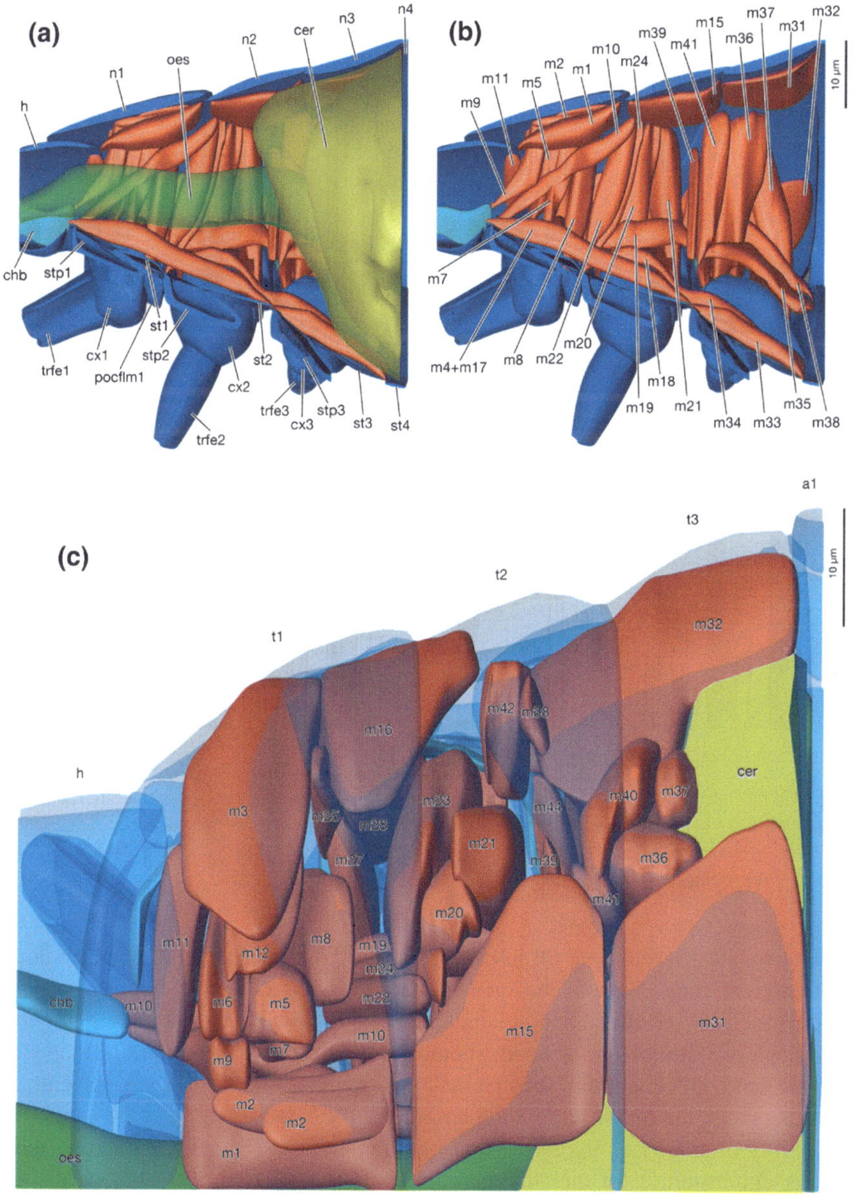

Fig. 11.3 First instar larva of *Mengenilla chobauti*, thorax, 3D reconstructions (modified after Osswald et al. (2010)). Skeleton *blue*; musculature *orange*; cerebrum *yellow*; gut *green*. **a**, **b** lateral view, **c** dorsal view. *a1* first abdominal segment; *cer* cerebrum; *chb* lateral 'Chitinbalken'; *cx1,2,3* pro-, meso-, metacoxa; *h* head; *m* muscles; *n1,2,3,4* notum of pro-, meso-, metathorax, first abdominal segment; *oes* oesophagus; *pocflm1* postcoxal 'Flankenmembran' prothorax; *st1,2,3,4* sternum of pro-, meso-, metathorax, first abdominal segment; *stp1,2,3* sternal plate of pro-, meso-, metathorax; *t1,2,3* pro-, meso-, metathorax; *trfe1,2,3* trochanterofemur of pro-, meso-, metathorax

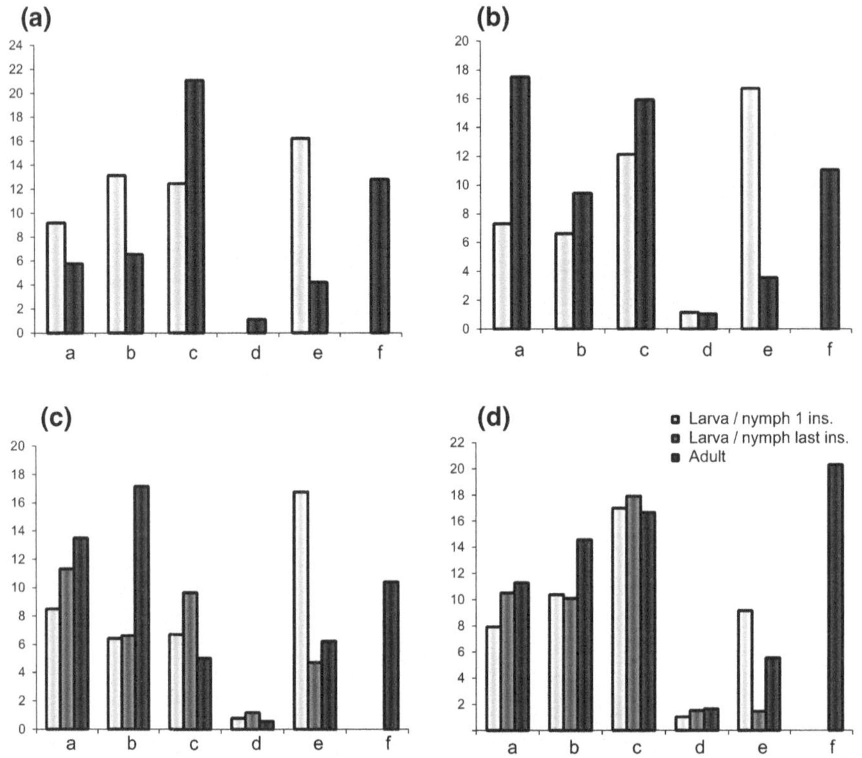

Fig. 11.4 Relative volume of organs during postnatal development: **a** psocopteran *Liposcelis bostrychophila*, **b** thrips *Heliothrips haemorrhoidalis*, **c** beetle *Mikado* sp., **d** beetle *Sericoderus lateralis*; *a* skeleton, *b* musculature, *c* digestive system, *d* excretory system, *e* central nervous system, *f* reproductive system

The smallest insects and all insects less than 0.4 mm long are holometabolous. The relatively larger size of the smallest hemimetabolous insects can be explained by the practically identical body plan of the nymphs and adults. In this case, the first instar nymph, rather than the adult, should be viewed as the main limiting stage: on the one hand, it is the size of the first instar nymph that limits the size of the egg; on the other hand, the first instar nymph has almost the same structure as the adult. The possibility of stepwise oligomerization in Holometabola (Dogel' 1954) gives them greater opportunities for miniaturization compared with Hemimetabola.

All insects with body length smaller than 0.3 mm are wasp parasitoids of eggs. This is largely determined by the considerably simplified structure of their larvae, strong desembryonization, and reduction of yolk in the egg, which is entirely alecithal in some species: these features are possible in larvae that develop inside host eggs (Ivanova-Kazas 1961).

References

Bate, C. M. (1976). Embryogenesis of an insect nervous system I. A map of the thoracic and abdominal neuroblasts in *Locusta migratoria*. *Journal of Embryology and Experimental Morphology, 35*, 107–123.

Beutel, R. G., Pohl, H., & Hunefeld, F. (2005). Strepsipteran brain and effect of miniaturization (Insecta). *Arthropod Structure & Development, 34*(3), 301–313.

Boivin, G. (2010). Reproduction and immature development of egg parasitoids. In F. L. Consoli, J. R. P. Parra, & R. A. Zucchi (Eds.), *Egg Parasitoids in Agroecosystems with Emphasis on Trichogramma* (pp. 1–23). Houten: Springer.

Doe, C. Q., & Goodman, C. S. (1985). Early events in insect neurogenesis. I. Development and segmental differences in the pattern of neuronal precursor cells. *Development Biology, 111*, 193–205.

Dogel', V. A. (1954). *Олигомеризация гомологичных органов как один из главных путей эволюции животных (Oligomerization of homologous organs as one of the main paths of evolution of animals)*. Leningrad: Leningrad Univ. Press.

Hinke, W. (1961). Das relative postembryonale wachstum der hirnteile von *Culex pipiens*, *Drosophila melanogaster* und *Drosophila*-mutanten. *Zeitschrift fuer Morphologie und Oekologie der Tiere, 50*(1), 81–118.

Ivanova-Kazas, O. M. (1952). Постэмбриональное развитие *Prestwichia aquatica* Lubb. (Hymenoptera) (The Postembryonic development of *Prestwichia aquatica* Lubb. (Hymenoptera)). *Trudy Leningradskogo obshchestva estestvoispytatelej, 71*(4), 165–213.

Ivanova-Kazas, O. M. (1954). Вопросы эволюции эмбрионального развития у перепончатокрылых (Hymenoptera) (On the evolution of postembryonic development of Hymenoptera). *Trudy Vsesoyuznogo ehntomologicheskogo obshchestva, 44*, 301–335.

Ivanova-Kazas, O. M. (1961). *Очерки по сравнительной эмбриологии перепончатокрылых (Essays on the comparative embryology of Hymenoptera)*. Leningrad: Leningrad Univ. Press.

Moritz, G. (1988a). Die Ontogenese der Thysanoptera unter besonderer Berücksichtigung des Fransenflüglers *Hercinothrips femoralis* (O.M. Reuter 1891) 1. Embryonalentwicklung. *Zoologische Jahrbuecher Abteilung fuer Anatomie und Ontogenie der Tiere, 117*, 1–64.

Moritz, G. (1988b). Die Ontogenese der Thysanoptera unter besonderer Berücksichtigung des Fransenflüglers *Hercinothrips femoralis* (O.M. Reuter 1891) 2. Erst- und Zweitlarve. *Zoologische Jahrbuecher Abteilung fuer Anatomie und Ontogenie der Tiere, 117*, 299–351.

Moritz, G. (1989a). Die Ontogenese der Thysanoptera unter besonderer Berücksichtigung des Fransenflüglers *Hercinothrips femoralis* (O.M. Reuter 1891) 3. Praepupa und Pupa. *Zoologische Jahrbuecher Abteilung fuer Anatomie und Ontogenie der Tiere, 118*, 15–54.

Moritz, G. (1989b). Die Ontogenese der Thysanoptera unter besonderer Berücksichtigung des Fransenflüglers *Hercinothrips femoralis* (O.M. Reuter 1891) 4. Imago-Kopf. *Zoologische Jahrbuecher Abteilung fuer Anatomie und Ontogenie der Tiere, 118*, 273–307.

Moritz, G. (1989c). Die Ontogenese der Thysanoptera unter besonderer Berücksichtigung des Fransenflüglers *Hercinothrips femoralis* (O.M. Reuter 1891) 5. Imago-Thorax. *Zoologische Jahrbuecher Abteilung fuer Anatomie und Ontogenie der Tiere, 118*, 391–427.

Moritz, G. (1989d). Die Ontogenese der Thysanoptera unter besonderer Berücksichtigung des Fransenflüglers *Hercinothrips femoralis* (O.M. Reuter 1891) 6. Imago-Abdomen. *Zoologische Jahrbuecher Abteilung fuer Anatomie und Ontogenie der Tiere, 119*, 157–217.

Osswald, J., Pohl, H., & Beutel, R. G. (2010). Extremely miniaturised and highly complex: the thoracic morphology of the first instar larva of *Mengenilla chobauti* (Insecta, Strepsiptera). *Arthropod Structure & Development, 39*(4), 287–304.

Panov, A. A. (1957). Строение головного мозга насекомых на последовательных этапах постэмбрионального развития (Structure of the insect cerebrum at consecutive stages of postembrynic development). *Entomologiceskoe Obozrenie, 36*(2), 269–284.

Panov, A. A. (1959). О формировании гломерулярной структуры нейропиля обонятельной доли мозга насекомых (On the formation of glomerular structure of the neuropil of the olfactory lobe of the insect cerebrum). *Zoologicheskii Zhurnal, 38*(5), 775–777.

Panov, A. A. (1960a). Строение головного мозга насекомых на последовательных этапах постэмбрионального развития: 2. Центральное тело (Structure of the cerebrum at consecutive stages of postembryonic development: 2. Central body). *Entomologiceskoe Obozrenie, 38*(2), 86–105.

Panov, A. A. (1960b). Строение головного мозга насекомых на последовательных этапах постэмбрионального развития: 3. Зрительные доли (Structure of the cerebrum at consecutive stages of postembryonic development: 3. Optic lobes). *Entomologiceskoe Obozrenie, 39*(1), 86–105.

Panov, A. A. (1961). Строение головного мозга насекомых на последовательных этапах постэмбрионального развития: 4. Обонятельный центр (Structure of the cerebrum in insects at consecutive stages of postembryonic development: 4. Olfactory center). *Entomologiceskoe Obozrenie*, *40*(2), 259–271.

Pohl, H. (2000). Die Primärlarven der Fächerflügler – evolutionäre Trends (Insecta, Strepsiptera). *Kaupia, 10*, 1–144.

Polilov, A. A. (2016). Anatomy of adult and first instar nymph of book lice *Liposcelis bostrychophila* (Psocoptera: Liposcelididae). *Entomological Review* (in press).

Polilov, A. A., & Beutel, R. G. (2009). Miniaturization effects in larvae and adults of *Mikado* sp. (Coleoptera: Ptiliidae), one of the smallest free-living insects. *Arthropod Structure & Development, 38*(3), 247–270.

Polilov, A. A., & Beutel, R. G. (2010). Developmental stages of the hooded beetle *Sericoderus lateralis* (Coleoptera: Corylophidae) with comments on the phylogenetic position and effects of miniaturization. *Arthropod Structure & Development, 39*(1), 52–69.

Polilov, A. A., & Shmakov, A. S. (2016). The anatomy of the thrips *Heliothrips haemorrhoidalis* (Thysanoptera, Thripidae) and its specific features caused by miniaturization. *Arthropod Structure & Development* (in press).

Wegerhoff, R., & Breidbach, O. (1992). Die entwicklung des zentral komplexes beim käfer. studie zur ontogenie eines zentralen hirnbereiches. *Verhandlungen der Deutschen Zoologischen Gesellschaft, 85*, 1–92.

Chapter 12
Analysis of Peculiar Miniaturization-Related Structural Features in Different Groups of Animals

12.1 Introduction

Miniaturization, or decrease in body size, is a widespread trend in animal evolution. Peculiar structural features related to miniaturization are classified into three groups: reductions and structural simplifications, novel structures, and increased variation (Hanken and Wake 1993). The third group has been analyzed only in amphibians (Hanken 1982, 1983, 1985, 1993). According to the geometrical approach to analyzing the influence of body size on structure in animals, scaling can follow three routes: isometric changes, allometric changes, and change to a new design (Schmidt-Nielsen 1984). All these variants of consequences of miniaturization have been found in insects.

Gorodkov (1984) set the threshold of miniaturization (pumilization) at approximately 1 mm and suggested that overcoming this limit should be accompanied by dramatic simplification, which he termed pumilistic degeneration. It should be noted that this is true of many invertebrate taxa (Swedmark 1964; Rundell and Leander 2010). However, microinsects go below this threshold without much simplification of structure, and some of them even display a number of novel structures. Thus, insects demonstrate an amazing ability to scale biological structures and processes: the smallest insects differ from other animals of comparable size in their principally higher levels of structural organization and considerably greater numbers of cells (Polilov 2015). Structural changes of microinsects related to miniaturization are associated with renunciation of certain vital functions (lack of wings, legs, mouthparts, sense organs) or considerable changes in proportional effects of physical forces and relative environmental parameters, such as capillary forces, air viscosity, or diffusion rate (reduction of the tracheal and circulatory systems, ptiloptery). The consequences of miniaturization displayed by different insect taxa include both common and unique changes.

A.A. Polilov, *At the Size Limit—Effects of Miniaturization in Insects*,
DOI 10.1007/978-3-319-39499-2_12

12.2 Morphological Consequences of Miniaturization in Insects

Comparison of morphological changes related to decreased body size shows that the majority of these changes are parallelisms evolved independently in different insect taxa, although some of them are unique for particular microinsects (Polilov 2015).

Key Parallelisms Typical of Most Microinsects:

- Reduction of sutures and fusions of sclerites in the head skeleton of adults and larvae.
- Simplification of the tentorium; fusion of thoracic sclerites in adults.
- Absence of most sclerites and extremely weak sclerotization of the integument in larvae.
- Simplification of the endoskeleton; ptiloptery.
- Reduction in the number of wing veins.
- Strongly flattened hypoderm cells.
- Simplification of alimentary canal differentiation.
- Reduction of gut musculature.
- Reduction in number of Malpighian tubules.
- Simplification of the circulatory system; simplification of the tracheal system in adults and larvae (reduction in number of spiracles).
- Oligomerization and concentration of the central nervous system.
- Asymmetry of the central nervous system.
- Increase in nuclear–cytoplasmic ratio of neurons; considerable reduction in size and number of neurons.
- Increase in relative volume of the central nervous system and brain.
- Dramatic reduction in number of ommatidia and sensilla.
- Reduction in number of ovarioles; increase in the relative volume of the reproductive system.

Parallelisms Typical of Some Microinsects:

- Reductions in numbers of antennomeres, palpomeres, and tarsomeres (Ptiliidae, Mymaridae, and Trichogrammatidae).
- Reduction in number of tergites and sternites (adult Mymaridae and Trichogrammatidae).
- Simplification of the male copulatory apparatus (Ptiliidae and Corylophidae).
- Absence of cuticle differentiation into exo- and endocuticle (Mymaridae and Trichogrammatidae).
- Reduction of the tracheal system and an accompanying shift to partly or entirely cutaneous respiration in larvae (Ptiliidae, Mymaridae, and Trichogrammatidae).
- Absence of a heart (adult and larval Ptiliidae, adult and larval Trichogrammatidae, and larval Mymaridae).

- Shift of the abdominal ganglia into the thorax (Liposcelididae, Ptiliidae, and Corylophidae).
- Considerable increase in the degree of chromatin compaction in nuclei of neurons (Ptiliidae, Trichogrammatidae, and Mymaridae).
- Reduction of particular muscles; desembryonization and strong simplification of structure in larvae (Mymaridae and Trichogrammatidae).

Unique Features of Particular Microinsects (known in only one family or order of microinsects):

- Absence of the metafurca in adults (Megaphragma).
- Absence of all elements of the thoracic endoskeleton (first instar strepsipteran larva of *Mengenilla chobauti*).
- Asymmetrical mouthparts (Thysanoptera).
- Mechanism of folding, fixation, and protection of wing (Ptiliidae).
- Replacement of hemolymph with parenchyma-like fat body (Ptiliidae).
- Partial shift of the midgut into the head (nymphal Liposcelis).
- Complete shift of the brain into the thorax in adults (Corylophidae).
- Lysis of cell bodies and nuclei of neurons (Megaphragma).
- Lysis of wing musculature during reproductive system maturation (Sericoderus).
- Unpaired gonads (Ptiliidae).
- Critical simplification of structure in males (*Dicopomorpha*).

12.3 Comparison of the Consequences of Miniaturization in Different Animal Taxa

Terrestrial arthropods

The smallest arthropods are mites of the family Microdispidae: *Cochlodispus minimus* Mahunka 1976, 79 μm long (Mahunka 1976), and *Microdispus australis* Mahunka 1969, 82 μm long (Mahunka 1969). The smallest representative of Eriophydae of comparable size is *Indosetacus rhinacanthi* Ghosh et Chakrabarti 1987, 86 μm long (Ghosh and Chakrabarti 1987). The smallest Collembola are 120 μm long (Bellinger et al. 1996–2013). The structure of Microdispidae and the smallest springtails has not been studied, but Eriophyidae and the smallest insects display many similar features related to miniaturization: absence of heart and replacement of hemolymph with the fat body, considerable decrease in neuron size, absence of neurilemma, unpaired gonads, absence of respiratory organs, and desmosome-like attachment of muscles to skeleton (Silvere and Shtein-Margolina 1976). However, some features of structure found in Eriophyidae, such as the morphologically smooth musculature, syncytial hypoderm, and lysis of gut during egg development (Silvere and Shtein-Margolina 1976), have not been found in insects. The respiratory system is simplified or absent also in many springtails

(Ruppel 1953). Features described in minute spiders—increased relative volume of the central nervous system and brain, decreased neuron size, and shift of some ganglia into the coxal area (Quesada et al. 2011)—are also found in microinsects as well.

Other Invertebrates

The smallest insects are among the smallest metazoans; only some representatives of several groups of invertebrates are even smaller. The very smallest metazoans are rotifers: the smallest of them, *Ascomorpha minima* Hofstein 1909, is only 43 μm long (von Hofstein 1909). In other phyla, adult animals smaller than 140 μm (the length of male *Dicopomorpha echmepterygis*) have been described among loriciferans, the smallest known representative of which is 108 μm long (Kristensen 2002), gastrotrichans, the smallest known representative of which is 70 μm long (McClain and Boyer 2009), annelids, which can be as small as 50 μm long (Westheide 1990), and crustaceans, which can be as small as 94 μm (Boxshall and Huys 1989).

Miniaturization of invertebrates is usually associated with interstitial (Swedmark 1964; Noodt 1974; Kirsteurer 1972; Westheide 1984, 1987; Rundell and Leander 2010; Worsaae et al. 2012) or parasitic modes of life (Spassky 1983; Petrunina and Kolbasov 2012), including parasitism of multicellular organisms on unicellular forms (Tchesunov et al. 2000). Miniaturization is accompanied by regressive evolution and multiple reductions in most cases (Swedmark 1964; Rundell and Leander 2010). Nevertheless, some minute invertebrates have novel structures as well, such as the original pharyngeal muscular organs and unpaired male copulatory organs found in polychaetes (Westheide 1987). In many cases, miniaturization in invertebrates is related to changes in ontogeny, such as heterochrony (Laurin and Garcia-Joral 1990; Snyder and Bretsky 1971), pedomorphosis (Serban 1960; Rundell and Leander 2010), or progenesis (Westheide 1987; Mooi 1990). Some invertebrate species have dwarf males (Turner and Yakovlev 1983; Schuchert and Rieger 1990; Rouse et al. 2004; Vortsepneva et al. 2008; Worsaae and Rouse 2010). One special phenomenon classified as miniaturization is the decrease in size in some individuals of colonial animals, but it is difficult to compare such cases of miniaturization with the others (Gorodkov 1984).

The consequences miniaturization shared by insects with other animals include the reduction of the vascular system and absence of the heart, as in the smallest archiannelids (Rensch 1948), mollusks (Brenzinger et al. 2013), crustaceans (Boxshall 1982), and tardigrades (Westheide and Rieger 2004). Many of the smaller invertebrates, e.g., tardigrades and rotifers, have unpaired gonads (Westheide and Rieger 2004); in some microinsects gonads also become unpaired. In the majority of miniature invertebrates, the respiratory system is simplified (Gorodkov 1984).

The majority of the smallest animals (Rotatoria, Loricifera, Nematoda) have approximately a thousand cells; many of them display eutely, constant number, and position of cells (Westheide and Rieger 2004). Thus, the body consists of 959 cells, 183 of them neurons, in the rotifer *Epihanes senta* (Müller 1773) (Martini 1912) and 900 cells, 196–200 of them neurons, in the rotifer *Asplanchna brightwellii*

Gosse 1950 (Ware and LoPresti 1975). The body consists of only 959–1031 cells, 302–381 of them neurons, even in the nematode *Caenorhabditis elegans* Gosse, 1850 (White et al. 1986; White 1988), which does not belong the smallest nematodes. At the same time, even the smallest flying insect *Megaphragma mymaripenne* (Hymenoptera: Trichogrammatidae, body length 220 μm) has as many as 8350–8600 cells in the body (without legs and other appendages), including 7400 cells of the central nervous system, 111 cells of the gut, 44 cells of Malpighian tubules, 585 cells of musculature, 160 cells of the reproductive system, and 139 cells of the fat body. Thus, the number of cells in the smallest insects is greater by one order of magnitude than in other animals of the microcosm. Insects are also distinguished by the absence of eutely. The number of cells in any organ differs between specimens; the position of cells also varies, and even paired organ of the same specimen consist of different numbers of cells with different positions of nuclei. For instance, one of the large flight muscles (IIdvm1) contains 27 nuclei on the left and 30 nuclei on the right in one specimen of *M. mymaripenne* and 18 and 17 nuclei, respectively, in another specimen of the same species. Surprisingly, the vital functions of the whole organism are supported by as few as 240–350 cells of the metabolic and circulatory systems, i.e., fewer than 4 % of the cells provide for digestion, excretion, transport, and storage of metabolic products.

Vertebrates
The principal consequences of miniaturization in vertebrate morphology include reductions and structural simplifications and are limited to fusion of skeletal elements in fishes (Te Winkel 1935; Myers and Bohlke 1956; Tyler 1970; Miller 1979; Springer 1983; Roberts 1986: Kottelat et al. 2006), amphibians (Hanken 1982, 1983, 1985, 1993; Wake 1986; Yeh 2002), reptiles (Rieppel 1984; Griffith 1990), and mammals (Shea 1992), and simplification of sense organs (Te Winkel 1935; Roth et al. 1988; 1990; Linke et al. 1986). Many small mammals also display allometry of the central nervous system (Roth et al. 1990, 1994, 1995, 1997; Rehkamper et al. 1991; Kaas 2000; Perge et al. 2012). Thus, it can be seen that the consequences of miniaturization in vertebrate morphology are largely similar to those found in insect morphology, but the degree of changes in insects is considerably greater, as should be expected, considering such a difference in degree of decrease in body size.

12.4 Steps of Miniaturization

As can be seen from the previous sections, microinsects share many common features. But at the same time there are some considerable differences. They can be explained by distinguishing steps of miniaturization.

I distinguish two such steps. The first is characterized by retention of the principal vital functions at all stages of the life cycle. It is typical mainly of free-living animals (e.g., Ptiliidae, Corylophydae, and Thripidae). The second is determined by loss of one or several functions at one or several stages of the life cycle. This loss of functions is accompanied by considerable reductions in systems responsible for those functions. For instance, the majority of mymarids and trichogrammatids retain all vital functions as adults but have no sense or locomotor organs as larvae, except for the first instar. The male of *Dicopomorpha echmepterygis* is the most vivid example of the second step in insects: it loses some vital functions at the larval stage, like all egg parasitoid wasps, and lacks wings, mouthparts, and digestive system at the adult stage.

These steps of miniaturization determine not only the number and type of morphological changes but also the factors that limit further decrease in body size. The limit to insect miniaturization at the first stage is 300 μm (the length of the smallest coleopteran). Only those insects that have lost some of the principal vital functions are even smaller than that.

12.5 Conclusions

Gorodkov (1984) set the threshold of miniaturization (pumilization) at about 1 mm and suggested that overcoming this limit should be accompanied by dramatic simplification, which he termed pumilistic degeneration. It should be noted that this is true of many invertebrate taxa. But microinsects go below this threshold without much simplification, and even the smallest of them not only are free of any considerable signs of simplification but also even display a number of novel structures. Thus, insects demonstrate an amazing ability to scale biological structures and processes: The smallest insects differ from other animals of comparable size in their principally higher levels of structural organization and considerably greater numbers of cells. Considerable structural changes of microinsects related to miniaturization are associated with renunciation of certain physiological functions (lack of wings, legs, mouthparts, sense organs) or considerable changes in proportional effects of physical forces and relative environmental parameters, such as capillary forces, air viscosity, or diffusion rate (reduction of the tracheal and circulatory systems, ptiloptery). The nervous and reproductive systems are especially intolerant of miniaturization. It is by these systems that the minimum body size of insects is limited.

References

Bellinger, P. F., Christiansen, K. A., & Janssens, F. (1996–2013). Checklist of the Collembola of the world. http://www.collembola.org. Accessed 2014, February 05.

Boxshall, G. A. (1982). On the anatomy of the misophirioid copepods, with special reference to *Benthomisophria palliata* Sars. *Philosophical Transactions of the Royal Society B, 297*(1086), 125–181.

Boxshall, G., & Huys, R. (1989). New tantulocarid, *Stygotantulus stocki,* parasitic on harpacticoid copepods, with an analysis of the phylogenetic relationships within the Maxillopoda. *Journal of Crustacean Biology, 91*(1), 126–140.

Brenzinger, B., Haszprunar, G., & Schrodl, M. (2013). At the limits of a successful body plan—3D microanatomy, histology and evolution of *Helminthope* (Mollusca: Heterobranchia: Rhodopemorpha), the most worm-like gastropod. *Frontiers in Zoology, 10*(1), 37.

Ghosh, N. K., & Chakrabarti, S. (1987). A new genus and three new species of eriophyid mites (Acarina: Eriophyoidea) from West Bengal India. *Entomon, 12*, 49–54.

Gorodkov, K. B. (1984). Олигомеризация и эволюция систем морфологических структур: 2. Олигомеризация и уменьшение размеров тела (Oligomerization and evolution of the morphological structure systems. 2. Oligomerization and body size decrease. *Zoologicheskii Zhurnal, 63*, 1765–1778.

Griffith, H. (1990). Miniaturization and elongation in *Eumeces* (Sauria: Scincidae). *Copeia, 1990* (3), 751–758.

Hanken, J. (1982). Appendicular skeletal morphology in minute salamanders, genus *Thorius* (Amphibia: Plethodontidae): growth regulation, adult size determination, and natural variation. *Journal of Morphology, 174*(1), 57–77.

Hanken, J. (1983). Miniaturization and its effect on cranial morphology in Plethodontid Salamanders, genus *Thorius* (Amphibia, Plethodontidae): The fate of the brain and sense organs and their role in skull morphogenesis and evolution. *Journal of Morphology, 177*(3), 255–268.

Hanken, J. (1985). Morphological novelty in the limb skeleton accompanies miniaturization in Salamanders. *Science, 229*(4716), 871–874.

Hanken, J., & Wake, D. B. (1993). Miniaturization of body size: organismal consequences and evolutionary significance. *Annual Reviews of Ecology and Systematics, 24*, 501–519.

Kaas, J. H. (2000). Why is brain size so important: design problems and solutions as neocortex gets bigger or smaller. *Brain and Mind, 1*(1), 7–23.

Kirsteurer, E. (1972). The interstitial nemertean fauna of marine sand. *Smithsonian Contributions to Zoology, 76*, 17–19.

Kottelat, M., Britz, R., Tan, H. H., & Witte, K. E. (2006). *Paedocypris*, a new genus of Southeast Asian cyprinid fish with a remarkable sexual dimorphism, comprises the world's smallest vertebrate. *Proceedings of the Royal Society B, 273*(1589), 895–899.

Kristensen, R. M. (2002). An introduction to Loricifera, Cycliophora and Micrognathozoa. *Integrative and Comparative Biology, 42*(3), 641–651.

Laurin, B., & Garcia-Joral, F. (1990). Miniaturization and heterochrony in *Homoeorhynchia meridionalis* and *H. cynocephala* (Brachiopoda, Rhynchonellidae) from the Jurassic of the Iberian range, Spain. *Paleobiology, 16*(1), 62–76.

Linke, R., Roth, G., & Rottluff, B. (1986). Comparative studies on the eye morphology of lungless salamanders, family Plethodontidae, and the effect of miniaturization. *Journal of Morphology, 189*(2), 131–143.

Mahunka, S. (1969). The scientific results of the Hungarian soil zoological expeditions to South America 9. Acari: Pyemotidae and Scutacaridae from the Guayaramerin Region in Bolivia. *Acta Zoologica Academiae Scientiarum Hungaricae, 15*, 63–90.

Mahunka, S. (1976). Äethiopische Tarsonemiden (Acari: Tarsonemida). II. *Acta Zoologica Academiae Scientiarum Hungaricae, 22*, 69–96.

Martini, E. (1912). Studien über die Konstanz histologischer Elemente. III. Hydatina senta. *Zeitschrift für wissenschaftliche Zoologie, 102*, 425–645.

McClain, C. R., & Boyer, A. G. (2009). Biodiversity and body size are linked across metazoans. *Proceedings of the Royal Society B, 276*(1665), 2209–2215.

Miller, P. J. (1979). Adaptiveness and implications of small size in teleosts. *Symposia of the Zoological Society of London, 44*, 263–306.

Mooi, R. (1990). Progenetic miniaturization in the sand dollar *Sinaechinocyamus*: Implications for clypeasteroid phylogeny. In C. De Ridder, P. Dubois, M.-C. Lahaye, et al. (Eds.), *Echinoderm research* (pp. 137–143). Rotterdam: Balkema.

Myers, G. S., & Bohlke, J. (1956). The Xenurobryconini, a group of minute South American characid fishes with teeth outside the mouth. *Stanford Ichthyological Bulletin, 7*, 6–12.

Noodt, W. (1974). Anpassung an interstitielle Bedingungen: Ein Faktor in der Evolution höherer Taxa der Crustacea. *Faunistisch-Ökologische Mitteilungen, 4*, 445–452.

Perge, J. A., Niven, J. E., Mugnaini, E., Balasubramanian, V., & Sterling, P. (2012). Why do axons differ in caliber? *Journal of Neuroscience, 32*(2), 626–638.

Petrunina, A., & Kolbasov, G. (2012). Morphology and ultrastructure of definitive males of *Arcticotantulus pertzovi* and *Microdajus tchesunovi* (Crustacea; Tantulocarida). *Zoologischer Anzeiger, 251*(3), 223–236.

Polilov, A. A. (2015). Small is beautiful: Features of the smallest insects and limits to miniaturization. *Annual Review of Entomology, 60*, 103–121.

Quesada, R., Triana, E., Vargas, G., Douglass, J. K., Seid, M. A., Niven, J. E., et al. (2011). The allometry of CNS size and consequences of miniaturization in orb-weaving and cleptoparasitic spiders. *Arthropod Structure & Development, 40*(6), 521–529.

Rehkamper, G., Schuchmann, K. L., Schleicher, A., & Zilles, K. (1991). Encephalization in hummingbirds (Trochilidae). *Brain, Behavior and Evolution, 37*, 85–91.

Rensch, B. (1948). Histological changes correlated with evolutionary changes in body size. *Evolution, 2*, 218–230.

Rieppel, O. (1984). Miniaturization of the lizard skull: its functional and evolutionary implications. *Symposia of the Zoological Society of London, 52*, 503–520.

Roberts, T. R. (1986). *Danionella translucida,* a new genus and species of cyprinid fish from Burma, one of the smallest living vertebrates. *Environmental Biology of Fishes, 16*, 231–241.

Roth, G., Blanke, J., & Wake, D. B. (1994). Cell size predicts morphological complexity in the brains of frogs and salamanders. *Proceedings of the National Academy of Sciences of the United States of America, 91*(11), 4796–4800.

Roth, G., Nishikawa, K. C., & Wake, D. B. (1997). Genome size, secondary simplification, and the evolution of the brain in salamanders. *Brain, Behavior and Evolution, 50*(1), 50–59.

Roth, G., Rotluff, B., Blanke, J., & Ohle, M. (1995). Brain size and morphology in miniaturized plethodontid salamanders. *Brain, Behavior and Evolution, 45*(2), 84–95.

Roth, G., Rotluff, B., Grunwald, W., Hanken, J., & Linke, R. (1990). Miniaturization in plethodontid salamanders (Caudata, Plethodontidae) and its consequences for the brain and visual system. *Biological Journal of the Linnean Society, 40*(2), 165–190.

Roth, G., Rottluff, G., & Linke, R. (1988). Miniaturization, genome size and the origin of functional constraints in the visual system of salamanders. *Naturwissenschaften, 75*(6), 297–304.

Rouse, G. W., Goffredi, S. K., & Vrijenhoek, R. C. (2004). *Osedax*: bone-eating marine worms with dwarf males. *Science, 305*(5684), 668–671.

Rundell, R. J., & Leander, B. S. (2010). Masters of miniaturization: Convergent evolution among interstitial eukaryotes. *BioEssays, 32*(5), 430–437.

Ruppel, H. (1953). Physiologische untersuchungen über die bedeutung des ventraltubus und die atmung der collembolen. *Zoologische Jahrbucher, 64*, 429–469.

Schmidt-Nielsen, K. (1984). *Scaling: Why is animal size so important?*. Cambridge: Cambridge Univ. Press.

Schuchert, P., & Rieger, R. M. (1990). Ultrastructural observations on the dwarf male of *Bonellia viridis* (Echiura). *Acta Zoologica, 71*(1), 5–16.

Serban, M. (1960). La neotenie et le probleme de la taille chez les Copepodes. *Crustaceana, 1*, 77–83.

Shea, B. T. (1992). Ontogenetic scaling of skeletal proportions in the talapoin monkey. *Journal of Human Evolution, 23*(3), 283–307.

Silvere, A. P., & Shtein-Margolina, V. V. (1976). Tetrapodili — четырехногие клещи: Электронномикроскопическая анатомия, проблемы эволюции и взаимоотношения с возбудителями болезней растений (Tetrapodili, four-legs mites: Electron microscopic anatomy, evolution problems, and mutual relations with plant pathogenic organisms). Tallinn: Valgus.

Snyder, J., & Bretsky, P. W. (1971). Life habits of diminutive bivalve mollusks in the Maquoketa formation (Upper Ordovician). *American Journal of Science, 271*(3), 227–251.

Spassky, A. A. (1983). Об условиях миниатюризации цепней — одного из направлений их эволюции (On the conditions for miniaturization in cyclophyllid cestodes, one of the trends of their evolution). *Izvestiye AN Moldavskoi SSR. Seria biologicheskih i chimicheskih nauk, 5*, 54–58.

Springer, V. G. (1983). *Tyson belos,* new genus and species of western Pacific fish (Gobiidae, Xenisthminae), with discussions of gobioid osteology and classification. *Smithsonian Contributions to Zoology, 390*, 1–40.

Swedmark, B. (1964). The interstitial fauna of marine sand. *Biological Reviews, 39*, 1–42.

Tchesunov, A., Milutin, D., & Evseev, A. (2000). Многоклеточные паразиты простейших (Multicellular parasites of protozoans). *Priroda, 3*(1015), 6–12.

Te Winkel, L. E. (1935). A study of *Mistichthys luzonensis* wilh special reference to conditions correlated with reduced size. *Journal of Morphology, 58*(2), 463–535.

Turner, R. D., & Yakovlev, Y. (1983). Dwarf males in the Teredinidae (Bivalvia, Pholodacea). *Science, 219*(4584), 1077–1078.

Tyler, J. C. (1970). An especially small, sexually dimorphic new species of filefish (Monacanthidae) from Australasian reefs. *Proceedings of the National Academy of Sciences of the United States of America, 122*, 273–290.

von Hofstein, N. (1909). Rotatorien aus dem Mastermyr (Gottland) und einigen anderen schwedischen Binnengewassern. *Arkiv för zoologi, 6*, 1–125.

Vortsepneva, E., Tzetlin, A., Purschke, G., Mugue, N., Hass-Cordes, E., & Zhadan, A. (2008). The parasitic polychaete known as *Asetocalamyzas laonicola* (Calamyzidae) is in fact the dwarf male of the spionid *Scolelepis laonicola* (comb. nov.). *Invertebrate Biology, 127*(4), 403–416.

Wake, M. H. (1986). The morphology of *Idiocranium russeli* (Amphibia: Gymnophiona), with comments on miniaturization through heterochrony. *Journal of Morphology, 189*(1), 1–16.

Ware, R. A., & LoPresti, V. (1975). Three-dimensional reconstruction from serial sections. *International Review of Cytology, 40*, 325–440.

Westheide, W. (1984). The concept of reproduction in polychaetes with small body size; adaptation in interstitial species. *Fortschritte der Zoologie, 29*, 265–287.

Westheide, W. (1987). Progenesis as a principle in meiofauna evolution. *Journal of Natural History, 21*(4), 843–854.

Westheide, W. (1990). *Polychaetes: Linterstitial families*. Oegstgeest: Universal Book Services.

Westheide, W., & Rieger, (2004). *Spezielle Zoologie*. Berlin: Spektrum Akademischer Verlag Heidelberg.

White, J. (1988). The anatomy. In W. B. Wood (Ed.), *The nematode* C. elegans (pp. 81–122). NY: Cold Spring Harbor Laboratory Press.

White, J. G., Southgate, E., Thomson, J. N., & Brenner, S. (1986). The structure of the nervous system of the nematode *Caenorhabditis elegans*. *Philosophical Transactions of the Royal Society B, 314*(1165), 1–340.

Worsaae, K., & Rouse, G. W. (2010). The simplicity of males: Dwarf males of four species of *Osedax* (Siboglinidae; Annelida) investigated by confocal laser scanning microscopy. *Journal of Morphology, 271*(2), 127–142.

Worsaae, K., Sterrer, W., Kaul-Strehlow, S., Hay-Schmidt, A., & Giribet, G. (2012). An anatomical description of a miniaturized acorn worm (Hemichordata, Enteropneusta) with asexual reproduction by paratomy. *PLoS ONE, 7*(11), e48529.

Yeh, J. (2002). The effect of miniaturized body size on skeletal morphology in frogs. *Evolution, 56*(3), 628–641.

Chapter 13
Limits to Insect Miniaturization

13.1 Introduction

One of the earliest authors who started the discussion of factors that limit the minimum size of animals was Bernhard Rensch. He put forward the idea that the size of animals is limited by the size of the egg, which is disproportionately large in many small animals; the smallest invertebrates (e.g., Gastrotricha, Rotatoria, and the smaller mollusks) have only one egg developing at a time, and the smaller amphibians and reptiles have at most two eggs developing at a time (Rensch 1947, 1948). The second idea proposed by Rensch is that the size of animals is limited by the maximum filling of the body with organs. This idea was based on the allometry of wing musculature in small dipterans and aphids and on the observation that the brain of the smaller insects occupies almost the entire head capsule (Rensch 1948). Studies of relative volumes of organs in the smallest insects and larger members of related taxa have shown that metabolic systems, circulatory systems, and musculature change isometrically or display the relative volume decreasing with decreasing body size in the majority of insect taxa, relative volume of the skeleton slightly increasing, and only the relative volume of the central nervous system (in all insects) and reproductive system (in most) increasing by a considerable factor with decreasing body size; hence, it is through analysis of these two systems that the factors limiting the minimum size of insects can be identified (Polilov 2005, 2008, 2014, Polilov and Beutel 2009).

13.2 Limits to Insect Miniaturization

The first hypothesis on the factors limiting the miniaturization of insects concerns the nervous system. The size and numbers of neurons in microinsects are considerably smaller than in other insects (Fig. 13.1). Nevertheless, the relative size of the

A.A. Polilov, *At the Size Limit—Effects of Miniaturization in Insects*,
DOI 10.1007/978-3-319-39499-2_13

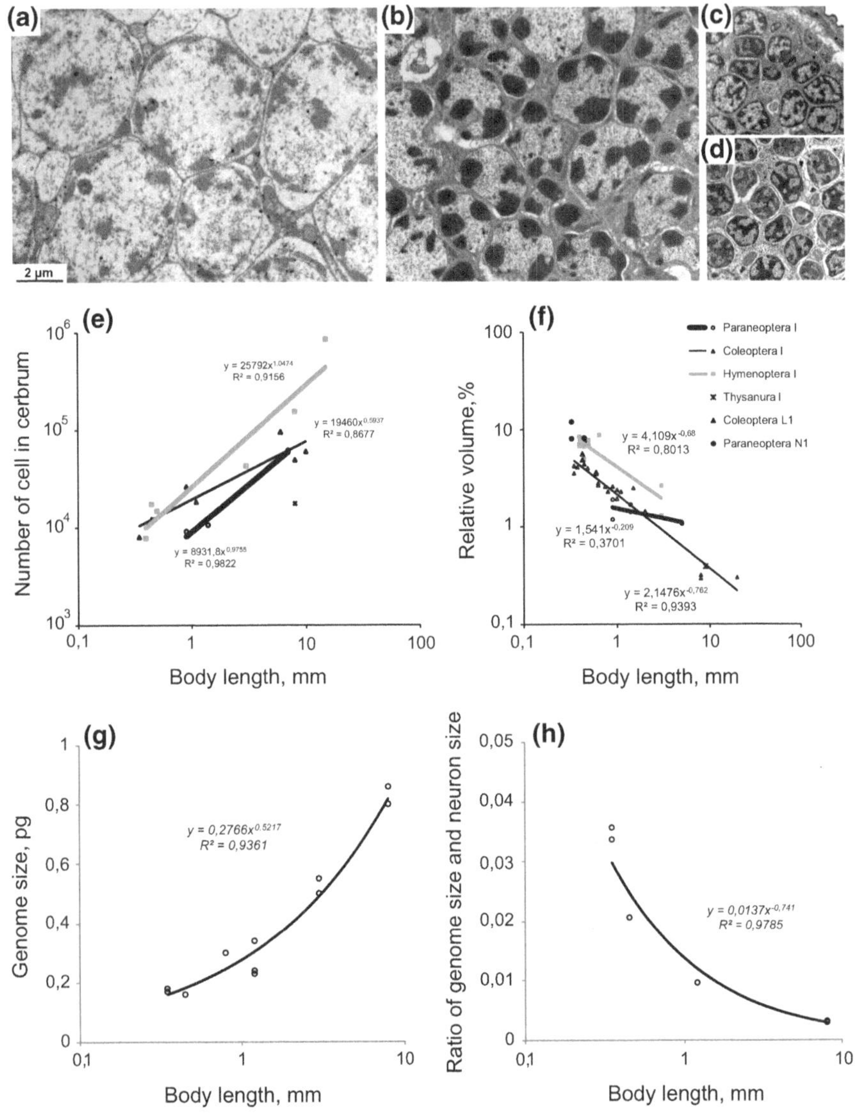

Fig. 13.1 Effects of miniaturization on insect nervous system: **a–d** neurons of beetles of different size (Coleoptera), TEM, **a** *Aleochara* sp. (Staphylinidae), **b** *Acrotrichis grandicollis*, **c** *Mikado* sp., **d** *Nanosella* sp. (Ptiliidae); **e** changes of number of cells in cerebrum; **f** changes of relative brain volume; **g**, **h** genome size in Staphylinoidae, **g** ratio of body size to genome size; **h** change in relative genome size (ratio of genome size and neuron size)

central nervous system in the smallest insects is greater by a considerable factor than in larger ones (Rensch 1948, Goossen 1949, Polilov 2005, 2007, 2008, 2015; Polilov and Beutel 2009, 2010). In extreme cases, the relative volume of the central nervous system can reach 11.8 % in adults (*Trichogramma*) and 16.75 % in first-instar larvae (*Liposcelis*) (Polilov 2015). Increasing relative size of the central nervous system with decreasing body size has been found also in other animals (Quesada et al. 2011). The relative brain size, or cerebral index, which is widely referred to in discussions of the evolution of neural activity in animals (Schmidt-Nielsen 1984), should be treated separately. The relative brain size in humans is 2.5 %; the highest previously known cerebral index among all animals, 8.33 %, was recorded in hummingbirds. However, in *Trichogramma* this index reaches 8.36 %, and in the first-instar nymph of *Liposcelis* it reaches 11.95 % (Makarova and Polilov 2013b; Polilov 2015). The principle of allometric changes of brain size with changing body size, known as Haller's rule or the brain–body allometry, has been shown to be true for many insects (Rensch 1948; Beutel and Haas 1998; Beutel et al. 2005; Mares et al. 2005; Polilov 2005, 2007, 2008, 2015; Wehner et al. 2007; Riveros and Gronenberg 2010; Eberhard and Wcislo 2011; Seid et al. 2011; Makarova and Polilov 2013a, b), spiders (Quesada et al. 2011), and other invertebrates (Eberhard and Wcislo 2011). The only known exceptions to Haller's rule among all animals are several lineages of laboratory cultures of *Trichogramma* (Woude et al. 2013). Not only does the dramatically increasing relative volume of the central nervous system in the smallest insects limit the body construction geometrically: it also leads to strongly increasing metabolic expenses determined by the increasing relative volume of the central nervous system and increasing relative surface area of one neuron, and consequently its energy cost (Chittka and Niven 2009).

The minimum size of the nervous system is limited by its conserved structure and ultrastructure, and by the number and size of neurons (Makarova and Polilov 2013a, b). Retaining a considerable number of neurons is required for performing the functions of the central nervous system, since it is by the number of neurons that the efficiency of the nervous system is determined (Kaas 2000). The nuclear-cytoplasmic ratio for neurons is considerably higher in microinsects than in other insects. In neurons of the smallest insects, the nucleus occupies up to 90 % of the cell volume (Polilov 2008). Hence, the size of the neurons in the smallest insects approaches the minimum value limited by the nucleus size. The size of the nucleus, in turn, is limited by the size of the genome (Gregory 2001). But it has been shown that in many animals (Gregory et al. 2000), including insects (Ferrari and Rai 1989; Finston et al. 1995), body size can correlate positively with genome size. Therefore, to test the hypothesis that the diminution of cell size can be limited by genome size, genome size in seven beetle species of the subfamily Staphylinoidea was determined in collaboration with the Ryan Gregory Laboratory (Department of Integrative Biology, University of Guelph, Ontario, Canada) by the Feulgen densitometry (DeSalle et al. 2005). The results have shown that genome size within the superfamily decreases to less than 0.2 of the initial value as body size decreases. At the same time, the relative genome size (ratio of genome size to cell volume), at least in neurons, considerably increases (Fig. 13.1g, h). Hence it can be concluded that in spite of the decreased genome size of microinsects, genome size does limit the diminution of neurons. This is also evidenced by the fact that the size of nuclei in the smallest insects decreases via decreasing the degree of chromatin compaction (Polilov 2008).

The second factor limiting the miniaturization of the nervous system is the diameter of the processes of neurons. According to mathematical models proposed by Faisal (2005), decreasing diameter of the processes increases the noise effects of ion channels and, as a result, disturbs the conduction of action potentials. Judging by these models, axons smaller than 80 nm in diameter are incapable of conducting the signal (Faisal et al. Faisal et al. 2005). Processes smaller in diameter than this theoretical limit have been found to occur sporadically in the smallest insects (Hustert 2012), but the majority of their axons are wider, and therefore the limits implied by this model actually apply to the size of neurons. Another consequence of these limits is that in spite of the obvious difficulties in limiting the cell body size of neurons (related to the size of the nucleus and principal organelles), the relative volume of the neuropil in all studied insects is practically constant (the average value for all studied insects is 59.4 %) and changes isometrically to the brain size with decreasing body size (Makarova and Polilov 2013a, b).

In the studied hymenopterans, the central nervous system clearly limits the size of the adult, because in the larvae it only primordial (Ivanova-Kazas 1961; Polilov 2014). In the beetles, both the first-instar larva and adult are important, since the former has the highest relative volume of the central nervous system during post-natal development (Polilov and Beutel 2009, 2010) and the latter has considerably more complex organization and greater number of neurons (Makarova and Polilov 2013a). In hemimetabolous insects, the critical stage is the first-instar nymph, which has the relative volume of the central nervous system considerably greater than at the other stages (3.8 times as great as in the adult in *Liposcelis* and 4.7 times as great in *Heliothrips*) and central nervous system and brain differentiation almost indistinguishable from those of the adult (Polilov 2014).

Therefore, it can be concluded that the size of the central nervous system does limit miniaturization in insects, and the conserved structure and ultrastructure on the one hand and the number and minimum size of neurons and the diameter of axons on the other hand limit the diminution of the nervous system.

The nervous systems of animals smaller than or comparable in size to the smallest insects are principally simpler and consist of two or three hundred neurons (Martini 1912; White 1988; Meinertzhagen 2010). Therefore, the size of these animals is less strongly limited by the central nervous system.

The size of the sense organs should also be considered among the factors that limit the minimum size of insects. One of the effects of miniaturization on sense organs is the dramatic reduction in the number of structural units (sensilla, ommatidia), but the size of particular sensilla and ommatidia changes only slightly (Makarova et al. 2015; Polilov 2014). The insect sensillum is a complex receptor structure that consists of the cuticular part and several cells (Snodgrass 1926). This construction is apparently sensitive to scaling and limits the further diminution of the sensillum (below the basal diameter of 0.6 μm). The limit to the miniaturization of the ommatidium is about 6 μm in diameter (Makarova et al. 2015). This limit is determined by several factors. One of them is structural. Even in the smallest insects, such as *Megaphragma*, each ommatidium consists of 21 cells (Makarova et al. 2015). All these cells have nuclei. Although almost all chromatin in these cells is compacted, the diameter of the majority of their nuclei is over 1 μm. The nucleus size clearly forbids further diminution of the ommatidium. The second factor limiting the ommatidium size is optical. The optical efficiency of the ommatidium is determined by the size of the lens and area of the rhabdom (Warrant and McIntyre 1993; Warrant and Nilsson 1998; Land and Nilsson 2012); as a result, these two parameters cannot be decreased ad infinitum. The length of the rhabdom considerably decreases with decreasing body size (Makarova et al. 2015). It is probably to retain the focal power of the ommatidium that the rhabdom diameter and lens size in *Megaphragma* is greater than in the larger *Trichogramma* (Makarova et al. 2015). It has also been shown that small insects approach the diffraction limit of vision (Fischer et al. 2011). In addition to constructive limits, there is probably also an energy limit to the size of receptors, since relative energy expenses increase with decreasing size (Niven et al. 2007; Niven and Laughlin 2008).

The second hypothesis concerns the size of the egg and volume of the reproductive system. It is based on my analysis of published data on the egg size in Staphylinoidea (Dybas 1966; De Coninck and Coessens 1981; Leschen 1988; Ashe and Newton 1993; Betz and Fuhrmann 2001; Staniec 2003; Staniec and Pietrykowska-Tudruj 2008, 2009; Pietrykowska-Tudruj and Staniec 2009) and Chalcidoidea (Balduf 1928; Ivanova-Kazas 1961; Vu Kuang Kon 1974; Sahad 1982, 1984; Saakyan-Baranova 1990; Boivin et al. 1993; Dahlan and Gordh 1996; Mockford 1997; Heraty et al. 2004; Hesami et al. 2004; Irvin and Hoddle 2009; Torrensa and Heraty 2013). In Coleoptera, by the example of Staphylinoidea, the relative egg size increases by a factor of more than 6 with decreasing body size (Fig. 13.2). In Ptiliidae, only one egg develops at a time, occupying more than half the body length of the female (Dybas 1966). It is probably the minimum size of the egg that limits the further miniaturization of beetles. The need to develop such an egg explains the strong increase in the relative volume of the reproductive system. In Chalcidoidea the relative egg size increases much less strongly, since the larva develops in the egg of the host, opening the possibility of strong desembryonization and depletion of yolk, up to completely alecithal eggs (Ivanova-Kazas 1961). This is also the explanation of the relatively small volume of the reproductive system found in Mymaridae and Trichogrammatidae. Increasing relative egg size with decreasing body size has also been recorded in other insects (Garcia-Barros 2000, 2002). In spiders, relative egg size and number of eggs also decrease with decreasing body size (Eberhard 2011). Therefore, Rensch's idea that the minimum size of animals is limited by the size of the egg (Rensch 1948) is also true for insects, except for those hymenopterans that have strongly simplified parasitic larvae.

The last possible hypothesis concerns feeding and the need to overcome the surface tension force and capillary forces. This is the decisive factor limiting the minimum size of sucking insects (Novotny and Wilson 1997). It is probably also quite important for other insects.

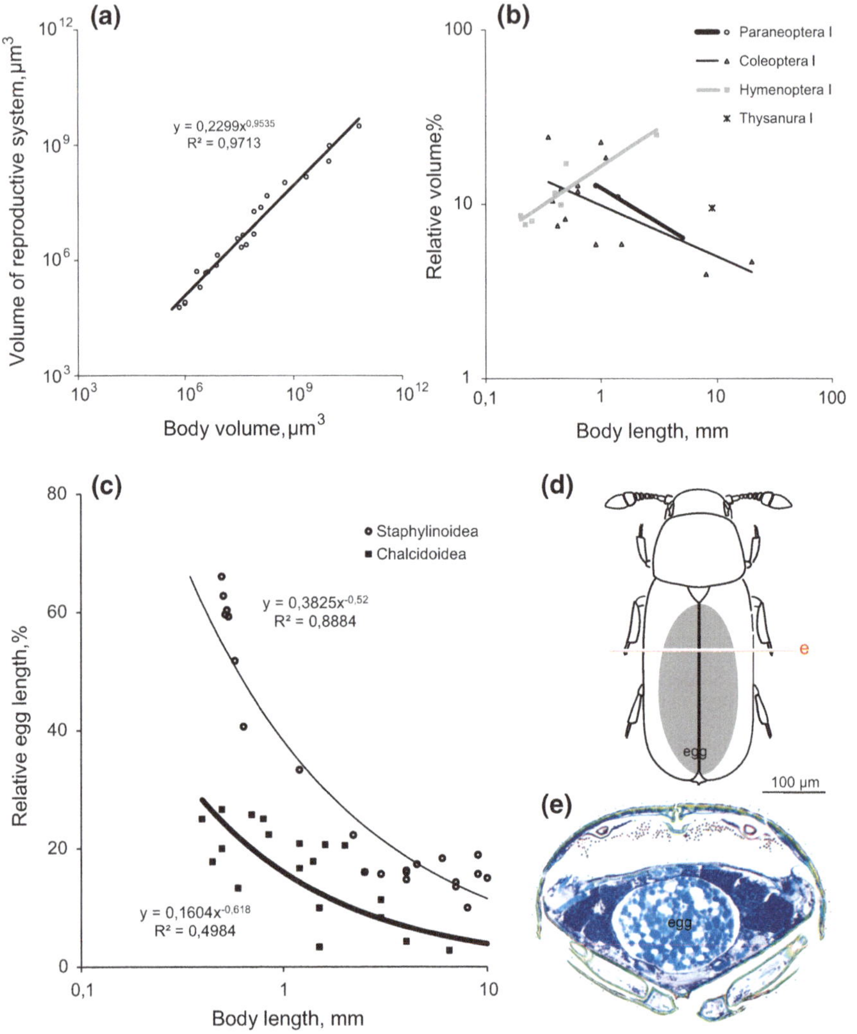

Fig. 13.2 Effects of miniaturization on reproductive system: **a** changes of volume of reproductive system; **b** changes of relative volume of reproductive system in different taxa; **c** changes of relative egg size; **d**, **e** egg of *Porophilla mystacea* (Coleoptera: Ptiliidae)

References

Ashe, J. S., & Newton, A. F. (1993). Larvae of *Trichophya* and phylogeny of the tachyporine group of subfamilies (Coleoptera: Staphylinidae) with a review, new species and characterization of the Trichophyinae. *Systematic Entomology, 18*, 267–286.

Balduf, W. V. (1928). Observations on the buffalo tree-hopper *Ceresa bubalus* Fabr. (Membracidae, Homoptera) and the bionomics of an egg parasite, *Polynema striaticorne*

Girault (Mymaridae, Hymenoptera). *Annals of the Entomological Society of America, 21*(3), 419–435.

Betz, O., & Fuhrmann, S. (2001). Life history traits in different life forms of predaceous Stenus beetles Coleoptera, Staphylinidae, living in waterside environments. *Netherlands Journal of Zoology, 51*(4), 371–393.

Beutel, R. G., & Haas, A. (1998). Larval head morphology of *Hydroscapha natans* LeConte 1874 (Coleoptera, Myxophaga, Hydroscaphidae) with special reference to miniaturization. *Zoomorphology, 118*(2), 103–116.

Beutel, R. G., Pohl, H., & Hunefeld, F. (2005). Strepsipteran brain and effect of miniaturization (Insecta). *Arthropod Structure & Development, 34*(3), 301–313.

Boivin, G., Picard, C., & Auclair, J. L. (1993). Preimaginal development of *Anaphes* n.sp. (Hymenoptera: Mymaridae), an egg parasitoid of the carrot weevil (Coleoptera: Curculionidae). *Biological Control, 3*(3), 176–181.

Chittka, L., & Niven, J. (2009). Are bigger brains better? *Current Biology, 19*, R995–R1008.

Dahlan, A. N., & Gordh, G. (1996). Development of *Trichogramma australicum* Girault (Hymenoptera: Trichogrammatidae) on *Helicoverpa armigera* (Hübner) eggs (Lepidoptera: Noctuidae). *Australian Journal of Entomology, 35*, 337–344.

De Coninck, E., & Coessens, R. (1981). Life cycle and reproductive pattern of *Acrotrichis intermedia* (Coleoptera: Ptiliidae) in experimental conditions. *Journal of Natural History, 15*(6), 1047–1055.

DeSalle, R., Gregory, T. R., & Johnston, J. S. (2005). Preparation of samples for comparative studies of arthropod chromosomes: visualization, in situ hybridization, and genome size estimation. *Methods in Enzymology, 395*, 460–488.

Dybas, H. S. (1966). Evidence for parthenogenesis in the featherwing beetles, with a taxonomic review of a new genus and eight new species (Coleoptera: Ptiliidae). *Fieldiana Zoology, 51*, 11–52.

Eberhard, W. G. (2011). Are smaller animals behaviourally limited? Lack of clear constraints in miniature spiders. *Animal Behaviour, 81*, 813–823.

Eberhard, W. G., & Wcislo, W. T. (2011). Grade changes in brain–body allometry: Morphological and behavioural correlates of brain size in miniature spiders, insects and other invertebrates. *Advances in Insect Physiology, 40*, 155–214.

Faisal, A. A., White, J. A., & Laughlin, S. B. (2005). Channel noise places limits on the miniaturization of the brain's wiring. *Current Biology, 12*(12), 1143–1149.

Ferrari, J. A., & Rai, K. S. (1989). Phenotypic correlates of genome size variation in *Aedes albopictus*. *Evolution, 43*, 895–899.

Finston, T. L., Hebert, P. D. N., & Foottit, R. (1995). Genome size variation in aphids. *Insect Biochemistry and Molecular Biology, 25*, 189–196.

Fischer, S., Müller, C. H. G., & Meyer-Rochow, V. B. (2011). How small can small be: the compound eye of the parasitoid wasp *Trichogramma evanescens* (Westwood, 1833) (Hymenoptera, Hexapoda), an insect of 0.3- to 0.4-mm total body size. *Visual Neuroscience, 28*(4), 295–308.

García-Barros, E. (2000). Body size, egg size, and their interspecific relationship with ecological and life history traits in butterflies (Lepidoptera: Papilionoidea, Hesperioidae). *Biological Journal of the Linnean Society, 70*, 251–284.

García-Barros, E. (2002). Taxonomic patterns in the egg to body size allometry of butterflies and skippers (Papilionoidea & Hesperiidae). *Nota Lepidopterologica, 25*(2/3), 161–75.

Goossen, H. (1949). Untersuchungen an gehirnen verschieden grosser, jeweils verwandter Coleopteren- und Hymenopteren. *Arten Zoologische Jahrbücher Abteilung fuer Allgemeine Zoologie, 62*, 1–64.

Gregory, T. R. (2001). Coincidence, coevolution, or causation? DNA content, cell size, and the C-value enigma. *Biological Reviews, 76*, 65–101.

Gregory, T. R., Hebert, P. D. N., & Kolasa, J. (2000). Evolutionary implications of the ralationship between genome size and body size in flatworms and copepods. *Heredity, 84*, 201–208.

Heraty, J., Hawks, D., Kostecki, J. S., & Carmichael, A. (2004). Phylogeny and behaviour of the Gollumiellinae, a new subfamily of the ant-parasitic Eucharitidae (Hymenoptera: Chalcidoidea). *Systematic Entomology, 29*(4), 544–559.
Hesami, S., Seyedoleslami, H., & Ebadi, R. (2004). Biology of *Anagrus atomus* (Hymenoptera: Mymaridae), an egg parasitoid of the grape leafhopper *Arboridia kermanshah* (Homoptera: Cicadellidae). *Entomological Science, 7*(3), 271–276.
Hustert, R. (2012). Giant and dwarf axons in a miniature insect, *Encarsia formosa* (Hymenoptera, Calcididae). *Arthropod Structure & Development, 41*(6), 535–543.
Irvin, N. A., & Hoddle, M. S. (2009). Egg maturation, oosorption, and wing wear in *Gonatocerus ashmeadi* (Hymenoptera: Mymaridae), an egg parasitoid of the glassy-winged sharpshooter, *Homalodisca vitripennis* (Hemiptera: Cicadellidae). *Biological Control, 48*(2), 125–132.
Ivanova-Kazas, O. M. (1961). *Очерки по сравнительной эмбриологии перепончатокрылых (Essays on the comparative embryology of Hymenoptera)*. Leningrad: Leningrad Univ. Press.
Kaas, J. H. (2000). Why is brain size so important: design problems and solutions as neocortex gets bigger or smaller. *Brain and Mind, 1*(1), 7–23.
Land, M. F., & Nilsson, D.-E. (2012). *Animal eyes* (2nd ed.). Oxford: Oxford University Press.
Leschen, R. A. B. (1988). The natural history and immatures of *Scaphisoma punctatum* (Coleoptera: Scaphidiidae). *Entomological News, 99*, 225–232.
Makarova, A. A., & Polilov, A. A. (2013a). Peculiarities of the brain organization and fine structure in small insects related to miniaturization. 1. The smallest Coleoptera (Ptiliidae). *Entomological Review, 93*(6), 703–713.
Makarova, A. A., & Polilov, A. A. (2013b). Peculiarities of the brain organization and fine structure in small insects related to miniaturization. 2. The smallest Hymenoptera (Mymaridae, Trichogrammatidae). *Entomological Review, 93*(6), 714–724.
Makarova, A. A., Polilov, A. A., & Fisher, S. (2015). Comparative morphological analysis of compound eye miniaturization in minute Hymenoptera. *Arthropod Structure & Development, 44*(1), 21–32.
Mares, S., Ash, L., & Gronenberg, W. (2005). Brain allometry in bumblebee and honeybee workers. *Brain, Behavior and Evolution, 66*, 50–61.
Martini, E. (1912). Studien über die Konstanz histologischer Elemente. III. Hydatina senta. *Zeitschrift für wissenschaftliche Zoologie, 102*, 425–645.
Meinertzhagen, I. A. (2010). The organisation of invertebrate brains: cells, synapses and circuits. *Acta Zoologica Stockholm, 91*(1), 64–71.
Mockford, E. L. (1997). A new species of *Dicopomorpha* (Hymenoptera: Mymaridae) with diminutive, apterous males. *Annals of the Entomological Society of America, 90*(2), 115–120.
Niven, J. E., Anderson, J. C., & Laughlin, S. B. (2007). Fly photoreceptors demonstrate energy-information trade-offs in neural coding. *PLoS Biology, 5*(4), e116.
Niven, J. E., & Laughlin, S. B. (2008). Energy limitation as a selective pressure on the evolution of sensory systems. *Journal of Experimental Biology, 211*, 1792–1804.
Novotny, V., & Wilson, M. R. (1997). Why are there no small species among xylem-sucking insects? *Evolutionary Ecology, 11*(4), 419–437.
Pietrykowska-Tudruj, E. (2009). Staniec B (2009) The egg and mature larva of *Anotylus insecatus* (Gravenhorst, 1806) (Coleoptera: Staphylinidae). *Genus, 20*(2), 209–223.
Polilov, A. A. (2005). Anatomy of the feather-winged beetles *Acrotrichis montandoni* and *Ptilium myrmecophilum* (Coleoptera, Ptiliidae). *Entomological Review, 85*(5), 467–475.
Polilov, A. A. (2007). Морфологические особенности Mymaridae связанные с миниатюризацией (Miniaturization-related structural features of Mymaridae). *Studies on hymenopterous insects: Collection of scientific papers* (pp. 50–64). Moscow: KMK.
Polilov, A. A. (2008). Anatomy of the smallest of the Coleoptera, feather-winged beetles from tribe Nanosellini (Coleoptera, Ptiliidae) and limits to insect miniaturization. *Entomological Review, 88*(1), 26–33.
Polilov, A. A. (2014). Морфологические особенности мельчайших насекомых (Morphological features of the smallest insects). Doctor Dissertation, Lomonosov Moscow State University.

Polilov, A. A. (2015). Small is beautiful: Features of the smallest insects and limits to miniaturization. *Annual Review of Entomology, 60*, 103–121.

Polilov, A. A., & Beutel, R. G. (2009). Miniaturization effects in larvae and adults of *Mikado* sp. (Coleoptera: Ptiliidae), one of the smallest free-living insects. *Arthropod Structure & Development, 38*(3), 247–270.

Polilov, A. A., & Beutel, R. G. (2010). Developmental stages of the hooded beetle Sericoderus lateralis (Coleoptera: Corylophidae) with comments on the phylogenetic position and effects of miniaturization. *Arthropod Structure & Development, 39*(1), 52–69.

Quesada, R., Triana, E., Vargas, G., Douglass, J. K., Seid, M. A., Niven, J. E., et al. (2011). The allometry of CNS size and consequences of miniaturization in orb-weaving and cleptoparasitic spiders. *Arthropod Structure & Development, 40*(6), 521–529.

Rensch, B. (1947). *Neuere Probleme der Abstammungslehre: Die transspezifische Evolution*. Stuttgart: Ferdinand Enke.

Rensch, B. (1948). Histological changes correlated with evolutionary changes in body size. *Evolution, 2*, 218–230.

Riveros, A. J., & Gronenberg, W. (2010). Brain allometry and neural plasticity in the bumblebee *Bombus terrestris*. *Brain, Behavior and Evolution, 75*, 138–148.

Saakyan-Baranova, A. A. (1990). Морфологическое исследование преимагинальных стадий развития шести видов рода *Trichogramma* Westwood (Hymenoptera, Trichogrammatidae) (Morphological study of pre-adult stages of development in six species of the genus *Trichogramma* Westwood (Hymenoptera, Trichogrammatidae)). *Entomologiceskoe Obozrenie, 69*(2), 257–263.

Sahad, K. A. (1982). Biology and morphology of *Gonatocerus* sp. (Hymenoptera, Mymaridae), an egg parsasitoid of the green rice leafhopper, *Nephotettix cincticeps* Uhler (Homoptera, Deltocephalidae). II. Morphology. *Kontyu, 50*(3), 467–476, Tokyo.

Sahad, K. A. (1984). Biology of *Anagrus optabilis* (Perkins) (Hymenoptera, Mymaridae), and egg parasitoid of delphacid planthoppers. *Esakia, 22*, 129–144.

Schmidt-Nielsen, K. (1984). *Scaling: Why is animal size so important?*. Cambridge: Cambridge Univ. Press.

Seid, M. A., Castillo, A., & Wcislo, W. T. (2011). The allometry of brain miniaturization in ants. *Brain, Behavior and Evolution, 77*(1), 5–13.

Snodgrass, R. E. (1926). The morphology of insect sense organs and the sensory nervous system. *Smithsonian Miscellaneous Collections, 77*(8), 1–80.

Staniec, B. (2003). Developmental stages of *Platystethus nitens* (C. Sahlberg, 1832) (Coleoptera; Staphylinidae). *Genus, 14*, 345–355.

Staniec, B., & Pietrykowska-Tudruj, E. (2008). Morphology of developmental stages of *Philonthus fumarius* (GravEnhorst, 1806) (Coleoptera, Staphylinidae) with notes on biology. *Acta Zoologica Academiae Scientiarum Hungaricae, 54*(3), 213–234.

Staniec, B., & Pietrykowska-Tudruj, E. (2009). Immature stages of *Rabigus tenuis* (FaBricius, 1792) (Coleoptera, Staphylinidae, Staphylininae) with observation on its biology and taksonomic comments. *The Belgian Journal of Zoology, 138*(1), 22–39.

Torrens, J., & Heraty, J. M. (2013). A new genus of Eucharitidae (Hymenoptera: Chalcidoidea), with notes on life history and immature stages. *Zootaxa, 3630*(2), 347–358.

Vu Kuang Kon. (1974). Морфологические и биологические особенности преимагинальных фаз и стадий развития *Encyrtus infidus* Rossi (Hymenoptera, Chalcidoidea) – паразита карагановой ложнощитовки (*Eulecanium caraganae* Borchs.) (Peculiar morphological and biological features of preadult phases and sgages of deelopment in *Encyrtus infidus* Rossi (Hymenoptera, Chalcidoidea), parasite of the caragana scale (*Eulecanium caraganae* Borchs.)). *Entomologiceskoe Obozrenie, 53*(4), 732–751.

Warrant, E. J., & McIntyre, P. D. (1993). Arthropod eye design and the physical limits to spatial resolving power. *Progress in Neurobiology, 40*, 413–461.

Warrant, E. J., & Nilsson, D. E. (1998). Absorption of white light in photoreceptors. *Vision Research, 38*, 195–207.

Wehner, R. T., Fukushi, T., & Isler, K. (2007). On being small: Brain allometry in ants. *Brain, Behavior and Evolution, 69*, 220–228.
White, J. (1988). The anatomy. In W. B. Wood (Ed.), *The nematode* C. elegans (pp. 81–122). NY: Cold Spring Harbor Laboratory Press.
Woude, E., Smid, H. M., Chittka, L., & Huigens, M. E. (2013). Breaking Haller's rule: Brain-body size isometry in a minute parasitic wasp. *Brain, Behavior and Evolution, 81*(2), 86–92.

Appendix

See Tables A.1 and A.2.

A.A. Polilov, *At the Size Limit—Effects of Miniaturization in Insects*,
DOI 10.1007/978-3-319-39499-2

Table A.1 Homology of head musculature in insects

Wipfler et al. (2011) (used in this study)		Kéler (1963)	Matsuda (1965)	Badonnel (1934) Psocoptera	Mickoleit (1963) Thysanoptera	Moritz (1982, 1989) Thysanoptera adult	Moritz (1982, 1989) Thysanoptera nymph	Snodgrass (1942) Hymenoptera	Youssef (1971) Hymenoptera
0an1	M. tentorioscapalis anterior	1	52	X				4	m. tent-scap I
0an2	M. tentorioscapalis posterior	2	54	Y	2?	4	4	2	m. tent-scap II
0an3	M. tentorioscapalis lateralis	3	51	Z	3?	3	3	3	m. tent-scap III
0an4	M. tentorioscapalis medialis	4	53	–				5	m. tent-scap VI
0an5	M. frontopedicellaris	–	–	–					
0an6	M. scapopedicellaris lateralis	5	+	1–3	4	2	2		
0an7	M. scapopedicellaris medialis	6	+	4–6	5	1	1		
0lb1	M. frontolabralis	8	62	b	6	5	12	1	m. fron-labr
0lb2	M. frontoepipharyngalis	9	61	a	11	17	5		
0lb3	M. epistoepipharyngealis	10		–					
0lb4	M. labralis transversalis		–	–					
0lb5	M. labroepipharyngealis	7	63	c	7	6	13		
0lb6	M. labrolabralis	–	–	–					
	M. tentoriolabralis				12				
0md1	M. craniomandibularis internus	11	21+24	M	8	18	6	9a,b	m. cran-intramand
0md2	M. craniomandibularis externus	–	22	–					
0md3	M. craniomandibularis externus	12	23	N				8	m. cran-extramand
0md4	M. hypopharyngomandibularis	13	26	O					
0md5	M. tentoriomandibularis lateralis	–	25a	–					m. tent-mand
0md6	M. tentoriomandibularis lateralis	14a	25b	Q	9	19 (Herc.)	7		m. tent-mand

(continued)

Table A.1 (continued)

Wipfler et al. (2011) (used in this study)		Kéler (1963)	Matsuda (1965)	Badonnel (1934) Psocoptera	Mickoleit (1963) Thysanoptera	Moritz (1982, 1989) Thysanoptera adult	Moritz (1982, 1989) Thysanoptera nymph	Snodgrass (1942) Hymenoptera	Youssef (1971) Hymenoptera
0md7	M. tentoriomandibularis medialis superior	–	25a	–					m. tent-mand
0md8	M. tentoriomandibularis medialis inferior	14b	25b	P					m. tent-mand
	M. laciniomandibularis				10	19			
0mx1	M. craniocardinalis	15	1	–				10	m. geno-card
0mx2	M. craniolacinialis	19	5	Rl	14, 15	21, 22	9		
0mx3	M. tentoriocardinalis	17	3	–	16			11	m. tent-card
0mx4	M. tentoriostipitalis anterior	18	4	R				12, 13	m.tent-stip I, II
0mx5	M. tentoriostipitalis posterior	18	4	–					
0mx6	M. stipitolacinialis	20	6	Lp1—Lp3	13	20	8		
0mx7	M. stipitogalealis	21	7	–					
0mx8	M. stipitopalpalis externus	22a	10	m1	17	23	10		
0mx9	M. stipitopalpalis medialis	22b	11	–					
0mx10	M. stipitopalpalis internus	23	9	m2?	18				
0mx11	M. stipitalis transversalis		12	–					
0mx12	M. palpopalpalis maxillae primus	24	+	m3?					
0mx13	M. palpopalpalis maxillae secundus	25	+	–					
0mx14	M. palpopalpalis maxillae tertius	26	+	–					

(continued)

Table A.1 (continued)

Wipfler et al. (2011) (used in this study)		Kéler (1963)	Matsuda (1965)	Badonnel (1934) Psocoptera	Mickoleit (1963) Thysanoptera	Moritz (1982, 1989) Thysanoptera adult	Moritz (1982, 1989) Thysanoptera nymph	Snodgrass (1942) Hymenoptera	Youssef (1971) Hymenoptera
0mx15	M. palpopalpalis maxillae quartus	27	+	–					
0la1	M. postoccipitoglossalis medialis	–	31	–					
0la2	M. postoccipitoglossalis lateralis	–	32	–					
0la3	M. postoccipitoparaglossalis	–	33	–					
0la4	M. postoccipitopraementalis	–	34	–				17	m. postoccip-prement
0la5	M. tentoriopraementalis	29	35	o				18	m. tent-prement
0la6	M. tentorioparaglossalis	30	36	–				18	m. tent-prement
0la7	M. tentorioglandularis	–	37	–					
0la8	M. submentopraementalis	28	38	p	19	26	11		
0la9	M. postmentomembranus	–	40	–					
0la10	M. submentomentalis	–	41	–					
0la11	M. praementoparaglossalis	31	43	j?	27				
0la12	M. praementoglossalis	32	42	i	20	27	22		
0la13	M. praementopalpalis internus	33	44	g	22				
0la14	M. praementopalpalis externus	34	45	l	21				
0la15	M. praementomembranus	–	–	–					
0la16	M. palpopalpalis labii primus	35	+	–					
0la17	M. palpopalpalis labii secundus	36	+	–					

(continued)

Table A.1 (continued)

Wipfler et al. (2011) (used in this study)		Kéler (1963)	Matsuda (1965)	Badonnel (1934) Psocoptera	Mickoleit (1963) Thysanoptera	Moritz (1982, 1989) Thysanoptera adult	Moritz (1982, 1989) Thysanoptera nymph	Snodgrass (1942) Hymenoptera	Youssef (1971) Hymenoptera
0hy1	M. frontooralis	41	74	C1				33	m. fron-susp
0hy2	M. tentoriooralis	47	75	C2?					
0hy3	M. craniohypopharyngealis	42	37	r	26?	13	19	38	m. tent-susp
0hy4	M. postoccipitalohypopharyngealis	–	78	–					
0hy5	M. tentoriosuspensorialis	48?	80	z					
0hy6	M. postmentoloralis	–	72	–	25	12	21		
0hy7	M. praementosalivaris anterior	38	76+46 +(39?)	k?	23	10	17		
0hy8	M. praementosalivaris posterior	39	77+47 +(39?)	h	24	11	18		
0hy9	M. oralis transversalis	67	–	+	39	15b	24		
0hy10	M. loroloralis	M b.	73	–					
0hy11	M. lorosalivarialis	–	–	–					
0hy12	M. hypopharyngosalivaris	37	71+79	–	28?	14	20		
0hy13	M. annularis salivarii	40?	–	–	29				
0te1	M. tentoriofrontalis posterior	–	–	–					
0te2	M. tentoriofrontalis anterior	–	–	–					
0te3	M. tentoriofrontalis dorsalis	–	–	–					
0te4	M. posterotentorialis	–	–	–					
0te5	M. tentoriotentorialis longis	–	–	–					
0te6	M. tentoriotentorialis brevis	–	–	–					

(continued)

Table A.1 (continued)

Wipfler et al. (2011) (used in this study)		Kéler (1963)	Matsuda (1965)	Badonnel (1934) Psocoptera	Mickoleit (1963) Thysanoptera	Moritz (1982, 1989) Thysanoptera adult	Moritz (1982, 1989) Thysanoptera nymph	Snodgrass (1942) Hymenoptera	Youssef (1971) Hymenoptera
0ci1	M. clypeopalatalis	43	81	A+A´	31–33	7–9	14–16	25–31	m. clyp-cibpariet, m. clyp-epiph
0bu1	M. clypeobuccalis	44	82	B					
0bu2	M. frontobuccalis anterior	45	83	E1+E2	40	16	23	34	m. fron-phar I
0bu3	M. frontobuccalis posterior	46	84	F				35	m. fron-phar II, ? III
0bu4	M. tentoriobuccalis lateralis	49	85	–					
0bu5	M. tentoriobuccalis anterior	48	87	x	35				
0bu6	M. tentoriobuccalis posterior	50	88	y					
0ph1	M. verticopharyngealis	51	86	H				36	m. cran-phar
0ph2	M. tentoriopharyngealis	52a	89+90	u+v					
0ph3	M. postoccipitopharyngealis	52b	–	–					
0st1	M. annularis stomodaei	68	93	+	36–37			36	m. circ.phar
0st2	M. longitudinalis stomodaei	69	91+92	+	38	15a	24		

Table A.2 Homology of thoracic musculature in insects

Friedrich and Beutel (2008) (used in this study)		General studies					Psocop.	Thysanoptera		
		Beutel and Haas (2000)	Snodgrass (1929)	Maki (1938) (Embioptera)	Kelér (1963)	Matsuda (1970)	Badonnel (1934)	Mickoleit (1961)	Moritz (1982, 1989) adult	Moritz (1982, 1989) nymph
Idlm1	M. prophragma-occipitalis	2	49	1	55 56	op–t 3	DPo	M.lev.cap.prphr.	29	25
Idlm2	M. pronoto-occipitalis	1	48	–	57	op-t 2	DVL1	M.lev.cap.terg.	30	26
Idlm3	M. prophragma-cervicalis	3	56	2	70	cv(d)-t 1 t14	–	–		
Idlm4	M. cervico-occipitalis dorsalis		47	5	–	op-t 1	–	–		
Idlm5	M. pronoto-phragmalis anterior	6	–	3 4	71 72?	t 12	DPr	M.dors.obl.	37	
Idlm6	M. pronoto-phragmalis posterior		–	–	72?	t 13		–		
Idvm1	M. cervico-occipitalis anterior	8?	–	–	61	op-cv 1	DVA	M.compr.occ.		
Idvm2	M. cervico-occipitalis medialis	15	50	11	61	op-cv 2	–	M.cerv.-postocc.	31 (Aelo)	27
Idvm3	M. cervico-occipitalis posterior	15	51	12	61	op-cv 3	DVL2?			
Idvm4	M. pronoto-cervicalis lateralis		–	14?	65	t-s(cv) 1?	T	–		
Idvm5	M. pronoto-cervicalis anterior	14	53	15 16?	60	t-cv 1	–	M.depr.cap.terg.	36	29
Idvm6	M. pronoto-cervicalis medialis	12	52	13	60	t-cv 2	S	M.depr.cap.terg.	36	29
Idvm7	M. pronoto-cervicalis posterior	5	–	–	–	t-cv 3	–	–		
Idvm8	M. prophragma-tentorialis	13	–	–	–	t-s(cv) 9	–	–		
Idvm9	M. profurca-occipitalis	11	–	–	–	op-s 2 p-s 3	–	–		
Idvm9	M. profurca-occipitalis	11		–		op-s 2	–			
Idvm10	M. profurca-phragmalis	16	59	17	87	t-s 1	U	M.prphr.-prfurc.	38, 38	30
Idvm11	M. prophragma-spinalis		–	–	–	t-s 8	–	–		
Idvm12	M. profurca-mesonotalis		–	–	–	t-s 2	–	–		

(continued)

Table A.2 (continued)

Friedrich and Beutel (2008) (used in this study)		General studies					Psocop.	Thysanoptera		
		Beutel and Haas (2000)	Snodgrass (1929)	Maki (1938) (Embioptera)	Kelér (1963)	Matsuda (1970)	Badonnel (1934)	Mickoleit (1961)	Moritz (1982, 1989) adult	Moritz (1982, 1989) nymph
Idvm13	M. pronoto-trochantinalis anterior	22	62	20	78	t-ti(cx) 2	–	–		
Idvm14	M. pronoto-trochantinalis posterior		–	21	78?	t-ti(cx) 3	–	M.prom.cox.terg. troch.	45	34
Idvm15	M. pronoto-trochantinocoxalis		–	27	79	t-ti(cx) 1	L	M.prom.cox.terg. bascox.	46, 47	
Idvm16	M. pronoto-coxalis anterior	23	63	–	81	t-cx 5	–	M.rem.cox.terg.	49	36
Idvm17	M. pronoto-coxalis posterior	23	64	23, 24?	81	t-cx 6 t-cx 7	C	M.rem.cox.terg.	50	
Idvm18	M. pronoto-coxalis lateralis	25	65	24?	81	t-cx 8	F			
Idvm19	M. pronoto-trochanteralis		71d	29	80	t-tr 1	A	M.depr.troch.terg.	54, 55	
Itpm1	M. pleurocrista-occipitalis	7	–	–	62	op-p 2	LL	–		
Itpm2	M. propleuro-occipitalis		57	–	63, 64	op-p 1 t-p 3	DVL2?	–	31, 32 (Herc)	
Itpm3	M. pronoto-pleuralis anterior	17?	–	19	–	–	G	–	33	
Itpm4	M. pronoto-apodemalis anterior		–	19	85	t-p 14?	G?	M.pl.-terg.	43, 44	39
Itpm5	M. pronoto-apodemalis posterior		–	19?	–	t-p 15?	–	–		
Itpm6	M. pronoto-intersegmentalis	18	–	–	–	t-p 1 t-p 2	–	–		
Ispm1	M. profurca-apodemalis		–	–	88	p-s 1	–	M. furc.-pl.		
Ispm2	M. prospina-mesopleuralis	20, 51	–	44	–	p-s 2	h2	–		
Ispm3	M. prospina-intersegmentalis		–	–	–	p-s 6	–	–		
Ispm4	M. prospina-intersegmentalis posterior		–	–	–	p-s 7	–	–		

(continued)

Table A.2 (continued)

Friedrich and Beutel (2008) (used in this study)		General studies					Psocop.	Thysanoptera		
		Beutel and Haas (2000)	Snodgrass (1929)	Maki (1938) (Embioptera)	Kelér (1963)	Matsuda (1970)	Badonnel (1934)	Mickoleit (1961)	Moritz (1982, 1989) adult	Moritz (1982, 1989) nymph
Ispm5	M. profurca-intersegmentalis posterior		–	18	–	p-s 5	–	–		
Ispm6	M. profurca-mesopleuralis transversalis	58	–	–	–	p-s 4	–	–		
Ispm7	M. prosterno-intersegmentalis posterior		–	–	–	p-s 10	–	–		
Ipcm1	M. procoxa-cervicalis	26	–	–	–	cv-cx 3	–			
Ipcm2	M. procoxa-cervicalis transversalis		54	–	59	cv-cx 1 cv-cx 2	X1	–		
Ipcm3	M. propleuro-trochantinalis		–	–	86	p-ti(cx) 1	–	–		
Ipcm4	M. propleuro-coxalis superior	24	68	28	82	p-cx 4	E	–		
Ipcm5	M. propleuro-coxalis inferior		68?	28?	83	p-cx 5	e	M.abd.cox.pl.	48	35
Ipcm6	M. propleuro-coxalis posterior		–	–	84	p-cx 6 p-cx 9	–	–		
Ipcm7	M. mesanepisterno-procoxalis transversalis	29	–	–	–	p-cx 7	–	–		
Ipcm8	M. propleuro-trochanteralis	31	71 b, c	30	80?	p-tr 1 p-tr 2	A'	M.depr.troch.pl.	56	37
Ivlm1	M. profurca-cervicalis	10	66	8	0vlm3	cv-s 1 cv-s 4?	–			
Ivlm2	M. submento-cervicalis		–	–	–	op-cv(v) 4	–	–		
Ivlm3	M. profurca-tentorialis	9	55	67	58	s 1 s 2	LV1	M.depr.cap.furc. lat	34, 35	28
Ivlm4	M. profurca-spinalis	21	61	10	75	s 14 s 16	d1d2	M.prfurc.-prsp.	41	32

(continued)

Table A.2 (continued)

Friedrich and Beutel (2008) (used in this study)		General studies					Psocop.	Thysanoptera		
		Beutel and Haas (2000)	Snodgrass (1929)	Maki (1938) (Embioptera)	Kelér (1963)	Matsuda (1970)	Badonnel (1934)	Mickoleit (1961)	Moritz (1982, 1989) adult	Moritz (1982, 1989) nymph
Ivlm5	M. profurca-sternalis		–	9	–	s 17	–	–		
Ivlm6	M. profurca-mesospinalis		–	35	76	s 15	–	–		
Ivlm7	M. profurca-mesofurcalis	42	60	34	74	s 13	LVL2LMI2	M.mesfurc.-profurc.	40	31
Ivlm8	M. prospina-mesospinalis		88	–	73	s 11	–	–		
Ivlm9	M. prospina-mesofurcalis	43	87	36	77	s 12	LVM2	M.mesfurc.-prsp.	42	33
Iscm1	M. profurca-coxalis anterior	27	–	22	89	s-cx 5	m	–		
Iscm2	M. profurca-coxalis posterior	30	66	25	91	s-cx 3	–	–		
Iscm3	M. profurca-coxalis medialis		69	26	90	s-cx 6	Y	M.add.cox.furc.	51	
Iscm4	M. profurca-coxalis lateralis		–	–	–	s-cx 2	–	M.rem.cox.furc.	53	
Iscm5	M. prospina-coxalis	28	67	–	93	s-cx 4	J	–		
Iscm6	M. profurca-trochanteralis		–	–	92	s-tr1	a	M.depr.troch.furc.	57	
Iscm7	M. prospina-mesocoxalis		–	–	–	s-cx 1 s-cx 7	X2	–		
IIdlm1	M. prophragma-mesophragmalis	39	81	31	95	t 14	DL	M.dors.rect.	61	
IIdlm2	M. mesonoto-phragmalis	40, 41	82	32, 33	96	t 12	DO	–		
IIdvm1	M. mesonoto-sternalis	44	83, 84	38 37	100 101	t-p 5 t-p 6	K	M.st.-not.	62	
IIdvm2	M. mesonoto-trochantinalis anterior	59	–	48?	103	t-ti 1 t-ti 2	–	–		
IIdvm3	M. mesonoto-trochantinalis posterior		89	48	–	t-ti 3	–	–		

(continued)

Table A.2 (continued)

Friedrich and Beutel (2008) (used in this study)		General studies					Psocop.	Thysanoptera		
		Beutel and Haas (2000)	Snodgrass (1929)	Maki (1938) (Embioptera)	Kelér (1963)	Matsuda (1970)	Badonnel (1934)	Mickoleit (1961)	Moritz (1982, 1989) adult	Moritz (1982, 1989) nymph
IIdvm4	M. mesonoto-coxalis anterior	60?	90	50	–	t-cx 5	–	–		
IIdvm5	M. mesonoto-coxalis posterior	60	91	51	105	t-cx 6 t-cx 7	I	M.rem.cox.terg.	82	58
IIdvm6	M. mesocoxa-subalaris	64	99	52	122	t-cx 8	F F'			
IIdvm7	M. mesonoto-trochanteralis	69	103	58	104	t-tr 1	K'	–		
IIdvm8	M. mesofurca-phragmalis	45	–	39	125	t-s 1	U	M.mesphr.-mesfurc.		
IIdvm9	M. mesospina-phragmalis		–	–	–	t-s 8 t-s 7?	–	–		
IItpm1	M. prophragma-mesanepisternalis		58?	40	106	t-p 3	–	M.pl.-terg.epist. caud.	73	
IItpm2	M. mesopleura-praealaris	47	–	–	106	t-p 4 t-p 20	Z	M.pl.-terg.condyl.	74	
IItpm3	M. mesonoto-basalaris	48	–	–	–	t-p 7, t-p 8, t-p 9	V?	–		
IItpm4	M. mesonoto-pleuralis anterior	49	–	–	108	t-p 10 t-p 11 t-p 18	W2W3	–	65	
IItpm5	M. mesonoto-pleuralis medialis		–	41?	–	t-p 12	–	–		
IItpm6	M. mesonoto-pleuralis posterior	46	–	41?	109	t-p 15	–	–		
IItpm7	M. mesanepisterno-axillaris	53	–	42	–	t-p 13	–	M.pl.-pt.basalar.?	67	
IItpm7	M. mesanepisterno-axillaris	53		42		t-p 13	–			
IItpm7	M. mesanepisterno-axillaris	53								
IItpm7	M. mesanepisterno-axillaris									

(continued)

Table A.2 (continued)

Friedrich and Beutel (2008) (used in this study)		General studies					Psocop.	Thysanoptera		
		Beutel and Haas (2000)	Snodgrass (1929)	Maki (1938) (Embioptera)	Kelér (1963)	Matsuda (1970)	Badonnel (1934)	Mickoleit (1961)	Moritz (1982, 1989) adult	Moritz (1982, 1989) nymph
IItpm8	M. mesepimero-axillaris secundus		–	–	111	–	–	–		
IItpm9	M. mesepimero-axillaris tertius	54	85	–	112	t-p 14	W1	M.pl.-pt.?	64	
IItpm10	M. mesepimero-subalaris	52	–	43	123	t-p 16	–	–		
IItpm11	M. mesopleura-subalaris		–	–	–	t-p 19	–	M.pl.-terg.crist.pl.	75	42, 43
IItpm12	M. prophragma-intersegmentalis	19?	–	–	–	t-p 17	–	M.pl.-terg.epist. cran.?	72	
IIppm1	M. mesotransanapleuralis	56	–	–	118	p 1	–	M.st.-pl.?	70	
IIppm2	M. mesobasalare-intersegmentalis		–	–	–	p 2	–			
IIspm1	M. mesopleura-sternalis	50	97	45, 46	115	p 3	–	M.praecox.-basalar.	63	
IIspm2	M. mesofurca-pleuralis	55	86	47	124	p-s 1	–	M.furc.-pl.	71	45
IIspm3	M. mesospina-metanepisternalis	87	–	74	–	p-s 2	–	–		
IIspm4	M. mesospina-intersegmentalis		–	–	–	p-s 6	–	–		
IIspm5	M. mesospina-intersegmentalis posterior		–	–	–	p-s 7	–	–		
IIspm6	M. mesofurca-metanepisternalis	88	–	–	–	p-s 3	h3	–	?66	
IIspm7	M. mesofurca-intersegmentalis anterior mesofurca		–	–	–	p-s 9	–	–		
IIspm8	M. mesofurca-intersegmentalis posterior		–	–	–	p-s 5	–	–		
IIpcm1	M. mesanepisterno-trochantinalis		94, 95	–	119	p-ti(cx) 1	e	M.prom.cox.pl. troch.	77	

(continued)

Table A.2 (continued)

Friedrich and Beutel (2008) (used in this study)		General studies					Psocop.	Thysanoptera		
		Beutel and Haas (2000)	Snodgrass (1929)	Maki (1938) (Embioptera)	Kelér (1963)	Matsuda (1970)	Badonnel (1934)	Mickoleit (1961)	Moritz (1982, 1989) adult	Moritz (1982, 1989) nymph
IIpcm2	M. mesobasalare-trochantinalis	68	98	–	116	p-ti(cx) 2 p-ti(cx) 3	–	–		
IIpcm3	M. mesanepisterno-coxalis anterior	62	96	56	117	p-cx 4 p-cx 6	E	–		
IIpcm4	M. mesanepisterno-coxalis posterior	61	–	57	120	p-cx 5	e'	M.abd.cox.pl.	84	56
IIpcm5	M. mesanepisterno-trochanteralis	71	103?	59	121	p-tr 2	–	–		
IIpcm6	M. mesopleura-trochanteralis	70	–	–	–	p-tr 1	–	–		
IIvlm1	M. mesofurca-spinalis		–	–	–	s 14 s 16	–	–		47
IIvlm2	M. mesofurca-metaspinalis		–	–	–	s 15	–	–		
IIvlm3	M. mesofurca-metafurcalis	82	116	65	98	s 13	LVL3LVI3	M.mesfurc.-metfurc.	68	46
IIvlm4	M. mesospina-metaspinalis		–	–	97	s 11	–	–		
IIvlm5	M. mesospina-metafurcalis	83	117	66	99	s 12	–	–		48
IIvlm6	M. mesospina-abdominosternalis		–	–	–	p-s 13	–	–		
IIvlm7	M. mesofurca-abdominosternalis		–	–	–	p-s 10	–	M.mesfurc.-abdst.	69	54
IIscm1	M. mesofurca-coxalis anterior	65, 63?	92	49	126	s-cx 5	m	M.prom.cox.furc.	78, 79	57
IIscm2	M. mesofurca-coxalis posterior	67	101	53	128	s-cx 3	J	M.rem.cox.furc. ventr.	83	59
IIscm3	M. mesofurca-coxalis medialis		100	55	127	s-cx 6	–	–		
IIscm4	M. mesofurca-coxalis lateralis		–	54	128?	s-cx 2	Y	M. rem.cox.furc.dors.	83	59
IIscm5	M. mesospina-coxalis		93	–	130	s-cx 4	–	M.rem.cox.furc. ventr.?		

(continued)

Table A.2 (continued)

Friedrich and Beutel (2008) (used in this study)		General studies					Psocop.	Thysanoptera		
		Beutel and Haas (2000)	Snodgrass (1929)	Maki (1938) (Embioptera)	Kelér (1963)	Matsuda (1970)	Badonnel (1934)	Mickoleit (1961)	Moritz (1982, 1989) adult	Moritz (1982, 1989) nymph
IIscm6	M. mesofurca-trochanteralis	72	103	60	129	s-tr1	–	M.depr.troch.furc.	85	60
IIscm7	M. mesospina-metacoxalis	106?	–	–	–	s-cx 1 s-cx 7	X3	M.mesfurc. metcox.	103	
IIIdlm1	M. mesophragma-metaphragmalis	79	112	62	131	t 14	DL	M.dors.rect.	88	52
IIIdlm2	M. metanoto-phragmalis	80, 81	112?	63 64	132	t 12 t 13	DO	M.metnot. abdterg. M.dors. obl.	89, 90	
IIIdvm1	M. metanoto-sternalis	84	113	67 68	135 136	t-p 5 t-p 6	K	M.st.-not.	91	
IIIdvm2	M. metanoto-trochantinalis anterior	100	–	78?	138	t-ti 1 t-ti 2	–	–		
IIIdvm3	M. metanoto-trochantinalis posterior		118	78	–	t-ti 3	–	–		
IIIdvm4	M. metanoto-coxalis anterior	101	119	80	140?	t-cx 5	–	–		
IIIdvm5	M. metanoto-coxalis posterior	102	120	81	140	t-cx 6 t-cx 7	I	M.rem.cox.terg.	100	64
IIIdvm6	M. metacoxa-subalaris	105	129	82	153	t-cx 8	F F'	M.cox.-subalar.	104 (Aelo)	
IIIdvm7	M. metanoto-trochanteralis	111	133c	88	139	t-tr 1	A	M.depr.troch.terg.		
IIIdvm8	M. metafurca-phragmalis	85, 86	–	69	156	t-s 1	U	M.metfurc.-metphr.	98	53
IIItpm1	M. mesophragma-metanepisternalis	89	–	70	142	t-p 3	p	–		

(continued)

Table A.2 (continued)

Friedrich and Beutel (2008) (used in this study)		General studies					Psocop.	Thysanoptera		
		Beutel and Haas (2000)	Snodgrass (1929)	Maki (1938) (Embioptera)	Kelér (1963)	Matsuda (1970)	Badonnel (1934)	Mickoleit (1961)	Moritz (1982, 1989) adult	Moritz (1982, 1989) nymph
IIItpm2	M. metapleura-praealaris	90	–	–	143	t-p 4 t-p 20	Z	M.pl.-terg.condyl.	95	
IIItpm3	M. metanoto-basalaris	93	–	–	–	t-p 7 t-p 8	V?	–		
IIItpm4	M. mesonoto-pleuralis anterior		–	–	144	t-p 10 t-p 11 t-p 18	W2W3	M.pl.-terg.crist.pl.	96	
IIItpm5	M. metanoto-pleuralis medialis	91	–	71?	–	t-p 12	–	M.pl.-terg.crist.pl.		
IIItpm6	M. metanoto-pleuralis posterior	92	–	71?	–	t-p 15	–	–		
IIItpm7	M. metanepisterno-axillaris	96	–	72	–	t-p 13	–	–		
IIItpm8	M. metepimero-axillaris secundus		–	–	145	–	–	–		
IIItpm9	M. metepimero-axillaris tertius	95	114	–	146	t-p 14	W1	M.pl.-pt.?	93	
IIItpm10	M. metepimero-subalaris	94	–	73	154	t-p 16	–	–		
IIItpm11	M. metapleura-subalaris		–	–	–	t-p 19	–	–		
IIItpm12	M. mesophragma-intersegmentalis		–	–	–	t-p 17	–	–		
IIIppm1	M. metatransanapleuralis	97	–	–	–	p 1	–	–		
IIppm2	M. metabasalare-intersegmentalis		–	–	–	p 2	–	M.pl.-basalar.		
IIIspm1	M. metapleura-sternalis	98	–	75, 76	–	p 3	–	M.praecox.-basalar.	92	
IIIspm2	M. metafurca-pleuralis	99	115	77	165	p-s 1	–	M.furc.-pl.	97	
IIIspm3	M. metaspina-intersegmentalis posterior		–	–	–	p-s 7	–	–		

(continued)

Table A.2 (continued)

Friedrich and Beutel (2008) (used in this study)		General studies					Psocop.	Thysanoptera		
		Beutel and Haas (2000)	Snodgrass (1929)	Maki (1938) (Embioptera)	Kelér (1963)	Matsuda (1970)	Badonnel (1934)	Mickoleit (1961)	Moritz (1982, 1989) adult	Moritz (1982, 1989) nymph
IIIspm4	M. metafurca-intersegmentalis anterior		–	–	–	p-s 9	–	–		
IIIspm5	M. metafurca-intersegmentalis posterior		–	–	–	p-s 5	–	–		
IIIpcm1	M. metanepisterno-trochantinalis		125?	–	149	p-ti(cx) 1	e	–		
IIIpcm2	M. metabasalare-trochantinalis	110	127	–	148	p-ti(cx) 2 p-ti(cx) 3	–	–		
IIIpcm3	M. metanepisterno-coxalis anterior	104	126	86	151	p-cx 4 p-cx 6	E	–		
IIIpcm4	M. metanepisterno-coxalis posterior	103	–	87	150	p-cx 5	e’	M.abd.cox.pl.	102	62
IIIpcm5	M. metanepisterno-trochanteralis	112	–	89	152	p-tr 2	A’	M.depr.troch.pl.	104 (Herc)	
IIIpcm6	M. metapleura-trochanteralis		–	–	–	p-tr 1	–	–		
IIIpcm7	M. metanepisterno-mesocoxalis		–	–	–	p-cx 8	H	–		
IIIvlm1	M. metafurca-spinalis		–	–	–	s 14 s 16	–			
IIIvlm2	M. metafurca-abdominosternalis	fam	–	94, 95	134	s 20	lv	M.metfurc.-abdst.?	94	54
IIIvlm3	M. metaspina-abdominosternalis		–	–	133	s 12	–	–		
IIIscm1	M. metafurca-coxalis anterior	107	121	79	157	s-cx 5	m	M.prom.cox.furc.	99	63
IIIscm2	M. metafurca-coxalis posterior	109	123, 124	83	159	s-cx 3	J	M.rem.cox.furc. ventr.	101	65

(continued)

Table A.2 (continued)

Friedrich and Beutel (2008) (used in this study)		General studies					Psocop.	Thysanoptera		
		Beutel and Haas (2000)	Snodgrass (1929)	Maki (1938) (Embioptera)	Kelér (1963)	Matsuda (1970)	Badonnel (1934)	Mickoleit (1961)	Moritz (1982, 1989) adult	Moritz (1982, 1989) nymph
IIIscm3	M. metafurca-coxalis medialis		130	85	159?	s-cx 6	–	–		
IIIscm4	M. metafurca-coxalis lateralis	108	122	84	159	s-cx 2	–	M. rem.cox.furc.dors.		
IIIscm5	M. metaspina-coxalis		–	–	–	s-cx 4	–	–		
IIIscm6	M. metafurca-trochanteralis	113	133d	90	160	s-tr1	a	M.depr.troch.furc.	105	

Friedrich and Beutel (2008) (used in this study)		Coleoptera		Hymenoptera					
		Larsen (1966)	Baehr (1975)	Duncan (1939)	Snodgrass (1942)	Alam (1951)	Gibson (1986)	Miko et al. (2007)	Vilhelmsen et al. (2010)
Idlm1	M. prophragma-occipitalis	M2	1	–	40, 41			*ph1(t1)-poc*	41
Idlm2	M. pronoto-occipitalis	M1	2	–	40, 41?	43, 44		*t1-poc*	2
Idlm3	M. prophragma-cervicalis	M3	3	–	40, 41			*ph1(t1)-poc*	25
Idlm4	M. cervico-occipitalis dorsalis					43, 44			
Idlm5	M. pronoto-phragmalis anterior	M4	4			47 (I)			5
Idlm6	M. pronoto-phragmalis posterior			l is1,2	45	46 (I)		*t1-ph1*	5
Idvm1	M. cervico-occipitalis anterior								15
Idvm2	M. cervico-occipitalis medialis	M10	9						15
Idvm3	M. cervico-occipitalis posterior	M10	9		42 b,c	41			15
Idvm4	M. pronoto-cervicalis lateralis								
Idvm5	M. pronoto-cervicalis anterior	M9	7						3
Idvm6	M. pronoto-cervicalis medialis	M7	6	lpm3, 4	48	50		*t1-pl1,? ph1 (t1)-pl1*	3

(continued)

Table A.2 (continued)

Friedrich and Beutel (2008) (used in this study)		Coleoptera		Hymenoptera					
		Larsen (1966)	Baehr (1975)	Duncan (1939)	Snodgrass (1942)	Alam (1951)	Gibson (1986)	Miko et al. (2007)	Vilhelmsen et al. (2010)
Idvm7	M. pronoto-cervicalis posterior			lpm1, 2	47			*t1-cv*	4
Idvm8	M. prophragma-tentorialis	M8			46	51			41
Idvm9	M. profurca-occipitalis		10	lfp	51	54		*fu1-cv*	26
Idvm9	M. profurca-occipitalis			Ois2	43	40+B		*fu1d-poc*	
Idvm10	M. profurca-phragmalis	M11	13	58	78	63			29
Idvm11	M. prophragma-spinalis								
Idvm12	M. profurca-mesonotalis								
Idvm13	M. pronoto-trochantinalis anterior	M14	14						
Idvm14	M. pronoto-trochantinalis posterior								
Idvm15	M. pronoto-trochantinocoxalis								
Idvm16	M. pronoto-coxalis anterior	M15	15						10, 11
Idvm17	M. pronoto-coxalis posterior	M15	15						10, 11
Idvm18	M. pronoto-coxalis lateralis	M17	17	Ilm6	55	57?		*?t1-cx1*	10, 11
Idvm19	M. pronoto-trochanteralis								
Itpm1	M. pleurocrista-occipitalis			Ois1	42	41a+b, 42a+b		*? pl1(m, l)-poc*	
Itpm2	M. propleuro-occipitalis			32	42a?	42			16, 17
Itpm3	M. pronoto-pleuralis anterior	M12	11						7
Itpm4	M. pronoto-apodemalis anterior								8, 9
Itpm5	M. pronoto-apodemalis posterior								8, 9
Itpm6	M. pronoto-intersegmentalis	M13	12	lpm5,6	49, 50	52, 53		*t1-fu1*	
Ispm1	M. profurca-apodemalis								
Ispm2	M. prospina-mesopleuralis		35						
Ispm3	M. prospina-intersegmentalis								

(continued)

Table A.2 (continued)

Friedrich and Beutel (2008) (used in this study)		Coleoptera		Hymenoptera					
		Larsen (1966)	Baehr (1975)	Duncan (1939)	Snodgrass (1942)	Alam (1951)	Gibson (1986)	Miko et al. (2007)	Vilhelmsen et al. (2010)
Ispm4	M. prospina-intersegmentalis posterior								
Ispm5	M. profurca-intersegmentalis posterior								
Ispm6	M. profurca-mesopleuralis transversalis	M38							
Ispm7	M. prosterno-intersegmentalis posterior								
Ipcm1	M. procoxa-cervicalis			llm2	53	55		*pl1-cx1*	
Ipcm2	M. procoxa-cervicalis transversalis			49	pl 1-cx1cX				1
Ipcm3	M. propleuro-trochantinalis								37
Ipcm4	M. propleuro-coxalis superior	M16	16						20, 32
Ipcm5	M. propleuro-coxalis inferior			llm5	57	59		*fu1-cx1l*	20, 32
Ipcm6	M. propleuro-coxalis posterior								
Ipcm7	M. mesanepisterno-procoxalis transversalis								
Ipcm8	M. propleuro-trochanteralis	M20	20	llm3	61	61		*pl1-tr1*	21, 22
Ivlm1	M. profurca-cervicalis	M6	8	lfp	51	54		? *fu1-cv*	28
Ivlm2	M. submento-cervicalis								
Ivlm3	M. profurca-tentorialis	M5	5	Oi s3,4	44	45		*fu1v-poc*	27
Ivlm4	M. profurca-spinalis								34
Ivlm5	M. profurca-sternalis								
Ivlm6	M. profurca-mesospinalis								
Ivlm7	M. profurca-mesofurcalis	M30	25	lis4	52	48	1	*fu2-fu1v*	24, 35, 36
Ivlm8	M. prospina-mesospinalis								
Ivlm9	M. prospina-mesofurcalis	M31	26						40
Iscm1	M. profurca-coxalis anterior	M18	18	Ilm1	54	56		*s1-cx1*	105
Iscm2	M. profurca-coxalis posterior	M19	19	l lm4	56	58		*fu1-cx1m*	30

(continued)

Table A.2 (continued)

Friedrich and Beutel (2008) (used in this study)		Coleoptera		Hymenoptera					
		Larsen (1966)	Baehr (1975)	Duncan (1939)	Snodgrass (1942)	Alam (1951)	Gibson (1986)	Miko et al. (2007)	Vilhelmsen et al. (2010)
Iscm3	M. profurca-coxalis medialis								31
Iscm4	M. profurca-coxalis lateralis								23
Iscm5	M. prospina-coxalis			lis3	58	60		*s2-cx1*	39
Iscm6	M. profurca-trochanteralis								33
Iscm7	M. prospina-mesocoxalis								
IIdlm1	M. prophragma-mesophragmalis	M28	24	IIdl1	71	62	2	*ph1-ph2*	42
IIdlm2	M. mesonoto-phragmalis	M29	24			70			49
IIdvm1	M. mesonoto-sternalis		27	IIdv1	72	111	3	*pl2-t2a*	56
IIdvm2	M. mesonoto-trochantinalis anterior	M39	39						67
IIdvm3	M. mesonoto-trochantinalis posterior								53
IIdvm4	M. mesonoto-coxalis anterior	M40	37						50
IIdvm5	M. mesonoto-coxalis posterior	M40	37						50
IIdvm6	M. mesocoxa-subalaris	M43	40	–	82	72		*cx2-sa2*	
IIdvm7	M. mesonoto-trochanteralis	M47	44				17		51, 52
IIdvm8	M. mesofurca-phragmalis	M32	32	IIdv2			11	*fu2-ph2*	75, 76
IIdvm9	M. mesospina-phragmalis								
IItpm1	M. prophragma-mesanepisternalis								43
IItpm2	M. mesopleura-praealaris	M33	29	–	74		5	*pl2-t2c*	58
IItpm3	M. mesonoto-basalaris								48
IItpm4	M. mesonoto-pleuralis anterior								57
IItpm5	M. mesonoto-pleuralis medialis			IIpm4	75	68?	4	*pl2-t2b*	57
IItpm6	M. mesonoto-pleuralis posterior		28			68 (II)			57, 67

(continued)

Table A.2 (continued)

Friedrich and Beutel (2008) (used in this study)		Coleoptera		Hymenoptera					
		Larsen (1966)	Baehr (1975)	Duncan (1939)	Snodgrass (1942)	Alam (1951)	Gibson (1986)	Miko et al. (2007)	Vilhelmsen et al. (2010)
IItpm7	M. mesanepisterno-axillaris	M36b	31	m3Ax?	76a?		7?	*t1-3ax2*	14
IItpm7	M. mesanepisterno-axillaris			I IIpm2a?			19	*pl2-3ax3*	14
IItpm7	M. mesanepisterno-axillaris			IIpm2	76b		8	*pl2-3ax2a*	14
IItpm7	M. mesanepisterno-axillaris								63
IItpm8	M. mesepimero-axillaris secundus								
IItpm9	M. mesepimero-axillaris tertius	M36a	30	IIpm3	76c		9	*pl2-3ax2p*	14, 64
IItpm10	M. mesepimero-subalaris	M35, M34?					16		
IItpm11	M. mesopleura-subalaris								66
IItpm12	M. prophragma-intersegmentalis								
IIppm1	M. mesotransanapleuralis								68
IIppm2	M. mesobasalare-intersegmentalis			–	–	65 (II)		*t1-ba2*	13, 38, 59
IIspm1	M. mesopleura-sternalis		33	IIpm1	77	64	10	*pl2-ba2*	62
IIspm2	M. mesofurca-pleuralis	M37	34	IIfpl1	79	69	12	*pl2-fu2*	69
IIspm3	M. mesospina-metanepisternalis		67						
IIspm4	M. mesospina-intersegmentalis								70
IIspm5	M. mesospina-intersegmentalis posterior								
IIspm6	M. mesofurca-metanepisternalis		68						
IIspm7	M. mesofurca-intersegmentalis anterior mesofurca								
IIspm8	M. mesofurca-intersegmentalis posterior								

(continued)

Table A.2 (continued)

Friedrich and Beutel (2008) (used in this study)		Coleoptera		Hymenoptera					
		Larsen (1966)	Baehr (1975)	Duncan (1939)	Snodgrass (1942)	Alam (1951)	Gibson (1986)	Miko et al. (2007)	Vilhelmsen et al. (2010)
IIpcm1	M. mesanepisterno-trochantinalis								
IIpcm2	M. mesobasalare-trochantinalis								60
IIpcm3	M. mesanepisterno-coxalis anterior	M42	39						60
IIpcm4	M. mesanepisterno-coxalis posterior	M41	38	IIlm1	80	73	13	*pl2-cx2*	71
IIpcm5	M. mesanepisterno-trochanteralis	M50	46						61
IIpcm6	M. mesopleura-trochanteralis	M48	45						
IIvlm1	M. mesofurca-spinalis								
IIvlm2	M. mesofurca-metaspinalis								
IIvlm3	M. mesofurca-metafurcalis	M62	52	IIis2	–	71		*fu3-fu2*	81, 82
IIvlm4	M. mesospina-metaspinalis								
IIvlm5	M. mesospina-metafurcalis	M63	53						83
IIvlm6	M. mesospina-abdominosternalis								
IIvlm7	M. mesofurca-abdominosternalis								
IIscm1	M. mesofurca-coxalis anterior	M44	41	IIlm2	81	74	14	*s2-cx2*	
IIscm2	M. mesofurca-coxalis posterior	M46	43	IIlm4	83	75	15	*fu2-cx2*	77
IIscm3	M. mesofurca-coxalis medialis				?83	?75 (II)			74
IIscm4	M. mesofurca-coxalis lateralis	M45	42						
IIscm5	M. mesospina-coxalis								
IIscm6	M. mesofurca-trochanteralis	M52	47	67	86	76			78, 79
IIscm7	M. mesospina-metacoxalis	M80							
IIIdlm1	M. mesophragma-metaphragmalis	M60	50	IIIdl	96		18	*ph3-ph2*	44, 45

(continued)

Table A.2 (continued)

Friedrich and Beutel (2008) (used in this study)		Coleoptera		Hymenoptera					
		Larsen (1966)	Baehr (1975)	Duncan (1939)	Snodgrass (1942)	Alam (1951)	Gibson (1986)	Miko et al. (2007)	Vilhelmsen et al. (2010)
IIIdlm2	M. metanoto-phragmalis	M61	51	IIis1	70	70	42 (III)	*t2-t3*	89
IIIdvm1	M. metanoto-sternalis	M64	55	IIIpm4	97–99	85-86		? *pl3-t3 (a, b)*	73, 86
IIIdvm2	M. metanoto-trochantinalis anterior	M74	69						92
IIIdvm3	M. metanoto-trochantinalis posterior								93
IIIdvm4	M. metanoto-coxalis anterior	M75	70						91
IIIdvm5	M. metanoto-coxalis posterior	M76	71						91
IIIdvm6	M. metacoxa-subalaris	M79	74	IIIp m5	105	81		*cx3-sa3*	101
IIIdvm7	M. metanoto-trochanteralis	M84	78	–	–			*t3-tr3*	92
IIIdvm8	M. metafurca-phragmalis	M65, 66	62						110
IIItpm1	M. mesophragma-metanepisternalis		56						
IIItpm2	M. metapleura-praealaris	M67	58						47
IIItpm3	M. metanoto-basalaris	M69	57						85
IIItpm4	M. mesonoto-pleuralis anterior								87, 88
IIItpm5	M. metanoto-pleuralis medialis			79		? 84-86			87, 88
IIItpm6	M. metanoto-pleuralis posterior	M68	59						87, 88
IIItpm7	M. metanepisterno-axillaris	M71b	61	IIIpm2b	100	88		*pl3-3ax3*	72, 98
IIItpm8	M. metepimero-axillaris secundus								
IIItpm9	M. metepimero-axillaris tertius	M71a	60	IIIpm2b	100	89		*pl3-3ax3*	99
IIItpm10	M. metepimero-subalaris	M70	66	IIIpm3a&b	102			*pl3-sa3*	
IIItpm11	M. metapleura-subalaris			IIIpm3a&b	102	82		*pl3-sa3*	100

(continued)

Table A.2 (continued)

Friedrich and Beutel (2008) (used in this study)		Coleoptera		Hymenoptera					
		Larsen (1966)	Baehr (1975)	Duncan (1939)	Snodgrass (1942)	Alam (1951)	Gibson (1986)	Miko et al. (2007)	Vilhelmsen et al. (2010)
IIItpm12	M. mesophragma-intersegmentalis								
IIIppm1	M. metatransanapleuralis	M72	64						95, 102
IIppm2	M. metabasalare-intersegmentalis								94, 96
IIIspm1	M. metapleura-sternalis	M73	65	IIIpm1	101	87		*pl3-ba3*	94, 96
IIIspm2	M. metafurca-pleuralis								103
IIIspm3	M. metaspina-intersegmentalis posterior								
IIIspm4	M. metafurca-intersegmentalis anterior								
IIIspm5	M. metafurca-intersegmentalis posterior								
IIIpcm1	M. metanepisterno-trochantinalis								
IIIpcm2	M. metabasalare-trochantinalis								97
IIIpcm3	M. metanepisterno-coxalis anterior	M78	73			77a (III)			
IIIpcm4	M. metanepisterno-coxalis posterior	M77	72	IIIlm4	103	77a+b		? *pl3-cx3l*	106
IIIpcm5	M. metanepisterno-trochanteralis		79						
IIIpcm6	M. metapleura-trochanteralis								113
IIIpcm7	M. metanepisterno-mesocoxalis								
IIIvlm1	M. metafurca-spinalis								
IIIvlm2	M. metafurca-abdominosternalis		54						115
IIIvlm3	M. metaspina-abdominosternalis								
IIIscm1	M. metafurca-coxalis anterior	M81	75	IIIl m1	104	78		*pl3-cx3m*	105

(continued)

Table A.2 (continued)

Friedrich and Beutel (2008) (used in this study)		Coleoptera		Hymenoptera					
		Larsen (1966)	Baehr (1975)	Duncan (1939)	Snodgrass (1942)	Alam (1951)	Gibson (1986)	Miko et al. (2007)	Vilhelmsen et al. (2010)
IIIscm2	M. metafurca-coxalis posterior	M83	77	IIIlm2	106	79		*fu3-cx3(m,l)*	111, 112
IIIscm3	M. metafurca-coxalis medialis				?106				
IIIscm4	M. metafurca-coxalis lateralis	M82	76			79 (III)			104
IIIscm5	M. metaspina-coxalis					80 (III)			
IIIscm6	M. metafurca-trochanteralis	M85	80	IIIl m3	109	83		*pl3-tr3* ? *fu3-tr3*	107

References

Alam, S. M. (1951). The skeleto-muscular mechanism of *Stenobracon deesae* Cameron (Braconidae, Hymenoptera) - An ectoparasite of sugarcane and juar borers of India Part I Head and thorax. *Aligarh Muslim University Publications on Indian Insect Types, 3*, 1–174.

Badonnel, A. (1934). Recherches sur l'anatomie des Psoques. *Bulletin biologique de la France et de la Belgique, Suppl. 18*, 1–241.

Baehr, M. (1975). Skelett und Muskulatur des Thorax von *Priacma serrata* Leconte (Coleoptera, Cupedidae). *Zeitschrift für Morphologie der Tiere, 81*, 55–101.

Beutel, R. G., & Haas, A. (2000). Phylogenetic releationships of the suborders of Coleoptera (Insecta). *Cladistics, 16*, 103–141.

Duncan, C. D. (1939). A contribution to the biology of North American vespine wasps. *Stanford University Publications Biological Sciences, 8*, 1–272.

Friedrich, F., & Beutel, R. G. (2008). The thorax of*Zorotypus*(Hexapoda, Zoraptera) and a new nomenclature for the musculature of Neoptera. *Arthropod Structure & Development, 37*(1), 29–54.

Gibson, G. A. P. (1986). Evidence for monophyly and relationships of Chalcidoidea, Mymaridae, and Mymarommatidae (Hymenoptera: Terebrantes). *Canadian Entomologist, 118*, 205–240.

Larsén, O. (1966). On the morphology and function of locomotor organs of the Gyrinidae and other Coleoptera. *Opuscula Entomologica, Suppl. 30*, 1–241.

Maki, T. (1938). Studies of the thoracic musculature of insects. *Memoirs of the Faculty of Science and Agriculture; Taihoku Imperial University, 24*, 1–343.

Matsuda, R. (1965). Morphology and evolution of the insect head. *Memoirs of the American Entomological Institute, 4*, 1–334.

Matsuda, R. (1970). Morphology and evolution of the insect thorax. *Memoirs of the Entomological Society of Canada, Suppl. 76*, 1–431.

Mickoleit, E. (1961). Zur Thoraxmorphologie der Thysanoptera. *Zoologische Jahrbuecher Abteilung fuer Anatomie und Ontogenie der Tiere, 79*, 1–92.

Mickoleit, E. (1963). Untersuchungen zur Kopfmorphologie der Thysanoptera. *Zoologische Jahrbuecher Abteilung fuer Anatomie und Ontogenie der Tiere, 81*, 101–150.

Mikó, I., Vilhelmsen, L., Johnson, N. F., Masner, L., & Penzes, Z. (2007). Skeletomusculature of Scelionidae (Hymenoptera:Platygastroidea): Head and mesosoma. *Zootaxa, 1571*, 1–78.

Moritz, G. (1982). Zur Morphologie und Anatomie des Fransenflüglers*Aeolothrips intermedius*Bagnall, 1934 (Aeolothripidae, Thysanoptera, Insecta) – 2. Mitteilung: Der Thorax. *Zoologische Jahrbuecher Abteilung fuer Anatomie und Ontogenie der Tiere, 108*, 55–106.

Moritz, G. (1989). Die Ontogenese der Thysanoptera unter besonderer Berücksichtigung des Fransenflüglers*Hercinothrips femoralis* (O.M.Reuter 1891) 3. Praepupa und Pupa. *Zoologische Jahrbuecher Abteilung fuer Anatomie und Ontogenie der Tiere, 118*, 15–54.

Snodgrass, R. E. (1929). The thoracic mechamism of a grasshopper, and its antecedents. *Smithsonian Miscellaneous Collections, 82*, 1–112.

Snodgrass, R. E. (1942). The skeleto-muscular mechanisms of the honey bee. *Smithsonian Miscellaneous Collections, 103*, 1–120.

Kéler, S. V. (1963). *Entomologisches Wörterbuch*. Berlin: Akademieverlag.

Vilhelmsen, L., Miko, I., & Krogmann, L. (2010). Beyond the wasp-waist: structural diversity and phylogenetic significance of the mesosoma in apocritan wasps (Insecta: Hymenoptera). *Zoological Journal of the Linnean Society, 159*(1), 22–194.

Wipfler, B., Machida, R., Uller, M., & Beutel, R. G. (2011). On the head morphology of Grylloblattodea (Insecta) and the systematic position of the order, with a new nomenclature for the head muscles of Dicondylia. *Systematic Entomology, 36*, 241–266.

Youssef, N. (1971). Topography of the cephalic musculature and nervous system of the honey bee *Apis mellifera* Linnaeus. *Smithsonian Contributions to Zoology, 99*, 1–54.

The manufacturer's authorised representative in the EU is Springer Nature Customer Service Centre GmbH, Europaplatz 3, 69115 Heidelberg, Germany. If you have any concerns regarding our products, please contact ProductSafety@springernature.com

Printed and bound by CPI Group (UK) Ltd, Croydon, CR0 4YY
30/09/2026
02225559-0014